PAVEMENTS AND SURFACINGS
FOR HIGHWAYS AND AIRPORTS

PAVEMENTS AND SURFACINGS FOR HIGHWAYS AND AIRPORTS

MICHEL SARGIOUS

Associate Professor of Civil Engineering,
The University of Calgary, Alberta, Canada

A HALSTED PRESS BOOK

JOHN WILEY & SONS
NEW YORK—TORONTO

PUBLISHED IN THE U.S.A. AND CANADA BY
HALSTED PRESS
A DIVISION OF JOHN WILEY & SONS, INC., NEW YORK

Library of Congress Cataloging in Publication Data

Sargious, Michel.
   Pavements and surfacings for highways and airports.

"A Halsted Press book."
Includes index.
1. Pavements.   I. Title.
TE250.S25   625.8   75–11891
ISBN 0–470–75418–4

WITH 48 TABLES AND 310 ILLUSTRATIONS

© APPLIED SCIENCE PUBLISHERS LTD 1975

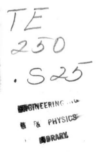

*TE
250
.S25*

ENGINEERING ...
& PHYSICS
LIBRARY

All rights reserved. No part of this publication may be reproduced,
stored in a retrieval system, or transmitted in any form or by any
means, electronic, mechanical, photocopying, recording, or otherwise,
without the prior written permission of the publishers, Applied
Science Publishers Ltd, Ripple Road, Barking, Essex, England

Printed Offset Litho in Great Britain by
Cox & Wyman Ltd, London, Fakenham and Reading

PREFACE

The book is intended to serve two purposes: first, to introduce the basic principles which must be known by people dealing with pavements; and secondly to present the theories and methods of pavement design that may be used by students, designers and researchers who have the prospect of designing and constructing highways or airfields. The book also includes design charts which can be used immediately by pavement designers. In addition, some of the new concepts developed in recent years to improve the methods of pavement systems analysis are explained.

This book is written in a relatively simple way so that it may be followed by people familiar with basic engineering courses in soil mechanics, concrete structures, and properties of concrete and asphalt mixes. Chapters 9 and 21 cover more advanced material which requires some knowledge of numerical methods of structural analysis and systems engineering, respectively.

The book consists of twenty-one chapters and is divided into four parts. The first part, which includes Chapters 1–4, covers pavement types, loads, climate, subgrades, sub-bases and stabilised soils. The second part, which includes Chapters 5–8, deals with flexible pavements for highways and airfields from the point of view of material characterisation, methods of design and construction, and cost analysis. The third part, consisting of Chapters 9–18, is concerned with methods of design and construction, and cost evaluation of the different types of concrete (rigid) pavements. Part four, which includes Chapters 19–21, explains the methods of pavement rehabilitation, design of overlays and pavement systems analysis.

Carefully selected solved examples are presented in the different chapters to illustrate the procedures and methods of design covered in the book.

In this book, various concepts of design, as well as several design charts,

v

have been brought together from many publications; each publication source is acknowledged in the text. The diversity of practices included in the book makes it possible to solve specific problems according to the amount of available data concerning the loading and environmental conditions. It also enables specific problems to be solved using different methods; the results may be compared before selecting an appropriate design.

The author wishes to thank Mr N. Finn of the University of Calgary for proofreading the manuscript. The typing of the manuscript was carefully done by Mrs M. May, Miss S. Leduchowski and Mrs M. Marsden of the University of Calgary, to all of whom the author's thanks are extended.

M. Sargious

CONTENTS

Chapter 11   PAVEMENT JOINTS, DOWEL BARS AND REINFORCING STEEL

Chapter 12   CONTINUOUSLY REINFORCED CONCRETE PAVEMENTS

**PART IV   PAVEMENT REHABILITATION AND
SYSTEMS ANALYSIS**

NOTATION

The following is a list of symbols which are common in the various chapters of the text. An attempt has been made to adhere as closely as possible to standard nomenclature used in soil mechanics and structural analysis. Some symbols are not included in this list but are explained where they appear in the text.

A	tyre contact area
A_s	cross-sectional area of steel per unit width of a concrete pavement
A_{sl}	cross-sectional area of slab per unit width of a concrete pavement
A_1	initial construction cost of travelled way per unit length
a	radius of circular area representing tyre imprint
a, b	semi-axes of an elliptical area representing tyre imprint
a_1, a_2, a_3	thickness coefficients for the different layers of a flexible pavement; unit costs of different pavement components
c	effective cohesion of soil; coefficient representing pavement condition for overlay design
C_1	annual cost of travelled way per unit length
CRF	capital recovery factor
d	clear distance between tyre contact areas; thickness of a pavement layer
D	average rut depth in a pavement, in both wheel paths
DI	distress index
E	modulus of elasticity of the soil

E_c modulus of elasticity of concrete

E_s dynamic subgrade modulus; modulus of elasticity of steel

f normal stress in a concrete pavement; frequency in cycles per second (c/s)

f_F stresses due to friction in a concrete pavement

f_F working stress for fibrous concrete pavements

f_L stress in a concrete pavement, due to load

f_P prestressing stress in a prestressed concrete pavement

f_s allowable working steel stress

f_t maximum tensile stress in a concrete pavement due to external load; allowable flexural concrete stress

f_t' tensile strength of concrete

$f_{\Delta t}$ curling stresses due to temperature differential in a concrete pavement

f_y yield strength of steel

F coefficient of subgrade friction; factor which depends upon subgrade class for overlay design

FS factor of safety

h thickness of concrete pavement

h_e thickness of a concrete pavement serving as a base

H anchor depth for a lug anchor at the end of a continuously reinforced concrete pavement

I moment of inertia of a concrete member

j a coefficient dependent on the load transfer characteristics of joints, or on the slab continuity

K modulus of subgrade reaction

L slab length; centre-to-centre spacing of anchor beams in a continuously reinforced concrete pavement

l radius of relative stiffness of a concrete pavement

$$= \sqrt[4]{\frac{E_c h^3}{12(1-v^2)K}}$$

M_R resilient modulus of soil

MR modulus of rupture of concrete

n	modular ratio, E_s/E_c; analysis period of a pavement, in years
N	number of load repetitions; number of lug anchors for a continuously reinforced concrete pavement
N_f	number of strain repetitions to failure in a pavement
P	wheel load
P_i	wheel load including impact effect
PSI	present serviceability index
p	present serviceability index; swell pressure of soil; tyre pressure
p_s	percentage of steel
q	intensity of a uniformly distributed load
R	pavement roughness
r	interest rate
S	centre-to-centre spacing of tyres; dynamic stiffness of bituminous bound materials
S_D	centre-to-centre spacing between a fore-tyre and a diagonally opposite aft-tyre
S_s	steel stress in continuously reinforced concrete pavement, due to temperature
S_T	centre-to-centre spacing between tandem axles
T	required thickness of asphalt concrete overlay on an existing flexible pavement
T_A	thickness of full-depth asphalt pavement
T_b	thickness of hot-mix sand asphalt base
T_e	effective thickness of asphalt concrete base
T_E	effective full-depth asphalt pavement thickness
T_s	thickness of pavement surface in asphalt concrete
t_f	thickness of flexible overlay on an existing concrete pavement
t_s	required thickness of surface coarse for a flexible overlay
Δt	difference in temperature between top and bottom surfaces of a pavement slab
V	volume of material
v	elastic deflection in a concrete pavement
w	weight of slab per unit area; pavement deflection; moisture content in soil

x	crack spacing
Δx	crack width; total movement at an expansion joint due to change in temperature
Z	cost of pavement per unit area
δ	end movement of a continuously reinforced concrete pavement
ε	strain in soil or pavement
ε_c	limiting compressive strain in soil
ε_t	allowable horizontal tensile strain in asphalt bound layers
σ	stress in soil or pavement
σ_r	radial horizontal stress in the soil at a distance Z below the centre of the load
$\sigma(T)$	thermal stress
σ_Z	vertical stress in the soil at a distance Z below the centre of the load
ϕ	angle of internal friction of subgrade or sub-base soil
v	Poisson's ratio of material
v_c	Poisson's ratio of concrete
γ	volume weight of material
γ_D	dry density of soil
γ_w	density of water

PART I

PAVEMENT COMPONENTS AND LOADS

CHAPTER 1

PAVEMENT TYPES, STRUCTURE AND WHEEL LOADS

1.1 INTRODUCTION

Pavements can be categorised, according to the purpose of their use, into two major groups:

1. Pavements for highways
   (a) freeways and expressways
   (b) rural highways
   (c) urban and suburban arterial streets
   (d) major streets in central business districts
2. Pavements for airports
   (a) general aviation airports for short-haul flights not exceeding 500 miles, such as those for pilot training and business, as well as for agricultural and industrial pursuits, etc.
   (b) air-carrier airports for scheduled continental and international passenger and cargo flights
   (c) military airports

A good pavement contributes to a large extent to the quality of a highway or airport. Good highway and air transportation is fundamental to our life. Highways link cities, towns and villages with the countryside, while airports link continents, nations and cities. A great deal of any modern nation's economy is geared to and depends on the speed and mobility of highway and air transportation.

A highway or airport cannot function properly unless its pavement provides an adequate, smooth, durable and serviceable support for the loads imposed by traffic at all times and in all weather conditions.

There are probably not too many engineering structures that are subjected to such extreme mechanical and climatic variations as the relatively

3

narrow highway or airfield ribbons. Both are at a disadvantage when compared with other structures in that the superstructure as well as the substructure, of both highways and airfields, are exposed to strong climatic influences which may in some situations change the properties of the subgrade supporting the pavement. For this reason, utmost care should be taken in designing pavements.

There are generally five groups of variables that should be considered in designing and constructing any pavement. These are:

1. Load variables
- Shape and value of the standard load representing the mixed traffic on the highway for which the pavement is to be designed
- Shape and value of the maximum load representing the different aircraft using the airfield for which the pavement is to be designed

2. Regional variables
- Types of existing soils, which will serve as a foundation for the pavement
- Characteristics of the environment in the region where the pavement is to be constructed
  - Temperature
  - Rainfall
  - Frost
  - Storms

3. Structural variables
- Types of subgrade
- Type of sub-base, which depends upon available material and the environmental conditions
- Type of pavement, which depends upon the loads, available materials, regional variables, and cost

4. Performance variables
- Safety of the pavement system
- Serviceability rating of the system
- Durability of the system

$$
5.\ \text{Cost variables} \begin{cases} \text{Construction cost} \begin{cases} \text{Materials} \\ \text{Equipment} \\ \text{Labour} \\ \text{Overheads} \end{cases} \\ \text{Annual maintenance cost} \end{cases}
$$

1.2 DIFFERENCE BETWEEN HIGHWAY AND AIRFIELD PAVEMENTS

Although the major factors, materials and theories to be considered in the design of highway and airfield pavements are generally the same, basic differences exist regarding the values assigned to the design factors. These are:

1. The total weight of an aircraft, excluding the private aircraft and those for general aviation purposes, is usually greater than that of a truck.

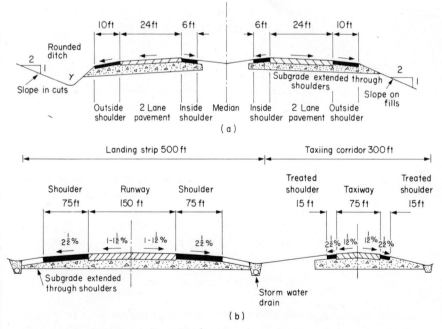

FIG. 1.1. Widths of highway and airfield pavements. (a) Typical cross-section of a highway. (b) Typical cross-section for the runway and taxiway of a national or international airport.

2. The wheel configuration for the main gears of an aircraft and the wheel spacings are generally different from those of a truck.

3. The tyre pressures and consequently the tyre imprints of an aircraft differ from those of a truck.

4. The repetition of load is much greater on highways than on airfields.

5. Lateral placement of traffic on highways and airfields is different. While vehicles generally travel within 1–4 ft of the pavement edge, aircraft movement is primarily concentrated in the airfield's centre with the outer wheels more than 15 ft from the pavement edge.

6. Airfield pavements are subjected to different types of impact during landing and take-off compared with highways.

7. Width of airfield pavements for air-carrier and military airports is generally much larger than that for a highway, as shown in Fig. 1.1.

If a highway is designed for the same wheel load and tyre pressure as an airport, the required thickness for the highway pavement will be larger than that of the airfield pavement because of the higher load repetition and because highway loads are applied closer to the pavement edge.

1.3 TYPES OF PAVEMENTS

Pavements can be classified into two general types:

1. Flexible pavements
   - Intermediate type, in which liquid bituminous materials are used as binders
   - High type, in which asphalt cements and the heaviest grades of tars are used

These pavements include all types of bituminous mats. They generally

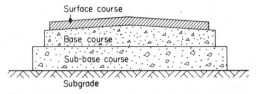

FIG. 1.2. Typical cross-section of a flexible pavement.

consist of a base course of suitable granular materials with or without bituminous binder, and a bituminous surface course. Figure 1.2 shows a typical cross-section of this type of pavement.

2. Rigid pavements
$\begin{cases} \text{Plain concrete} \\ \text{Lightly reinforced concrete} \\ \text{Continuously reinforced concrete} \\ \text{Prestressed concrete} \\ \text{Fibrous concrete} \end{cases}$

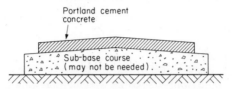

FIG. 1.3. Typical cross-section of a rigid pavement.

These pavements are constructed of Portland cement concrete as slabs resting on a continuous soil foundation. Fig. 1.3 shows a typical cross-section of this type of pavement.

1.4 PAVEMENT STRUCTURE

The structure of a pavement for main highways and airports generally consists of three significantly different layers:

1. Subgrade. The subgrade is the prepared and compacted soil forming the foundation of the pavement system. It is the uppermost material placed in an embankment or unmoved from cuts in the normal grading of the road or airfield bed.

2. Sub-base. The sub-base is defined as the layer between the subgrade and the pavement. The sub-base may consist of one or more layers of stabilised or granular material, properly compacted.

3. Pavement. The pavement is the top layer, constructed to provide adequate support for the loads imposed by traffic. It should produce a firm, stable, smooth, all-year surface in all climates, and should withstand without damage the abrasive action of traffic and other deteriorating influences.

In flexible pavements, a surface or wearing course and a base course are generally needed. In some cases where heavy loads are expected, a binder course may be used between the surface and base courses.

If the material of the subgrade, which acts as the foundation of the pavement system, is of high quality, a sub-base layer may not be necessary.

1.5 LOADS

In recent years, the volumes and weights of vehicles and aircraft using highways and airfields have been increasing steadily. Since a pavement should be designed to carry these loads safely under all climatic conditions, a knowledge of truck and aircraft-wheel arrangements, spacings and loads is essential.

1.5.1 Loads for Highways

The great variety of traffic using highways makes it necessary to replace the traffic loads by a simple loading system that can be easily used in pavement design. Although passenger cars form the highest percentage of vehicles using the highway system, their loads are small compared with those of trucks. The loads produced by the movement of heavy trucks are critical to pavements and so will be discussed in some detail in this chapter.

The live highway loads that generally affect pavement design consist of a combination of single-unit and multiple-unit vehicles. Single units include all 2-, 3- and 4-axle-unit trucks and all buses. The axles are either single or tandem, and are provided with single or dual tyres. Multiple units include 3-, 4- and 5-axle tractor trucks with semi-trailers, and all full trailer combinations. Figure 1.4 shows the different axle types for rigid chassis and articulated commercial vehicles.

The over-all length of single-powered commercial vehicles, in North America, varies between 35 ft (10·7 m) and 40 ft (12·2 m) and the length of combinations, consisting of a tractor, semi-trailer and/or full trailer varies between 65 ft (19·8 m) and 70 ft (21·4 m). Maximum width is about $8\frac{1}{2}$ ft (2·6 m) and maximum height is between $13\frac{1}{2}$ ft (4·1 m) and $14\frac{1}{2}$ ft (4·4 m). Table 1.1 shows the relationship between stowage capacity and lengths

TABLE 1.1 STOWAGE CAPACITY RELATED TO LENGTHS OF VAN–TRAILER COMBINATIONS

Combination type	Length (ft) Over-all	Total cargo body	Total cargo stowage capacity (ft^3)
Tractor and semi-trailer	45	35	1880
Tractor and semi-trailer	50	40	2150
Truck and full trailer	60	50	2660
Tractor semi-trailer and full trailer	60	50	2660
Tractor semi-trailer and full trailer	65	54	2870

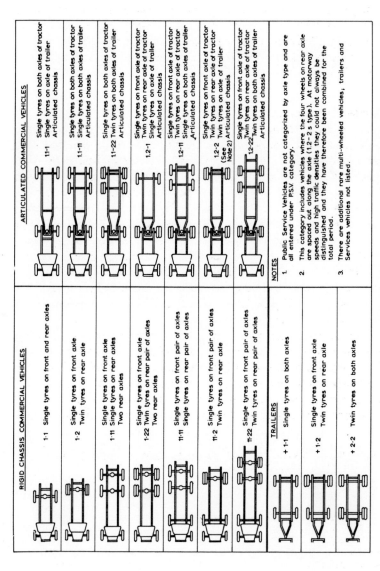

FIG. 1.4. Classification of axle types for rigid chassis and articulated commercial vehicles. (From [2].)

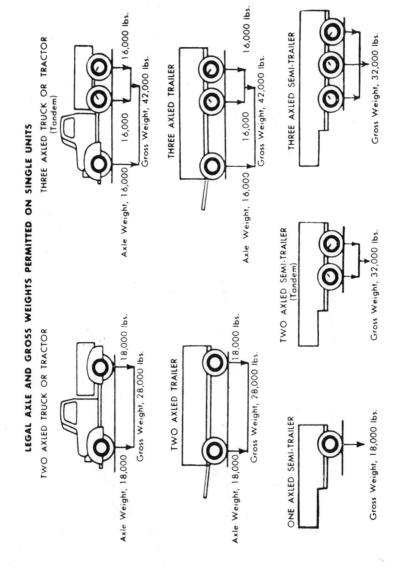

FIG. 1.5. Permissible weights on motor vehicles in Ontario.
(From Ontario Department of Transport Form CO-2 (1969).)

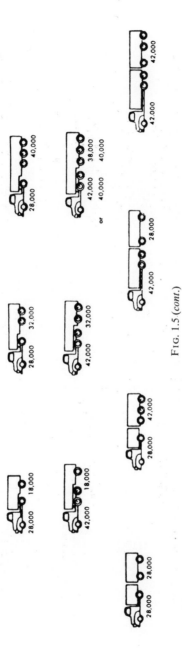

LEGAL GROSS WEIGHTS PERMITTED ON FOLLOWING COMBINATIONS

FIG. 1.5 (*cont.*)

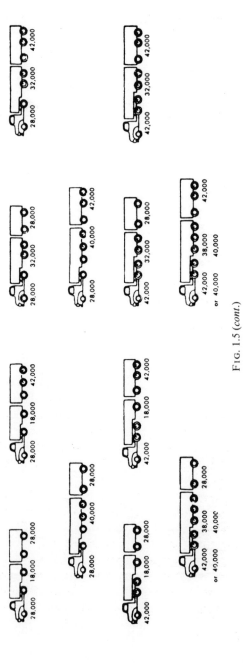

FIG. 1.5 (*cont.*)

of van–trailer combinations, while Table 1.2 relates payload weight to storage capacity [3].

There are different combinations for commercial vehicles, as well as different allowable axle and gross weights. An example of these standards, for the Province of Ontario in Canada, is shown in Fig. 1.5.

TABLE 1.2 PAYLOAD WEIGHTS RELATED TO STOWAGE CAPACITY OF VAN–TRAILER COMBINATIONS[a]

Commodity density (pcf)	Payload weight (lb)				
	Ta-S 35 ft	Ta-S 40 ft	Tr-F 50 ft	Ta-S-F 50 ft	Ta-S-F 54 ft
10	18 800	21 500	26 600	26 600	28 700
15	28 200	32 250	39 900	39 900	43 050
20	37 600	43 000	53 200	53 200	57 400
25	47 000	53 750	66 500	66 500	71 750
30	56 400	64 500	79 800	79 800	86 100
35	65 800	75 250	93 100	93 100	100 450
40	75 200	86 000	106 400	106 400	114 800
45	84 600	96 750	119 700	119 700	129 150
50	94 000	107 500	133 000	133 000	143 500

[a] Ta = tractor, Tr = truck, S = semi-trailer, and F = full trailer.

It should be noted that truck-axle spacing and weight limits are variable quantities in different countries. Even more than that, these values vary from one state to another in North America. The maximum allowable load for single-unit trucks, for example, is between 28 kips (12·6 t) and 60 kips (27·3 t), according to the number of axles and wheels, and according to the specifications for pavement design in each state. Similarly, for a tractor with a semi-trailer and/or a full trailer, the maximum allowable load varies between 60 kips (27·3 t) and 135 kips (61 t). The maximum single-axle load ranges between 16 000 lb (7·1 t) and 22 400 lb (10 t). The maximum tandem-axle load ranges between 28 000 lb (12·5 t) and 40 000 lb (18 t). Tandem spacings range between 40 in (100 cm) and 53 in (135 cm). Spacing centreline to centreline between dual wheels is about 13·5 in (34 cm) and tyre pressures are 60–90 psi (4·2–6·3 kgf/cm^2).

In Europe, higher axle loads are generally used. The maximum single-axle load is 29 000 lb (13 ton) and the maximum tandem-axle load is 45 000 lb (20 ton). Values of the axle or wheel loads that are specified for highway pavement design in the different European countries are given in Appendix A [11].

Roads in Europe are categorised into three groups from the point of view of traffic loads. These are:
1. Roads with light traffic: opposing vehicles with a width sum more than 13·30 ft (4·0 m) are met rarely.
2. Roads with heavy traffic: opposing vehicles with a width sum more than 13·30 ft (4·0 m) are met frequently.
3. Roads with heavy and fast traffic: same as (2) except that the speeds exceed 65 miles/hour (100 km/h).

In North America, roads are categorised according to location, as follows:
1. Local city streets: percentage of heavy trucks of average gross weight 15 000–25 000 lb (6800–11 300 kg) is less than or equal to 5%.
2. Urban highways: percentage of heavy trucks of average gross weight 20 000–45 000 lb (9000–20 500 kg) is 5–15%.
3. Inter-urban highways: percentage of heavy trucks of average gross weight 30 000–45 000 lb (13 500–20 500 kg) is 5–25%.

In many design procedures used in North America, a standard load is chosen for the design. The standard load commonly used is the 18 000 lb (8100 kg) single-axle load with dual wheels on each side. Other standard loads are sometimes used, for instance, the 22 400 lb (10 100 kg) single-axle load, 32 000 lb (14 500 kg) tandem-axle load or 40 000 lb (18 100 kg) tandem-axle load.

Extensive studies on both flexible and rigid pavements have provided factors for converting various axle loads to an equivalent number of applications of the 18 000 lb single-axle load. These equivalency factors, for any given single- or tandem-axle load, express the number of 18 000 lb single-axle load applications that will have the same effect on pavement performance as that produced by one application of the given axle load.

The load equivalency factors for flexible pavements are different from those for rigid pavements.

1.5.2 Load Equivalency Factors for Highway Flexible Pavements

The chart presented in Fig. 1.6 gives load equivalency factors for loads equal to or greater than 10 000 lb. These factors represent the equivalent number of applications of the 18 000 lb standard single-axle load.

The findings of the road tests conducted by the American Association of State Highway Officials in Northern Illinois, completed in the early 1960s and known as the A A S H O road tests, have established equivalency between tandem-axle loads and single-axle loads, as have other road tests. Insofar as general performance of pavement structure and subgrade soil is concerned, a 32 000 lb tandem-axle load is basically equivalent to an 18 000 lb single-axle load. Furthermore, the relationship holds for other

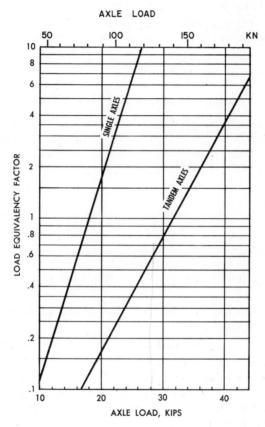

FIG. 1.6. Load equivalency factors for loads equal to or greater than 10 000 lb. (From *Asphalt Institute Manual*, M S-1, August 1970.)

loads, so that a tandem-axle load when multiplied by 0·57 gives the equivalent single-axle load.

1.5.3 Load Equivalency Factors for Highway Rigid Pavements
For preliminary calculations, the following values can be used in determining the equivalent 18 000 lb single-axle load applications for different vehicles [1]:

0·0004   for passenger cars

0·123   for single-unit vehicles, which include all 2-axle and 3-axle single-unit trucks and all buses

1·134   for multiple units, which include 3-, 4- and 5-axle tractor trucks with semi-trailer, and all full trailer combinations

For more accurate calculations, Table 1.3 can be used [1]. In this table, the equivalency factors are given for single axles in load increments of 2000 lb and for tandem axles in load increments of 4000 lb. These factors are tabulated for present serviceability levels of 2·0 and 2·5, which are commonly used in the design.

TABLE 1.3 18 000 LB SINGLE-AXLE EQUIVALENCY FACTORS

Single-axle load (lb)	18 000 lb single-axle equivalency factor		Tandem-axle load (lb)	18 000 lb single-axle equivalency factor	
	$p^a = 2 \cdot 0$	$p = 2 \cdot 5$		$p = 2 \cdot 0$	$p = 2 \cdot 5$
2 000	0·000 2	0·000 2	4 000	0·000 5	—
4 000	0·002	0·002	8 000	0·005	—
6 000	0·010	0·010	12 000	0·030	0·030
8 000	0·030	0·030	16 000	0·082	0·085
10 000	0·082	0·085	20 000	0·207	0·212
12 000	0·178	0·183	24 000	0·443	0·452
14 000	0·343	0·352	28 000	0·850	0·850
16 000	0·603	0·610	32 000	1·490	1·473
18 000	1·000	1·000	36 000	2·467	2·388
20 000	1·572	1·552	40 000	3·858	3·673
22 000	2·363	2·302	44 000	5·797	5·430
24 000	3·437	3·300	48 000	8·412	7·760

[a] Present Serviceability Index, p, represents the degree to which the public considers itself to be served by the pavement system.

Equations developed from the AASHO road tests on rigid pavements are used in obtaining the values given in Table 1.3. These equations have been derived from results on test pavements that were subjected to over one million repetitions of single-axle loads ranging from 3000 to 30 000 lb and tandem-axle loads ranging from 24 000 to 58 000 lb. Analysis of the results produced rigid pavement performance equations which made it possible to convert the mixed traffic into equivalent standard single-axle load repetitions.

If the average daily volume during the life period of the pavement is determined per lane for each type of vehicle, the total number of equivalent

18 000 lb single-axle load applications can be obtained using the equivalency factors for passenger cars, single-unit trucks, and multiple-unit trucks, or using Table 1.3 for accurate calculations.

Example
The rigid pavement of a highway is designed for a life period of 25 years. Average daily volume per lane during this period is 5000 vehicles distributed as follows:

Passenger cars	$=96\%$	$=4800$ vehicles per day
Single-unit trucks	$= 3\cdot75\% =$	187 vehicles per day
Multiple-unit semi-trailers and trailers $=$	$0\cdot25\% =$	13 vehicles per day

Determine the equivalent number of 18 000 lb single-axle load applications using the approximate equivalency factors for rigid pavements.

TABLE 1.4 EQUIVALENT 18 000 LB SINGLE-AXLE LOAD
APPLICATIONS FOR THE EXAMPLE

(1)	*(2)*	*(3)*	*(4)*	*(5)*
Vehicle type	*Average operations per day per lane*	*25 years operations per lane = $(2)^a \times 365 \times 25$*	*Equivalency factors*	*Number of equivalent load applications = $(3) \times (4)$*
Passenger cars	4 800	$43\cdot5 \times 10^6$	$0\cdot000\,4$	$1\cdot74 \times 10^4$
Single units	187	$1\cdot71 \times 10^6$	$0\cdot123$	$21\cdot10 \times 10^4$
Multiple units	13	$0\cdot118 \times 10^6$	$1\cdot134$	$13\cdot40 \times 10^4$
				$36\cdot24 \times 10^4$

^{a} (2) means the value in column 2.

Solution. Table 1.4 gives the required value, which is 362 400 equivalent repetitions of the 18 000 lb single-axle load.

It should be noted, in this example, that the effect of $0\cdot25\%$ multiple-unit combinations is much higher than the effect of 96% passenger cars.

1.5.4 Loads for Airports
The number of aircraft using an airfield is generally much less than the number of vehicles using a highway. Under ideal conditions a runway can

accommodate up to 50 landings and take-offs per hour, whereas a highway lane can serve up to 2000 vehicles per hour.

The distribution of traffic across airfield pavements is very important. Large aircraft are provided with modern equipment which permits the pilots to operate their planes within narrow paths during landing and taxiing operations. This results in channelised movement of aircraft on the airfields, which in turn produces a high number of load repetitions on the central lanes. Some studies have indicated that over 75% of heavy jet aircraft movements take place within the central 35 ft strip of the runway.

On the other hand, not all aircraft using an airport are of the same size or configuration. A large percentage of these aircraft are relatively small, and are not critical to the airfield pavements.

In the design of airfield pavements, the gross weight of the heaviest aircraft that will use the airport is used. The 'heaviest aircraft' does not necessarily mean the largest aircraft. Tyre pressure, number of main undercarriage assemblies, number of wheels per assembly, spacing between the wheels, distribution of the weight between the nose wheel and the main gears, and the aircraft gross weight, are the major factors in determining the effect of aircraft loading on the pavement system. For this reason, it may be important to show some basic types of aircraft wheel configurations as well as some aircraft data illustrating the load information generally needed in the design of airfield pavements.

More detailed aircraft assembly configurations are given in Appendix B.

Since most modern aircraft have nose wheels, the wheel configuration of aircraft with tail wheels will not be shown. In general, 5–10% of the gross aircraft weight is carried by the nose or wing wheels and the remaining weight is distributed on the main gears.

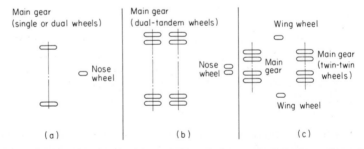

FIG. 1.7. Basic types of aircraft wheel configurations. (a) Tricycle single-tyres landing gear. (b) Twin-tandem landing gear. (c) Twin-twin bicycle gear.

There are three basic aircraft wheel arrangements, as shown in Fig. 1.7. These are: (1) single and dual wheels, (2) single and tandem axles and (3) tricycle and bicycle landing gears.

General aviation airports are designed to receive small aircraft of the single-engine and twin-engine types serving on local routes in the short-haul category, which normally does not exceed 500 miles. They accommodate inter-city flights for business and industry, instructional flights and flights for aerial photography, fire patrol and crop dusting. In general, these airports operate under visual flight rules (VFR) according to the Federal Aviation Administration (FAA) standards and regulations. Maximum weight of aircraft operating under these rules is 12 500 lb, with a seating capacity up to 12.

Figure 1.8 shows the wheel configuration and pertinent data for two types of aircraft belonging to the general avaiation fleet. The Evcoupe Model is one of the smallest of this fleet. Aircraft of the air-carrier fleet

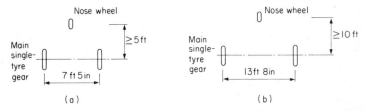

FIG. 1.8. Aircraft of the general aviation fleet. (a) Single-engine, tricycle-landing-gear aircraft. Evcoupe, Model 415 D, E, G Type certificate holder: Air Products. Single engine, two seats. Maximum gross weight: 1400 lb. (b) Twin-engine, tricycle-landing-gear aircraft. Dove, Model 104, Series 1 Type certificate holder: Havilland. Twin engine, ten seats. Maximum gross weight: 8500 lb.

with short- to medium-range capabilities up to 2000 miles include the DC-3, Convair 600, DC-9-10, DC-6B, and the B-737-200. These have a normal take-off weight of 25 000, 46 200, 77 700, 103 000 and 107 000 lb, respectively.

Aircraft with long-range capabilities operating out of international airports include the Boeing 727, 707, 747 and the Douglas DC-8, which have maximum take-off gross weights of 170 000, 316 000, 778 000 and 318 000 lb, respectively. The largest of the air-carrier fleet operating now is the Boeing 747, which has a passenger capacity of 490 and a maximum still-air range of 8000 miles.

Figure 1.9 shows the wheel configuration and pertinent data for three types of aircraft used in the air-carrier fleet. Military aircraft are generally more critical to airfield pavements than civilian aircraft. This is because

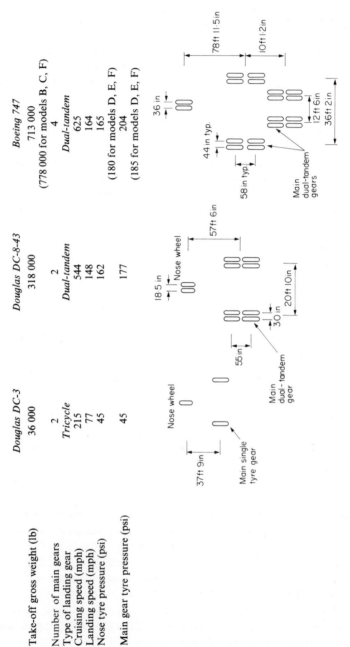

Fig. 1.9. Wheel configuration for aircraft of the air-carrier fleet.

of the large weights, smaller number of wheels and higher tyre pressure. The B-52 Bomber, which weighs over 500 000 lb, has two twin–twin main gears, four wheels each, and the tyre pressure is 260 psi. Till now, no standard loads have been established for the design of airfield pavements, since there are no equivalency factors available as in the case of highways. Accordingly, the pavement design for each airport depends upon the types of aircraft operating from that airport. The gross weight of the heaviest aircraft using the airfields, generally called the 'Design Aircraft', is employed in the design.

For very busy airports, the pavements are designed for an unlimited number of operations of the 'Design Aircraft'. Otherwise a number of load repetitions ranging between 30 000 and 60 000, according to the volume of traffic in the airport, can be used.

Example

The runway of a new airport is to be designed for a life period of 20 years. The combination of aircraft using the runway can be categorised as 30% heavy and 70% medium and light. Due to the channelised movement of the heavy aircraft, about 64% of this movement takes place within the central 36 ft strip of the runway. The runway consists of 10 lanes each 15 ft wide. The aircraft average peak-hour movement is 10 and forms 0·05% of the total annual movement.

Required: The number of repetitions of heavy aircraft movement per lane during the 20 years.

Solution. Sixty-four per cent of the movement of heavy aircraft takes place within the central 36 ft, i.e. within the four central lanes. This counts for 32% per lane for each of the two main undercarriage assemblies.

Total number of aircraft movements during the 20 years =

$$\frac{10}{0·0005} \times 20 = 400\,000$$

Number of heavy aircraft movements per lane during the 20 years =

$$\frac{32}{100} \times \frac{30}{100} \times 400\,000 = 38\,400$$

REFERENCES AND BIBLIOGRAPHY

1. Chastain, W. E., Sr, Beanblossom, J. A. and Chastain, W. E., Jr, 'AASHO Road Test Equations, Applied to the Design of Portland Cement Concrete Pavements in Illinois', *Highway Research Record*, No. 90, 1965, 26–41.

2. Currer, E. W. H. and Thompson, P. D., 'The Classification of Traffic for Pavement Design Purposes', *Proc. 3rd International Conf. on the Structural Design of Asphalt Pavements, London, September 1972*, The University of Michigan, **1**, 72–79.
3. 'Economic Effects of Changes in Legal Vehicle Weights and Dimensions on Highways', National Cooperative Highway Research Program, Final Report, Project No. 19-3, June 1972.
4. Federal Aviation Agency, 'Airport Paving', US Government Printing Office, November 1962.
5. Hennes, R. G. and Ekse, M., *Fundamentals of Transportation Engineering*, Second Edition, McGraw-Hill, 1969.
6. *Highway Research Board Special Report 87*, 'Highway Capacity Manual', Highway Research Board, 2101 Constitution Avenue, Washington, D.C., 1965.
7. *Highway Research Board Special Report 95*, 'Rigid Pavement Design', Highway Research Board, Washington, D.C., 1968.
8. Yoder, E. J., *Principles of Pavement Design*, Wiley, 1959.
9. Shook, J. F. and Lepp, T. Y., 'Method for Calculating Equivalent 18-kip Load Applications', *Highway Research Record*, No. 362, 1971.
10. Skok, E. L. and Root, R. E., 'Use of Traffic Data for Calculating Equivalent 18,000 lb Single-Axle Loads', *Highway Research Record*, No. 407, 1972.
11. *Synoptic Table of European Concrete Road Standards and Practices, PIARC Technical Committee for Concrete Roads, Berne, 1973*, compiled by D. Raymond Sharp, The Cement and Concrete Association, 52 Grosvenor Gardens, London.
12. The Asphalt Institute Manual Series No. 1, 'Thickness Design of Full-Depth Asphalt Pavement Structures for Highways and Streets', Revised Eighth Edition, August 1970.

CHAPTER 2

EQUIVALENT SINGLE-WHEEL LOAD CONCEPT AND APPLICATION

2.1 INTRODUCTION

The trend towards heavier trucks and aircraft, equipped with multiple-wheel assemblies, has emphasised the need in pavement design for a method whereby multiple-wheel loads can be related to a common standard, irrespective of arrangement or configuration of the wheels. In many cases, the conversion of multiple-wheel loads to equivalent single-wheel loads offers the designers of flexible and rigid pavements a convenient means for both the design and evaluation of highway and airfield pavements.

The interrelation of loading effects produced by single- and multiple-wheel assemblies has been determined through theoretical analysis, laboratory investigations, accelerated traffic tests on prototype pavements and investigations of pavement performance. Although further investigation along these lines is needed, the conversion methods presented herein will provide satisfactory solutions to the problem of converting multiple-wheel loads to equivalent single-wheel loads. The conversion for flexible pavements is naturally different from that for rigid pavements.

2.2 EQUIVALENT LOADS FOR DESIGN OF FLEXIBLE PAVEMENTS

2.2.1 Relationship Between Wheel Arrangements and Loading Effects on Pavements

A relationship has been established between wheel spacing and the depth of action and interaction of the forces induced in the pavement by the load. The depths of load reaction that control the conversion to equivalent single wheel loads are:

23

1. The depth at which each wheel acts independently as a single wheel, bearing its proportionate weight of the load. This depth is equal to $d/2$, where d is the clear distance between the contact areas.

2. The depth at which the loads on all adjacent wheels interact so that the combined effect produced in the subgrade is the same as that obtained by a single wheel having the same tyre pressure. For dual wheels this depth is approximately equal to twice the centre-to-centre spacing of the tyres.

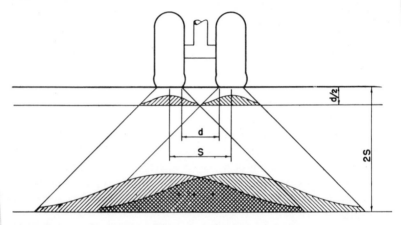

FIG. 2.1. Distribution of dual-wheel loads through flexible pavements.

Figure 2.1 represents the relationship of wheel spacing and the depth of action and interaction of forces for this case.

In the case of dual-tandem wheel assemblies, the depth at which the interaction of forces from the different wheels becomes approximately

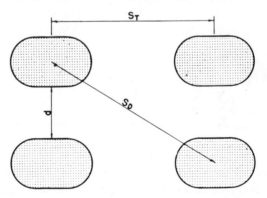

FIG. 2.2. Tyre imprint of a dual-tandem assembly.

equivalent to that obtained by a single wheel, is 2 S_D. S_D is the centre-to-centre spacing between a fore-tyre and a diagonally opposite aft-tyre, as shown in Fig. 2.2.

2.2.2 Equivalent Single-Wheel Loads (ESWL)

The thickness of a flexible pavement varies according to the qualities of the subgrade upon which it rests. A subgrade having a high supporting strength will require a thin pavement and the interaction of the loads on the subgrade caused by several wheels will be slight. On the other hand, a poor subgrade will require a thick flexible pavement section, and a considerable interaction of the several loads from the multi-wheel assemblies will take place.

As mentioned before, the critical controlling depths are $d/2$ and $2S$ for dual wheels, and $d/2$ and $2S_D$ for dual-tandem wheels.

The case of dual wheels can be represented by a truck having an 18 000 lb axle load, distributed equally between a pair of dual wheels on each side (9000 lb on each side). If the clear distance between contact areas, d, is 6 in, and the centre-to-centre spacing of the tyres S is 14 in, then $d/2$ is 3 in and $2S$ is 28 in. At pavement depths of 3 in or less, the dual-wheel loading of 9000 lb produces the same effect as a single-wheel load of 4500 lb. At depths of 28 in or more, the overlapping of stresses is such that the effects of the dual-wheel load are similar to those produced by a 9000 lb load on a single wheel.

The dual-tandem wheel loading can be represented by one main undercarriage assembly of a large aircraft having a gear loading of, say, 160 000 lb. If the centre-to-centre spacing of a pair of wheels, S, is 30 in, the clear distance between the tyre imprints is 18 in and the gear-axle spacing S_T is 60 in, then S_D is equal to 67 in. Under these conditions, the total dual-tandem wheel load of 160 000 lb produces the same effect at a depth of 9 in ($d/2$) as does a single-wheel loading of 40 000 lb; that is, one-quarter of the total gear load. At depths of 134 in ($2S_D$) or greater, the effects are similar to those which would be produced by a 160 000 lb load on a single wheel.

If these general principles are followed, it is possible to determine by simple graphic means the equivalent single-wheel load for any dual or dual-tandem wheel assembly of a truck or an aircraft, as will be shown later. Terzaghi [11], when referring to Boussinesq's work, has shown that the maximum vertical stress at any point below an earth mass due to a circular loaded area is dependent only on the depth Z, radius a of contact area and tyre pressure p (Fig. 2.3).

$$\sigma_z = p\left[1 - \frac{Z^3}{(a^2+Z^2)^{\frac{3}{2}}}\right] = p\left[1 - \frac{1}{\left(\frac{a^2}{Z^2}+1\right)^{\frac{3}{2}}}\right] \tag{2.1}$$

and

$$\sigma_r = \frac{p}{2}\left[1 + 2v - \frac{2(1+v)Z}{(a^2+Z^2)^{\frac{1}{2}}} + \frac{Z^3}{(a^2+Z^2)^{\frac{3}{2}}}\right] \tag{2.2}$$

where σ_z and σ_r are, respectively, the vertical and radial horizontal stresses at a distance Z below the centre of the loaded area. v is Poisson's ratio of the soil and is generally assumed to be equal to 0·5.

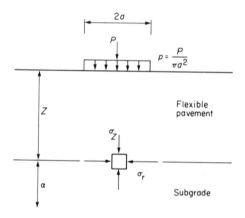

FIG. 2.3. Stresses acting on an element.

For geometrically similar points below the pavement surface and for a constant tyre pressure, vertical stresses are equal regardless of the value of total load. For a constant contact pressure and for a given allowable stress, the following equation should then apply:

$$\frac{Z}{a} = \text{constant} = c \tag{2.3}$$

where $a = \sqrt{P/p\pi}$. Substituting,

$$Z = c\sqrt{\frac{P}{p\pi}} = \frac{c}{\sqrt{p\pi}}\sqrt{P} \tag{2.4}$$

For a constant tyre pressure p,

$$Z = c'\sqrt{P} \tag{2.5}$$

This equation indicates that the required depth, Z, of a flexible pavement is proportional to the square root of the total load. Z is the total thickness of the flexible pavement layers above the subgrade. c' depends upon the allowable stress in the subgrade soil and the tyre pressure.

For v equal to 0.5, eqn. (2.2) can be written in the form:

$$\sigma_r = \frac{p}{2}\left[2 - \frac{3Z}{(a^2+Z^2)^{\frac{1}{2}}} + \frac{Z^3}{(a^2+Z^2)^{\frac{3}{2}}}\right] \tag{2.6}$$

Computations for determining equivalent single-wheel loads can be based on either equal-deflection or equal-stress criteria. The single-wheel load resulting in a deflection or a stress of the same amount as that produced by dual or dual-tandem assemblies is said to be the equivalent load. The equal-deflection criterion will be considered in the computation of equivalent single-wheel loads for flexible pavements.

The deflection in a subgrade layer of infinite depth starting at a depth Z below the surface can be obtained by integrating the equation of the strain over the depth of the layer:

$$\text{strain} \quad \delta = \frac{1}{E}\left[\sigma_z - 2v\sigma_r\right] \tag{2.7}$$

where E = modulus of deformation of the soil.

Substituting eqns. (2.1) and (2.6) in eqn. (2.7), with $v = 0.5$, gives

$$\delta = \frac{1\cdot5p}{E}\left[\frac{Z}{(a^2+Z^2)^{\frac{1}{2}}} - \frac{Z^3}{(a^2+Z^2)^{\frac{3}{2}}}\right] = \frac{1\cdot5pa^2}{E}\cdot\frac{Z}{(a^2+Z^2)^{\frac{3}{2}}} \tag{2.8}$$

Integrating between Z and $Z = \infty$, the total deflection is:

$$\Delta = \frac{1\cdot5pa}{E}\cdot\frac{a}{(a^2+Z^2)^{\frac{1}{2}}} \tag{2.9}$$

or

$$\Delta = \frac{pa}{E}K \tag{2.10}$$

where

$$K = \frac{1\cdot5}{\left[1+\left(\dfrac{Z}{a}\right)^2\right]^{\frac{1}{2}}} \tag{2.11}$$

Equation (2.10) is the Boussinesq settlement equation for deflections under the centre of a flexible plate.

The factor K is dimensionless and depends only upon the ratio of depth to radius.

For constant tyre pressure, the deflections under a single tyre and under a set of duals are as follows:

$$\text{single} \quad \Delta_1 = \frac{pa_1}{E} K_1 \tag{2.12}$$

$$\text{dual} \quad \Delta_2 = \frac{pa_2}{E} (K'_1 + K'_2) \tag{2.13}$$

where a_1 = radius of contact area of single tyre, corresponding to the tyre pressure, p;

a_2 = radius of contact area for each tyre of a set of duals;

K_1 = settlement factor for single tyre;

K'_1 = settlement factor contributed by one tyre of duals;

K'_2 = settlement factor contributed by remaining tyre of duals;

E = modulus of deformation of the soil.

For equal deflections:

$$\frac{pa_1}{E} K_1 = \frac{pa_2}{E} (K'_1 + K'_2)$$

or

$$\sqrt{P_1} K_1 = \sqrt{P_2} (K'_1 + K'_2) \tag{2.14}$$

where P_1 = gross load on single tyre;

P_2 = gross load on each of the dual tyres.

K_1 is the settlement factor under the centre of the single tyre and depends upon the depth-over-radius ratio only. The factors K'_1 and K'_2 depends on the depth-over-radius ratio as well as the offset distances. Their values can be obtained from the chart shown in Fig. 2.4.

The values of K'_1 and K'_2 can be calculated at the centre of the dual tyres as well as at the centre of one of the two tyres. The case which gives the larger value for $K'_1 + K'_2$ is the one that should be used. Once $K'_1 + K'_2$ and K_1 are determined and P_2 is known, the value of P_1 can be calculated for an assumed value of a_1. If the calculated value of a_1 differs greatly from the assumed one, calculations should be repeated until both radii are compatible.

Another procedure for determining equivalent single-wheel loads is the graphical method presented by Boyd and Foster.

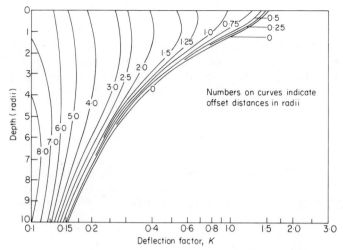

FIG. 2.4. Vertical deflection (Poisson's ratio = 0·5). Deflection = $(p \times a/E)K$.
(From C. R. Foster and R. G. Ahlvin, *Proceedings*, Highway Research Board, 1954.)

Assuming a straight-line relationship between the depths $d/2$ and $2S$, a convenient relationship can be derived to determine the equivalent loads.

Equation (2.5) can be written in the following form:

$$\log Z = \log c' + \tfrac{1}{2} \log P \tag{2.15}$$

If the depth of pavement is plotted as the abscissa and P as the ordinate

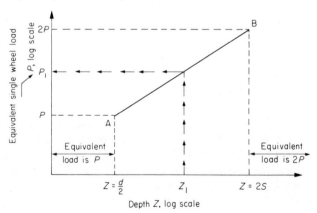

FIG. 2.5. Determination of equivalent single-wheel load for any dual-wheel system (flexible pavement).

(Fig. 2.5), then the point A (P, $d/2$) represents the case where each wheel is acting separately. Similarly, point B ($2P$, $2S$) represents the case where both wheel loads are acting together. The straight line connecting A and B is the locus of points where any single-wheel load is equivalent to a certain dual-wheel loading. Once a reasonable thickness Z is assumed for the pavement, the equivalent single-wheel load corresponding to the dual-wheel configuration used in plotting points A and B can be found. Enter the chart at the assumed depth Z, then proceed vertically to line A B and horizontally to the P axis, where the required value will be read.

A similar procedure can be used for converting dual-tandem loads to single-wheel loads. The only difference is that the diagonal distance S_D should be used instead of S, as shown in Fig. 2.6.

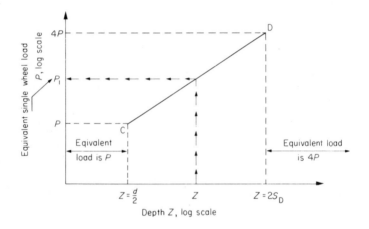

FIG. 2.6. Determination of equivalent single-wheel load for any dual-tandem wheel system (flexible pavement).

The procedures outlined above are very useful when design techniques are based on single-wheel loads, while the critical traffic load is caused by multiple-wheel assemblies. If the design thickness calculated is different from the thickness assumed in determining the equivalent single-wheel load, the process should be repeated until both the thicknesses become compatible.

2.3 EQUIVALENT LOADS FOR DESIGN OF RIGID PAVEMENTS

2.3.1 Equivalent Single-Wheel Loads for Central Loading Conditions

The concept of equivalent single-wheel load, for the central loading case, is based on a theory developed by Westergaard. It consists of determining the magnitude and contact area of a single-wheel load which, when placed centrally on a concrete slab, will produce the same maximum stress as that produced by two or four closely spaced loads. These loads may represent the axle load of a truck or the dual-tandem load of an aircraft undercarriage.

Several variables are involved in the calculation of the equivalent single-wheel load. Some of these, such as the load on the wheel assembly, spacing of wheels and tyre pressure, are related to the loading system. The other

TABLE 2.1 / VALUES (PSI/IN) FOR $E_c = 5 \times 10^6$ PSI

Thickness of pavement 'h' (in)	K = 50	K = 100	K = 150	K = 200	K = 250	K = 300	K = 350	K = 400	K = 500	K = 1 000
6	36·84	30·98	27·99	26·04	24·63	23·54	22·64	21·91	20·71	17·42
6½	39·11	32·89	29·72	27·66	26·16	25·00	24·04	23·26	21·99	18·50
7	41·35	34·78	31·42	29·23	27·65	26·42	25·42	24·58	23·25	19·55
7½	43·55	36·62	33·08	30·79	29·12	27·83	26·77	25·89	24·49	20·59
8	45·71	38·43	34·73	32·32	30·57	29·20	28·10	27·17	25·70	21·61
8½	47·83	40·22	36·34	33·82	31·98	30·57	29·40	28·44	26·90	22·62
9	49·93	41·99	37·94	35·30	33·39	31·90	30·69	29·69	28·07	23·61
9½	51·99	43·72	39·50	36·76	34·78	33·22	31·96	30·92	29·24	24·59
10	54·03	45·43	41·06	38·21	36·13	34·52	33·22	32·13	30·39	25·55
10½	56·05	47·13	42·59	39·63	37·48	35·81	34·46	33·33	31·52	26·50
11	58·04	48·81	44·10	41·04	38·82	37·08	35·68	34·51	32·64	27·44
11½	60·00	50·46	45·59	42·43	40·13	38·34	36·89	35·67	33·74	28·36
12	61·95	52·10	47·07	43·81	41·43	39·59	38·09	36·84	34·84	29·29
12½	63·87	53·71	48·53	45·17	42·72	40·81	39·27	37·98	35·92	30·19
13	65·79	55·32	49·98	46·51	44·00	42·03	40·44	39·11	37·00	31·12
13½	67·67	56·91	51·42	47·86	45·25	43·23	41·61	40·24	38·05	31·99
14	69·54	58·48	52·85	49·18	46·50	44·43	42·76	41·35	39·11	32·88
14½	71·40	60·04	54·25	50·49	47·75	45·62	43·89	42·45	40·15	33·75
15	73·24	61·59	55·65	51·79	48·98	46·80	45·02	43·55	41·18	34·62
15½	75·06	63·12	57·03	53·08	50·19	47·96	46·14	44·83	42·21	35·49
16	76·87	64·64	58·41	54·36	51·41	49·11	47·26	45·71	43·22	36·34

For values of E_c $\begin{Bmatrix} 4\ 000\ 000 \\ 3\ 000\ 000 \\ 2\ 000\ 000 \end{Bmatrix}$ multiply the value of l given in table by $\begin{Bmatrix} 0·95 \\ 0·88 \\ 0·80 \end{Bmatrix}$

variables, such as the thickness of the pavement slab, the modulus of elasticity and Poisson's ratio for the concrete used, and the modulus of subgrade reaction, are related to the pavement system. Westergaard chose the radius of relative stiffness, l, as a basis for setting out the results of the calculations, because it combines all the pavement variables in one expression:

$$l = \sqrt[4]{\frac{E_c h^3}{12(1 - v^2)K}} \qquad (2.16)$$

where h = thickness of the concrete pavement;
$\quad E_c$ = modulus of elasticity of the concrete;
$\quad v$ = Poisson's ratio of the concrete; and
$\quad K$ = modulus of subgrade reaction.

Table 2.1 gives values of l for various values of h and K.

In the case of dual-wheel loads, once the value of l is calculated and the spacing between the centrelines of the two wheels is known together with the tyre pressure, the value of the equivalent single-wheel load can be obtained from the curves shown in Fig. 2.7. In the construction of these conversion curves, it was found convenient to adopt the two factors S/l and A/l^2 in order to obtain the ratio of the total gear load to the equivalent single-wheel load. S is the distance between the centrelines of the dual

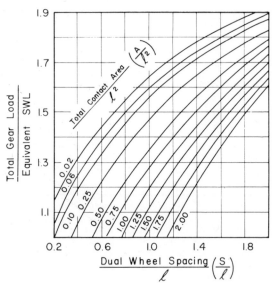

FIG. 2.7. Chart for determination of equivalent single-wheel load–rigid pavements–dual-wheel gear. (*Courtesy: Federal Aviation Agency.*)

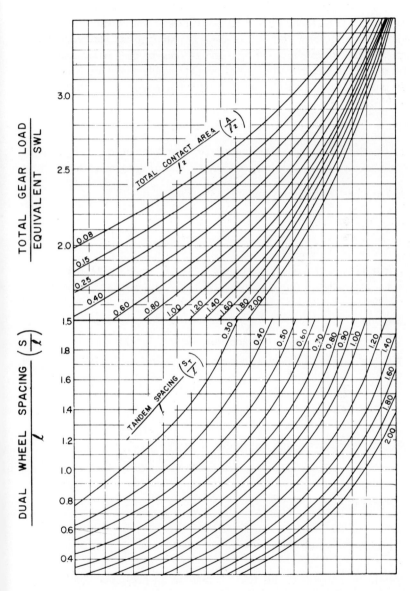

FIG. 2.8. Chart for determination of equivalent single-wheel load–rigid pavements–dual-tandem gear. *(Courtesy: Federal Aviation Agency.)*

wheels and A is the total contact area of the tyres of the assembly, which can be obtained by dividing the total gear load by the tyre pressure p.

In the case of dual-tandem wheel assemblies, it is necessary to introduce an additional factor S_T/l, in which S_T is the spacing of axles as shown in Fig. 2.2. The curves shown in Fig. 2.8 can then be used to determine the ratio of the total gear load to the equivalent single-wheel load.

In plotting the curves of Figs. 2.7 and 2.8, it was assumed that the tyre imprints are circular.

Examples

1. *Dual-wheel assembly.* It is required to determine the equivalent single-wheel load for a truck-axle load of 18 000 lb, if the following data are given:

Thickness of rigid pavement	$h = 8$ in
Spacing between centrelines of dual wheels	$S = 14$ in
Modulus of subgrade reaction	$K = 200$ pci
Modulus of elasticity of concrete	$E_c = 4\,000\,000$ psi
Tyre pressure of the wheels	$p = 60$ psi

Solution.

Load per dual wheels on each side $= 9000$ lb

From Table 2.1, the value of l $\qquad l = 32\cdot32 \times 0\cdot95 = 30\cdot7$ in

Total contact area $\qquad A = \dfrac{P}{p} = \dfrac{9000}{60} = 150$ in^2

$$\frac{S}{l} = \frac{14}{30\cdot7} = 0\cdot456$$

$$\frac{A}{l^2} = \frac{150}{30\cdot7^2} = 0\cdot16$$

From Fig. 2.7 $\dfrac{\text{total gear load}}{\text{ESWL}}$ $\qquad = 1\cdot17$

Equivalent single-wheel load $\qquad = \dfrac{9000}{1\cdot17} = 7700$ lb

It should be noted that this ESWL acts on an area such that the tyre pressure is 60 psi, the same as that of the dual wheels.

2. *Dual-tandem wheel assembly.* It is required to determine the equivalent

single-wheel load for an aircraft having a dual-tandem wheel load of 170 000 lb if the following data are given:

Thickness of rigid pavement	$h = 12$ in
Spacing between centrelines of dual wheels	$S = 32$ in
Spacing of axles	$S_T = 60$ in
Modulus of subgrade reaction	$K = 500$ pci
Modulus of elasticity of concrete	$E_c = 5\,000\,000$ psi
Tyre pressure of the wheels	$p = 170$ psi

Solution.
From Table 2.1 the value of l

$$l = 34\cdot84 \text{ in}$$

Total contact area

$$A = \frac{P}{p} = \frac{170\,000}{170} = 1000 \text{ in}^2$$

$$\frac{S}{l} = \frac{32}{34\cdot84} = 0\cdot92$$

$$\frac{S_T}{l} = \frac{60}{34\cdot84} = 1\cdot73$$

$$\frac{A}{l^2} = \frac{1000}{34\cdot84^2} = 0\cdot83$$

To determine the ratio of the total gear load to the ESWL, enter the ordinate of the lower group of curves in Fig. 2.8 at $S/l = 0\cdot92$. Carry a line horizontally until it intersects the curve representing $S_T/l = 1\cdot73$, interpolating when necessary. From this point of intersection, proceed vertically to intersect the curve representing $A/l^2 = 0\cdot83$ in the upper group of curves. A horizontal line from this point will intercept the ordinate of the upper group of curves at $2\cdot94$.

Therefore, the equivalent single-wheel load $= 170\,000/2\cdot94 = 58\,000$ lb.

2.3.2 Equivalent Single-Wheel Loads for Edge Conditions

In some situations, especially in highways, the edge loading condition may govern the pavement design. Accordingly charts similar to those shown in Figs. 2.7 and 2.8, but for the case of dual or dual-tandem wheel loads placed parallel to a free edge and with their outer wheels adjacent to the edge, are needed.

For this reason, the finite element method has been used, in a recent study [10], to determine the magnitude of the single-wheel load which, when placed at the edge of a rigid pavement, will produce the same maximum stress as that produced by two or four closely spaced edge loads representing dual and dual-tandem wheel loading systems. Elliptical tyre imprints were used in the analysis. The results are plotted in the form of curves and are shown in Figs. 2.9 and 2.10 for the cases of dual and dual-tandem wheel assemblies, respectively.

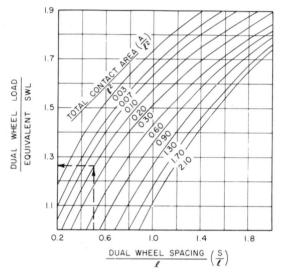

FIG. 2.9. Chart for determination of equivalent single-wheel edge load–rigid pavements–dual-wheel assemblies. (From [10].)

Chart 2.10(a) is for l values between 20 and 30, while Chart 2.10(b) is for l values between 30 and 50. Two examples are given to illustrate the use of these charts.

In the case of dual-tandem wheel assemblies, the equivalent single-wheel load should be determined for one axle as well as for the tandem axles. The larger value of the two is the one to be used in the design.

Examples
  1. Dual-wheel loads.
  Given: Truck-axle load   = 18 kips
        Load on dual wheels on each side   = 9 kips
        Tyre pressure   = 60 psi

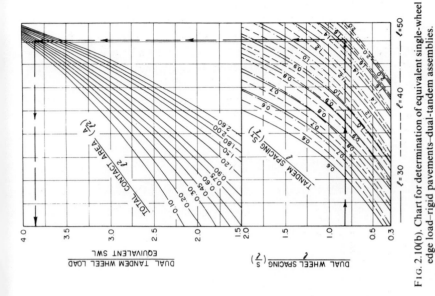

FIG. 2.10(b). Chart for determination of equivalent single-wheel edge load–rigid pavements–dual-tandem assemblies.

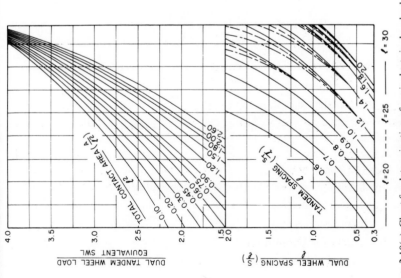

FIG. 2.10(a). Chart for determination of equivalent single-wheel edge load–rigid pavements–dual-tandem assemblies.

Spacing between dual wheels on each side $S = 14$ in

Slab thickness $h = 7 \cdot 0$ in

Modulus of subgrade reaction $K = 200$ pci

Modulus of elasticity of concrete $E_c = 4\,000\,000$ psi

Poisson's ratio of the concrete $v = 0 \cdot 15$

Required: Value of the equivalent single-wheel load.

Solution.

$$l = 27 \cdot 65 \text{ (from Table 2.1)}$$

$$\frac{S}{l} = \frac{14}{27 \cdot 65} = 0 \cdot 506$$

$$A = \frac{9000}{60} = 150 \text{ in}^2$$

$$\frac{A}{l^2} = \frac{150}{27 \cdot 65^2} = 0 \cdot 196$$

From the chart in Fig. 2.9:

$$\frac{\text{Dual-wheel load}}{\text{Equivalent single-wheel load}} = 1 \cdot 27$$

$$\text{Equivalent single-wheel load} = \frac{9000}{1 \cdot 27} = 7090 \text{ lb}$$

If the chart of Fig. 2.7 for the central loading case is used, then:

$$\frac{\text{Dual-wheel load}}{\text{Equivalent single-wheel load}} = 1 \cdot 19$$

$$\text{Equivalent single-wheel load} = \frac{9000}{1 \cdot 19} = 7560 \text{ lb}$$

2. Dual-tandem wheel loads.

Given: Load on one main undercarriage assembly

of a large aircraft $= 161\,000$ lb

Tyre pressure $= 195$ psi

Spacing between dual wheels $S = 31 \cdot 25$ in

Spacing between axles $S_T = 61 \cdot 25$ in

Slab thickness $h = 12 \cdot 0$ in

Modulus of subgrade reaction $\qquad$ $K = 300$ pci

Modulus of elasticity of concrete $\qquad$ $E_c = 4\,000\,000$ psi

Poisson's ratio of the concrete $\qquad$ $v = 0.15$

Required: Value of the equivalent single-wheel load.

Solution.

$l = 37.44$ in (from Table 2.1)

$$\frac{S}{l} = \frac{31.25}{37.44} = 0.83$$

$$\frac{S_T}{l} = \frac{61.25}{37.44} = 1.63$$

$$A = \frac{161\,000}{195} = 825 \text{ in}^2$$

$$\frac{A}{l^2} = \frac{825}{37.44^2} = 0.59$$

From the chart in Fig. 2.10(b), by interpolation:

$$\frac{\text{Dual-tandem wheel load}}{\text{Equivalent single-wheel load}} = 3.87$$

$$\text{Equivalent single-wheel load} = \frac{161\,000}{3.87} = 41\,600 \text{ lb}$$

If Fig. 2.8 is used, for the central loading case, then:

$$\frac{\text{Dual-tandem wheel load}}{\text{Equivalent single-wheel load}} = 2.87$$

$$\text{Equivalent single-wheel load} = \frac{161\,000}{2.87} = 56\,000 \text{ lb}$$

These two examples indicate that the equivalent single-wheel load, for the edge loading condition, is smaller than that for the central loading condition.

2.4 TYRE CONTACT AREA

If the dimensions of the imprint areas of the tyres are not available from the manufacturer, the tyre contact area can be approximately calculated by dividing the gross weight supported on the tyre, by the tyre pressure.

In reality, the contact pressure between the pavement and tyre differs from the tyre pressure, the difference depending on the magnitude of the tyre pressure. The walls of high-pressure tyres are in tension, so that contact pressure is less than the tyre pressure. However, for low-pressure tyres, the contact pressure is greater than the tyre pressure. If the tyre pressure in the latter case is used to determine the tyre contact area, this area will be larger than the actual tyre imprint, and will produce lower stresses in the pavement system than the actual case.

Some investigations have shown that, for the case of low-pressure tyres, the contact pressure is about 10% higher than the tyre pressure. This means that the contact area is equal to the gross weight divided by 1·1 times the tyre pressure.

Since the case of tyres with low pressures is not common with the heavy loads that are generally used in the design of highway and airfield pavements, the actual value of the tyre pressure is used in most calculations.

Although many design procedures consider the load on the tyre to be distributed over a circular area, the actual tyre imprint takes the shape of an ellipse. Impressions taken of actual tyre imprints indicate that the shape of the contact area can be closely approximated by a geometric

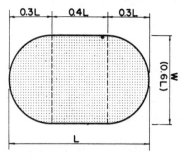

FIG. 2.11. Tyre imprint data.

figure composed of a rectangle and two semicircular ends. The width of the imprint W is 0·6 times its length L. Thus, the area enclosed by the tyre imprint consists of the central rectangular area with width 0·6L and length 0·4L and two equal semicircular areas having radii 0·3L, as shown in Fig. 2.11.

The contact area $A = (0.60L \times 0.4L) + [3.14 \times (0.3L)^2]$

$$= 0.24L^2 + 0.2826L^2 = 0.5226L^2$$

$$L = \sqrt{\frac{A}{0.5226}} \quad \text{and} \quad W = 0.6L$$

2.5 EFFECT OF LOAD REPETITION

It is not only the values of the load and the tyre pressure which govern the load variables used in the design of pavements, but also the repetition of the load. Both the cement concrete of rigid pavements and the asphalt concrete of flexible pavements can be subjected to fatigue failures if the number of load repetitions is high and the safety factor is low. The stress causing failure in both materials bears a linear relationship to the logarithm to the base 10 of the number of load applications. For this reason, it is important to determine the number of repetitions of the load during the life period of the pavement. The procedure for determining the equivalent number of repetitions of the standard design load which represents the mixed traffic on highways and city streets, has been given in Chapter 1, subsections 1.5.2 and 1.5.3. Also, an approximate procedure for determining the number of repetitions of the aircraft load used in the design of airfield pavements has been given earlier.

2.6 EFFECT OF TRANSIENT LOADS

The speed of moving vehicles has an effect on the stress–strain characteristic of both the pavement and underlying soil material. Early observations showed that on highways constructed on relatively steep grades the uphill portions generally exhibited more distress than the downhill parts. This means that the parts of the highway subjected to slow-moving traffic are affected more than those subjected to fast-moving vehicles.

Past observations have indicated that the static load has more influence on the pavement system than the transient load. This fact has also been supported by previous research data, which indicated a high increase in the modulus of deformation of some pavement materials and subgrade soils from static to transient load. Thus, smaller deflections are expected under transient loads than under stationary loads, simply because the deflection of a pavement is inversely proportional to the modulus of deformation of its materials.

Furthermore, previous data on test roads in the USA and Europe have shown that measured stresses and deflections decrease with increasing speeds for speed ranges between 0 and 40 mph.

However, more recent studies have indicated that pavement deflections and stresses tend to increase again at speeds greater than 40 or 50 mph because of vibrations set up by the vehicle as it moves at higher speeds. Recent research work in the USA [8] has indicated that curling, loss of subgrade support and the viscoelastic properties of the pavement materials have a large effect on the values of stresses and deflections in pavements under moving loads. The main conclusion from these studies is that, contrary to common opinion, an increase in velocity can produce an increase in deflection and stress.

2.7 IMPACT OF MOVING VEHICLES

Loads travelling over a paved surface may create impact at the transverse joints and cracks of highway and airfield rigid pavements. The amount of impact depends upon the amount of differential settlement between adjacent slabs and the amount of temperature steel which will affect the faulting at the cracks.

Tests on flexible pavements have indicated that wheel loads may change several times a second by as much as 1000 lb. However, as discussed before, the increased resistance to deformation resulting from transient loads may reduce or even cancel the impact effect.

2.7.1 Impact Factors

(a) *For highways*. Due to lack of definite information regarding the dynamic response of the pavement to moving loads, an impact effect ranging between zero and 20% of the value of the load is generally considered in the design of highway pavements.

In the design of flexible pavements, a zero impact effect is generally considered.

In the design of rigid pavements, the impact factors are more accurately described as load safety factors (LSF). The Portland Cement Association recommends the following load safety factors for highways:

1. Freeways and multi-lane expressways carrying high volumes of truck traffic, $LSF = 1.2$.
2. Highways and arterial streets serving moderate volumes of truck traffic, $LSF = 1.1$.
3. Highways, residential streets, and other streets carrying small volumes of truck traffic, $LSF = 1.0$.

(b) *For airfields*. The US Corps of Engineers' tests for the Airforce on rigid and flexible pavements during the period 1941–1956 gave the following basic findings:

1. The repetitive traffic of slow-moving aircraft is the most severe loading to which airfield pavements are subjected. This occurs on taxiways, runway ends and aprons.
2. Impact on normal landings can be ignored due to lift on the wing surfaces at aircraft landing speeds.

Accordingly, a zero impact effect is generally considered in the design of airfield pavements.

REFERENCES AND BIBLIOGRAPHY

1. *Aerodrome Physical Characteristics*, International Civil Aviation Organization, Part 2, Second Edition, 1965.
2. Burmister, D. M., 'Evaluation of Pavement Systems of the WASHO Road Test by Layered Systems Methods', *Highway Research Board Bulletin 177*, 1958.
3. Deacon, J. A. and Deen, R. C., 'Equivalent Axle Loads for Pavement Design', *Highway Research Record*, No. 291, 1969.
4. Drennon, C. B. and Kenis, W. J., 'Response of a Flexible Pavement to Repetitive and Static Loads', *Highway Research Record*, No. 337, 1970.
5. Federal Aviation Agency, 'Airport Paving', US Government Printing Office, November 1963.
6. Huang, Y. H., 'Computation of Equivalent Single-Wheel Loads Using Layered Theory', *Highway Research Record*, No. 291, 1969.
7. Yoder, E. J., *Principles of Pavement Design*, Wiley, 1959.
8. Lewis, K. H. and Harr, M. E., 'Analysis of Concrete Slabs on Ground Subjected to Warping and Moving Loads', *Highway Research Record*, No. 291, 1969.
9. Newmark, N. M., 'Influence Charts for Computation of Vertical Displacements in Elastic Foundations', Bulletin 367, University of Illinois Experiment Station, 1947.
10. Sargious, M. A. and Wang, S. K., 'Equivalent Single-Wheel Edge Load on Rigid Pavements', Department of Civil Engineering, The University of Calgary, Research Report No. CE 73-12, August 1973.
11. Terzaghi, K., *Theoretical Soil Mechanics*, Wiley, 1943.

CHAPTER 3

REGIONAL VARIABLES

3.1 INTRODUCTION

Before starting the design and construction of any highway or airfield pavement, two important regional variables must be carefully studied in the location proposed for the construction. These are the climate and soil conditions, and should be considered as two of the most important factors affecting pavement performance.

3.2 CLIMATE

Both the pavement and the underlying supporting layers are exposed to strong climatic influences. For this reason, statistical data and general information on the following climatic variables should be gathered before starting the design and construction of the pavement system.

3.2.1 Temperature

Air temperature has a direct influence on the performance of the pavement and supporting medium in both flexible and rigid pavements. Sudden temperature variations accompanied by fluctuations in soil moisture cause deformations in the pavement system, which may result in cracking, spalling or even the blow out of some slabs. Low temperatures cause soil shrinkage, particularly in cohesive soil. This may be accompanied by cracks which are filled with water during the next period of rain, resulting in a decrease in the bearing capacity of the soil. If the water in the cracks is frozen due to depressed air temperatures, then a break-up in the soil mass may result.

A temperature potential applied to soil will cause the movement of soil moisture from warmer regions to colder ones, thus changing the moisture distribution in the soil. If the colder region is then exposed to freezing temperatures, the migrating moisture acts as a supply, which causes the growth of ice lenses under the pavement and may contribute to frost damage.

It is the values of air temperature that affects, to a large extent, the type and amount of bitumen that should be used in flexible pavements and the spacing between the joints of rigid pavements. Also, the variation in temperature between the top and bottom surfaces of the pavement affects its deflection and load-bearing capacity.

3.2.2 Rainfall

Rain has an influence on the stability and strength of the supporting medium because it affects the moisture content of the subgrade and sub-base. Also, rainfall is well established as a factor affecting the elevation of the water table, the intensity of frost action, erosion, pumping and infiltration. The extent of the detrimental effects of frost action—in particular, loss of strength during the thaw period—are dependent on the combined effect of rainfall and air temperature. Where the frost problem is absent, the moisture content of the pavement will also vary with rainfall, and this will in turn affect the expansion and contraction of the pavement. Long periods of rainfall of low intensity can be more adverse than short periods of high intensity, because the amount of moisture absorbed by the soil is greatest under the former conditions.

3.2.3 Frost

The effect of frost action on pavements is the most severe of the environmental factors. In its broadest sense, frost action includes both frost heave and loss of subgrade support during the frost-melt period. Frost heave may cause a portion of the pavement to rise, due to ice crystal formation in a frost susceptible subgrade or base course. Thawing of the frozen soil and ice crystals during the spring period may cause pavement damage under loads. This damage usually results in high maintenance costs. The loss of strength under the pavement due to thawing may be of such a magnitude as to require prohibition of heavy loads during the critical period. The economic loss resulting when trucks and aircraft operate during this period while carrying much smaller loads than their capacity, may be very high.

The term 'frost' generally involves two concepts: (1) The existence of freezing temperatures below $32°F$; (2) The action of the freezing temperatures upon the soil, which leads to the state of frozen soils.

When the soil is subjected to freezing temperatures, several phenomena occur, the intensity of which depends upon the intensity of freezing temperatures. These phenomena are:

1. The penetration of frost into the soil occurs at a relatively low rate.
2. As a layer of water-bearing soil is frozen underneath the pavement, it becomes essentially dry, and as such has a high affinity for free water.
3. Free water in the underlying layers is attracted to the undersurface of the frozen soil layer and is consequently frozen as freezing temperature proceeds downward during cold weather.
4. Layers or lenses of clear ice, several inches in thickness, are built up in this way, as shown in Fig. 3.1. When water freezes, it increases by 9% in volume.

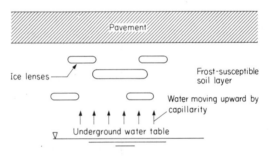

FIG. 3.1. Mechanics of formation of ice lenses.

The expansion of water as it freezes within the soil mass is not the main reason for heave and break-up of pavements due to freezing conditions. If a quick freeze process were to occur in the moist soil mass, total vertical expansion might amount to less than 1 in even though freezing were carried to a considerable depth. It is mainly the non-uniform formation of the above-mentioned ice lenses that actually causes pavement heave.

It is important for the pavement designer to recognise that for frost heave to take place all of the following factors must be present: (1) a frost-susceptible soil, (2) slowly depressed air temperatures and (3) a supply of water.

If any of the above factors are not present, there will be no frost heave. Also, in the cases where the heave extends fairly uniformly over a considerable length of the pavement, the seriousness of the heaving problem can be greatly reduced. Unfortunately, this is not often the case because of the variations in soil characteristics and conditions even within a fairly restricted pavement area.

Soil freezing depends to a large extent upon the duration of depressed air temperatures. Usually, time and temperature are measured by degree days. One degree day represents 1 day with a mean air temperature 1° below freezing. Thus, 30 degree days may mean either that the air temperature is 31°F for 1 month or that the air temperature is 2°F for 1 day. If a cumulative plot of degree days versus time is determined, as shown in Fig. 3.2, the difference between the maximum and minimum points on

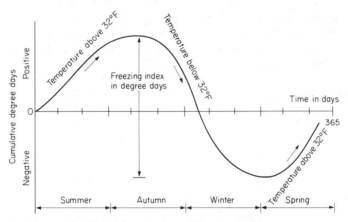

Fig. 3.2. Determination of the freezing index.

this curve is termed the freezing index. In plotting this curve, any convenient day can be used as a start for the plot, since the data are cumulative.

Since the design of the pavement structure in frost-affected regions depends to a large degree on the depth of frost penetration, formulae and curves giving the relationship between frost penetration and freezing index have been in use for a long time.

Aldrich [1] developed the following expression for frost penetration:

$$Z = \lambda\sqrt{\frac{48\,kF}{L}} \tag{3.1}$$

where Z = depth of penetration, in feet, in a homogeneous mass;

k = thermal conductivity in Btus per square foot per hour per degree Fahrenheit per foot;

F = freezing index in degree days;

L = volumetric heat of latent fusion in Btus per cubic foot

$= 1.434\,\omega\gamma$, where ω is moisture content and γ is dry density of soil in pounds per cubic foot; and

λ = dimensionless coefficient which depends upon the factors shown in Fig. 3.3.

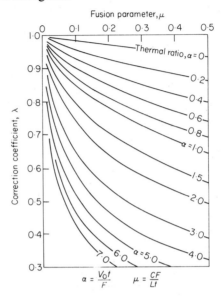

FIG. 3.3. Correction coefficient for Aldrich formula. V_0 = difference between mean annual temperature and freezing point of soil; t = duration of freeze (days); F = freezing index (degree days); L = latent heat (Btus/ft^3); C = volumetric heat (Btus/ft^3°F) (bituminous concrete 28, Portland cement concrete 30); $C = \gamma(0.17 + w/100)$ for unfrozen soil, $= \gamma(0.17 + 0.5w/100)$ for frozen soil; w = moisture content in percent (without dividing by 100). (From [1].)

In the case of layered systems an effective L/k can be used in eqn. (3.1) and may be computed as follows:

$$\frac{L}{k} = \frac{2}{Z^2}\left[\frac{h_1}{k_1}\left(\frac{L_1 h_1}{2} + L_2 h_2 + \ldots + L_n h_n\right) + \right.$$

$$\left. + \frac{h_2}{k_2}\left(\frac{L_2 h_2}{2} + L_3 h_3 + \ldots + L_n h_n\right) + \ldots + \frac{h_n}{k_n}\left(\frac{L_n h_n}{2}\right)\right] \qquad (3.2)$$

where Z is the estimated value for the frost penetration depth and h_1, $h_2, \ldots, h_n$ are the thicknesses of the different layers.

The value of the thermal conductivity, k, of soils depends on soil type, density and moisture content, and can be obtained from the chart shown in Fig. 3.4. This chart has been developed by Kersten [6] from laboratory tests on sands, gravel and fine-grained soils.

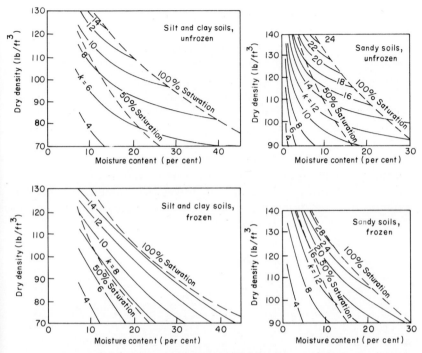

FIG. 3.4. Thermal conductivity of soil (k is in Btus per square foot per hour per degree Fahrenheit per inch). (From [6].)

The value of thermal conductivity for paving materials ranges from 0·84 Btu per square foot per hour per degree Fahrenheit per foot, for bituminous concrete, to 0·54 for Portland cement concrete.

The curve shown in Fig. 3.5 has been developed by the Corps of Engineers [2] to relate depth of penetration to freezing index for a well-drained non-frost-susceptible base course.

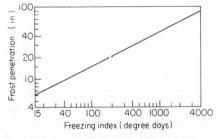

FIG. 3.5. Freezing index and frost penetration in a well-drained, non-frost-susceptible base course. *(Courtesy: Corps of Engineers.)*

Frost thawing during the springtime can cause pavement distress due to loss in subgrade supporting capacity, even though frost heaving may not have taken place. This phenomenon is sometimes referred to as spring break-up, and is particularly apparent immediately after a quick spring thaw.

During the thawing period, frozen soil generally thaws out from above and below. Excessive amounts of water freed by melting soil and ice lenses may soften the material of the layer immediately underneath the pavement. Unless this layer consists of non-frost-susceptible materials, a great reduction in its bearing strength may result, causing a decrease in pavement-supporting capacity. Factors which accentuate severe spring break-up include periods of high rainfall during the Autumn and Winter, resulting in a high degree of saturation of the subgrade. Still more critical is a high rainfall during the frost-melting period.

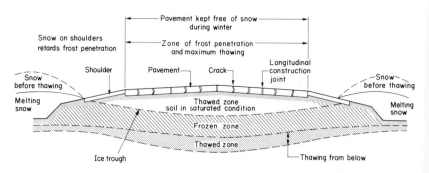

FIG. 3.6. Mechanism of thawing during the spring period.

In Winter, pavements are generally kept free of snow, which is deposited on the shoulder or in side ditches. Snow, though a good insulator, is undesirable in these locations in the early part of the thawing season. The reason for this is that when the pavement is exposed to the warming sun in the Spring, the soil underneath the pavement will thaw out while the drainage facilities and base-drainage course are still frozen. This may result in the accumulation of large quantities of water beneath the pavement between the relatively non-porous pavement surface and the still frozen layer underneath, as in a trough (Fig. 3.6). When heavy traffic moves over the pavement at such locations, it may cause failure in the pavement. This condition may prevail for a considerable period of time, because the excess water cannot evaporate through the pavement and will not immediately drain laterally. The seasonal variation of load-bearing capacity of a pavement subjected to freezing and thawing conditions is presented in Fig. 3.7 [9]. This figure shows that loss of subgrade support can be of considerable magnitude and may exist for relatively long periods

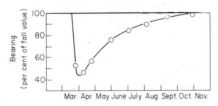

FIG. 3.7. Variation of bearing with seasons. (After [9].)

of time after thaw has taken place. Loss of strength is particularly apparent in areas where alternate freezing and thawing take place throughout the winter months, because each time the soil freezes, a loss of soil density results. This in turn results in higher potential moisture absorption. After several cycles of freezing and thawing, a large portion of the subgrade supporting capacity may be lost.

3.2.4 Winds

The type of winds blowing in the region should be carefully studied. Winds generally contribute to moisture evaporation from soils and set water in motion, in the form of waves which may wash and erode embankments and earthworks.

Winds causing sandstorms may have a severe effect on the pavement. If fine particles of sand penetrate through the cracks or joints of a pavement, they may cause widening of the cracks and spalling at the joints.

3.3 PERMAFROST

3.3.1 Definitions

The term 'permafrost' refers to areas of permanently frozen ground in the Arctic regions in the northern parts of Canada, Europe, Siberia and Alaska.

There are both continuous and discontinuous permafrost zones, as illustrated schematically in Fig. 3.8. The permafrost may be from several hundred to more than 1000 ft thick in the continuous zone, the active surface layer being generally 1–3 ft thick. The active layer is defined as the surface layer subjected to freezing and thawing according to the seasons of the year. In the discontinuous zones, the permafrost is interspersed with areas of unfrozen ground and can range in thickness from a few feet to 200 ft. The active layer is thicker than in the continuous zone and does not necessarily extend everywhere to the permafrost table. There is, naturally, some seasonal variation in the position of the permafrost table.

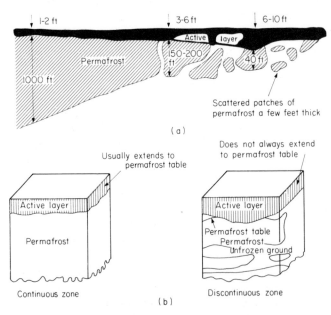

FIG. 3.8. Schematic representation of permafrost zones. (a) Typical vertical distribution and thickness of permafrost. (b) Typical profiles in permafrost region.

The effect of the seasonal changes in surface temperatures on the thickness of the active layer varies with the nature of the soil, vegetative cover, and snow cover.

In general, it is considered practice in permafrost areas to avoid cuts as much as possible and construct the pavements on fill.

Before discussing in some detail the problems associated with highway and airfield construction in the permafrost regions, and proposing solutions, definitions of some commonly used terms are summarised as follows.

Permafrost table. This is the upper surface of the permafrost; the ground above this surface is called the suprapermafrost layer.

Active layer. This layer is the part of the suprapermafrost layer which freezes in the Winter and thaws in the Summer. Although seasonal frost penetrates down to the permafrost table in most places, sometimes an unfrozen zone, called a talik, remains between the bottom of the seasonal frost and the permafrost table.

Frost mound. This is a heap or bank of ice-rich soil in a frozen condition, occurring most commonly in the upper 25 ft of the soil profile. When ice-rich soil thaws, the soil mass may change to a semi-liquid condition and large settlements may occur as the melted ice drains away.

Pingos. These are large mounds up to 150 ft in height originating from the arching of an impervious sheet of permanently frozen ground forced up by the intrusion of water under pressure. Pingos are widely spaced and should simply be avoided by any construction.

Frost blisters. These blisters are developed when suprapermafrost water is trapped below the freezing active layer. The hydrostatic pressure and resulting ice build-up in the closed space between the permafrost table and the active layer cause heaving of the ground surface, tilting of trees (drunken forest) and, possibly, icing on the surface adjacent to the mound. These are usually seasonal features which may or may not develop in the same location each year.

Icing mounds. These are frost blisters which have a greater available supply of groundwater and cause large mounds of ice to form on top of that ground which has heaved.

Peat mounds and hummocky terrain. These are the result of differential ice lensing or soil sorting due to frost action. The extent of ice in this terrain is less than in the larger features noted above.

Ice wedges and polygon terrain. Ice wedges are vertical ice formations that extend downward from the base of the active layer. They may attain a width of 6 ft and a height of 30 ft and are linked together in the horizontal plan to form patterns of polygon terrain. The individual wedge is developed in the continuous permafrost zone where very deep freezing of the surface ground causes contraction, resulting in tension cracks that develop in a polygonal pattern. These cracks are then filled with spring meltwater which freezes. The annual cyclic cracking and infilling can cause wedges of large dimensions to develop. When actively developing, the wedges are topped with a clefted ridge, resulting in a polygonal pattern on the surface.

3.3.2 Problems in Permafrost Regions

The problems generally encountered in pavement construction in the permafrost regions can mainly be related to the following two reasons:

Deformations Caused by Seasonal Freezing and Thawing
Processes that take place during seasonal freezing and thawing are reflected in heaving and settlement of roadbeds, in icing and in the appearance of frost fissures.

Differential deformations, which are caused primarily by the heterogeneity of the ground, and its differential moisture conditions and freezing, present the major danger to roadbeds.

For roads with flexible pavements, the most dangerous period is the Spring, or the so-called active period, when the roadbed begins to thaw. Thawing is greater under the middle of the road than under the shoulders. With the melting of ice inclusions formed in the upper part of the roadbed during the Winter, slurry forms in a closed cavity, bound by frozen ground below and at the sides, and by the pavement above. As traffic moves, the pavement may break and this slurry is squeezed to the surface.

Winter heaving, during which fissures form, is more dangerous for rigid pavements.

Roadbeds built on slopes containing streams of suprapermafrost water are most often subjected to icing formation. Spring icing feeding on subpermafrost water, may cause the stoppage of traffic by flooding the roadbed.

Deformations Caused by Changes in the Position of the Permafrost Table
Deformation caused by lowering of the permafrost table manifests itself primarily as settlement of the subgrade, which often upsets the stability of the roadbed and costs a lot to repair.

3.3.3 Methods of Pavement Construction in the Permafrost Regions
To ensure stability of the roadbed, it is recommended that one of the three following methods be employed.

Method of Preserving the Subgrade in a Frozen State
This method is used primarily in regions with thin active layers, containing ice in the form of lenses, veins, etc. The subgrade may be kept frozen in the following ways:

(a) By preserving the permafrost table at its natural level. This is recommended only when there is no ice-rich soil in the upper horizons.

(b) By raising the permafrost table to the level of the foot of the embankment, i.e. by freezing the entire active layer.

(c) By raising the permafrost table to a lower level than the natural depth of seasonal thawing, i.e. freezing only part of the active layer.

Methods (b) and (c) are applicable under unfavourable geocryological conditions and to important structures, such as airfields, that require an even surface.

The method of preserving a frozen subgrade is reliable in ensuring the stability of roadbeds. However, the most important and complicated problem in the design is the selection of embankment height to ensure that the permafrost table is at the required level, and also that the roadbed is protected from snowdrifts. In addition, the type of material used in roadbed construction is of great importance. Well-graded granular draining materials are recommended for this purpose.

The use of multi-layer embankment, each layer serving a specific function, is recommended when methods (b) and (c) are used. The lower layer serves to level the microrelief of the natural surface and to protect the sod-moss cover from damage in case large stones are used to fill the body of the embankment. An embankment may consist of layers of materials having better thermal insulation properties than the local natural ground. The materials used must be resistant to heaving, have a high compressive strength and be very resistant to frost weathering. Special attention should be paid to the connection of the embankment with the surrounding ground.

The preservation of vegetation in the foundation of the embankment is imperative when constructing according to the method of preserving a frozen subgrade. In addition to improving the thermal insulation properties of the embankment foundation, vegetation also interferes with the penetration of fines into the body of an embankment made of coarse material. Furthermore, the presence of vegetation under an embankment reduces heaving.

In installing drainage systems, the natural sod-moss cover must be disturbed as little as possible. Therefore, the construction of upslope channels, especially in silty, ice-saturated ground, can be allowed only if they are protected against washout and thawing by a paving layer over the moss or peat. These channels should be located at least 35 ft from the shoulder, should have a narrow section to prevent early freezing and should have a depth less than the depth of the active layer, particularly if they are upslope of the grade.

The timing and method of construction are of great importance for projects that involve the preservation of a frozen subgrade. Embankments are best erected when the active layer is at least partially frozen (to a depth of no less than $0.5 \rightarrow 0.7$ ft). This will ensure that the vegetation cover is not damaged significantly during the work. If an embankment is placed in Summer, it must be built up to its full height, and provision made for additional material to compensate for settlement of the thawed layer of the natural subgrade.

Method of Improving the Subgrade Prior to Construction

This method can be used in situations where the nature of the subgrade soil and climatic conditions allows of the use of procedures that will increase the modulus of deformation of the subgrade soil prior to erection of the roadbed. The improvement of subgrades is achieved primarily by decreasing their moisture content and lowering the permafrost table. Therefore, this method of construction is generally used in areas where permafrost has a relatively high temperature and where the active layer consists of sandy or sandy loam soils of low cohesion, that can easily drain any accumulated water. The presence of a relief, allowing of the discharge of run-off water from the drained areas, is very important.

When this method is used, run-off channels in the site of a road or an airfield are drained by removing the vegetation cover immediately after the snow has disappeared. Attention is required to ensure draining the water from all depressions and subsequently backfilling them. This drainage creates conditions which ensure the lowering of the permafrost table, the reduction of moisture in the active layer and the melting of ice lenses. All of these factors increase the stability of a roadbed and reduce its subsequent settlement under traffic.

When a permanent drainage system is installed, careful attention should be paid to maintaining the natural drainage pattern. In some sections, several years of preliminary drainage of meltwater may also be needed before a main highway or an airfield is constructed.

Both embankments and cuts can be used in locations where construction will involve subgrade improvement. However, if fine-grained or silty soils are present together with a shallow permafrost table, cuts must be avoided and embankment should mainly be used.

Method of Gradual Thawing of Permafrost in the Subgrade

Frozen ground serving as a subgrade may be thawed during and after construction in areas with a deep permafrost table. This applies specifically to situations where sufficient suprapermafrost water is present. Such thawing does not lead to catastrophic disturbances of the roadbed stability when the subgrade consists of sandy soils. These soils have high filtration capability after thawing, and generally do not contain ice lenses. In this case, a well-developed drainage system must be provided, so that the subgrade soil will consolidate as rapidly as possible. Rigid road pavements can be installed only 2–3 years after the construction of the roadbed; during this time, sufficient meltwater is drained from the subgrade soil and the permafrost table approaches a state of stability.

3.4 SOIL

3.4.1 Soil Identification

The importance of accurate identification and evaluation of the pavement foundation, i.e. the underlying soil, cannot be overemphasised. Soil condition and availability of suitable construction materials are important items affecting the safety and cost of the pavement structure.

An investigation of soil conditions at a highway or airport site will include:

1. A soil survey to determine the arrangement of the different layers, the soil profile and the physical properties of the various soils. The survey should also disclose the groundwater level and the drainability of the different layers as well as the composition of the soil as a whole.
2. Sampling of the soil layers.
3. Testing of soil samples to determine the physical properties of various soil materials with respect to stability and subgrade support.
4. A survey to determine the availability of materials for use in construction of subgrade and pavement.

The practice of determining data on soils by use of aerial photographs is becoming more and more widespread. By employing this method of investigation, it is possible to expedite soil studies and reduce the amount of effort required to gather data. Naturally, an experienced interpreter is essential if this method is to be attempted.

The principal soil properties influencing subgrade performance are stability, compressibility and permeability.

Soil is not a homogeneous fabricated material with controlled and defined properties. Rather, the soil material exists in nature under many different conditions, and constitutes a very complex variable system. Because of the complex nature of soil, every engineering problem involving soil is an individual one and must be treated as such.

It is not only the knowledge of the physical properties and performance of soil as a material that is needed, but also the knowledge of the behaviour of soils with varying moisture conditions under static and dynamic loads and under different climatic conditions.

Textural classification of soils depends upon the size of their constituent particles. The grain size of silty soils ranges between 0·005 mm and 0·05 mm. Soils finer than silt are called clay and those coarser than silt, up to 2 mm, are sand. Soils with larger grain size can be classified as gravel.

Also, sand can be classified as fine, medium and coarse. A sieve analysis is necessary to determine the grading of sand and gravel. The sieves are generally with square openings, the finest sieve commonly used being No. 200, with opening width equal to 0·074 mm.

There are several soil classifications made by different agencies. One of the effective methods for classification of subgrade soils is the Highway Research Board (HRB) system presented in the *Highway Research Board Proceeding 25*, 1945 [5].

3.4.2 HRB Classification of Subgrade Soils

According to this system, soils are divided into two main groups. The first group consists of granular materials containing 35% or less material passing sieve No. 200. This group covers three group classifications, referred to as A-1, A-2 and A-3, which have a general rating as a subgrade ranging between excellent and good, and have the following properties.

Group A-1. The typical material of this group is a well-graded mixture of gravel or stone fragments, coarse sand, fine sand and a non-plastic or slightly plastic soil binder. This group also includes gravel, stone fragments, coarse sand, volcanic cinders, etc., without soil binder.

Group A-2. This group includes a variety of granular materials which are borderline between the materials falling in A-1 and A-3 groups and the silt-clay materials of groups A-4, A-5, A-6 and A-7. It includes all materials containing 35% or less passing sieve No. 200 which cannot be classified as A-1 or A-3 because of fine content or plasticity, or both, that are in excess of the limitations for these groups.

Each of the group classifications A-1 and A-2 has been divided in subgroups as shown in Table 3.1.

Group A-3. The typical material of this group is fine desert blown sand or fine beach sand without silty or clay fines or with a very small percentage of non-plastic silt. This group also includes stream-deposited mixtures of poorly graded fine sand and limited amounts of coarse sand and gravel.

The second group consists of clay and silt-clay materials, containing more than 35% passing sieve No. 200. This group covers four group classifications, referred to as A-4, A-5, A-6 and A-7, which have a general

	Group 1: Granular materials, 35% or less passing sieve No. 200							Group 2: Silt-clay materials, more than 35% passing sieve No. 200				
General classification												
Group classification	A-1	A-1	A-3	A-2	A-2	A-2	A-2	A-4	A-5	A-6	A-7	A-7
Subgroup classification	A-1-a	A-1-b	—	A-2-4	A-2-5	A-2-6	A-2-7	A-4	A-5	A-6	A-7-5	A-7-6
Sieve analysis, percent passing sieve:												
No. 10	50 max											
No. 40	30 max	50 max	51 min									
No. 200	15 max	25 max	10 max	35 max	35 max	35 max	35 max	36 min	36 min	36 min	36 min	36 min
Characteristics of fraction passing sieve No. 40:												
Liquid limit				40 max	41 min	40 max	41 min	40 max	41 min	40 max	41 min	41 min
Plasticity index	6 max	6 max	NP	10 max	10 max	11 min	11 min	10 max	10 max	11 min	11 min	11 min
Group index	0	0	0	0	0	4 max	4 max	8 max	12 max	16 max	20 max	20 max
Usual types of significant constituent materials	Stone fragments and gravel with or without binder	Coarse sand with or without binder	Fine sand	Silty gravel and silt	Silty gravel and silt	Clayey gravel and sand	Clayey gravel and sand	Silty soils	Silty soils	Silty soils	Clayey soils	Clayey soils
Other remarkable characteristics of the material											*	†
General rating as subgrade	Excellent to good							Fair to poor				

* Highly elastic and subject to considerable volume change.
† Highly plastic and subject to extremely high volume changes.

rating as subgrade ranging between fair and poor, and have the following properties.

Group A-4. The typical material of this group is a non-plastic or moderately plastic silty soil usually having 75% or more passing sieve No. 200. This group also includes mixtures of fine silty soil and up to 64% of sand and gravel retained on sieve No. 200. The group index (G I) values for the materials of this group range from 1 to 8. Increasing percentages of coarse material are reflected by decreasing group index values.

Group A-5. The typical material of this group is similar to that of Group A-4, except that it is usually of micaceous or diatomaceous character and may be highly elastic, as indicated by the high liquid limit. The group index values for the materials of this group range from 1 to 12, with increasing values indicating the combined effect of increasing liquid limits and decreasing percentages of coarse materials.

Group A-6. The typical material of this group is a plastic clay soil usually having 75% or more passing sieve No. 200. This group also includes mixtures of fine clayey soil and up to 64% of sand and gravel retained on sieve No. 200. The group index values for the materials of this group range from 1 to 16, with increasing values indicating the combined effect of increasing plasticity indices and decreasing percentages of coarse material. Materials of this group usually have high volume change between wet and dry conditions.

Group A-7. The typical material of this group is similar to that of Group A-6, except that it has the high liquid limit characteristic of Group A-5. The range of group index values for the materials of this group is 1 to 20, with increasing values indicating the combined effect of increasing liquid limits and plasticity indices, and decreasing percentages of coarse material. This group's materials may be elastic as well as subject to high volume change.

Group classification A-7 has, in addition, been divided in subgroups as shown in Table 3.1.

The group index for a soil, mentioned in Table 3.1, can be calculated from eqn. (3.3):

$$\mathrm{G\,I} = 0\cdot2a + 0\cdot005ac + 0\cdot01bd \qquad (3.3)$$

where a = that portion of the percentage passing sieve No. 200 greater than 35% and not exceeding 75%, expressed as a positive whole number (0–40);

b = that portion of the percentage passing No. 200 sieve greater than 15% and not exceeding 55%, expressed as a positive whole number (0–40);

c = that portion of the numerical liquid limit greater than 40 and not exceeding 60, expressed as a positive whole number (0–20); and

d = that portion of the numerical plasticity index greater than 10 and not exceeding 30, expressed as a positive whole number (0–20).

Example

An A-5 material has 60% passing sieve No. 200. The liquid limit of the material is 45 and the plasticity index is 9. Calculate the group index for this material.

Solution.

$a = 60–35 = 25$

$b = 55–15 = 40$ (55 is substituted for 60 since the critical range is between 15 and 55)

$c = 45–40 = 5$

$d = 0$, since plasticity index is below 10

$GI = 0.2 \times 25 + 0.005 \times 25 \times 5 + 0.01 \times 40 \times 0$

$\quad = 5 + 0.625 + 0$

Since the group index should be recorded to the nearest number, $GI = 6$ for this example.

Definitions of the other factors contained in Table 3.1 are as follows:

Plastic limit (AASHO Method T-90). Plastic limit is the minimum moisture content at which the soil becomes plastic. When the moisture content exceeds the plastic limit, a sharp drop in the stability of the soil will occur.

Liquid limit (AASHO Method T-89). The liquid limit of a soil is the water content at which the soil passes from a plastic to a liquid state. The liquid state is defined as the condition in which the shear resistance of the soil is so slight that a small force will cause it to flow.

Plasticity index (AASHO Method T-91). The plasticity index of a soil is the numerical difference between the plastic limit and the liquid limit. In other words, it indicates the increase in moisture content required to change a soil from a plastic to a liquid condition.

Shrinkage limit (AASHO Method T-92). The shrinkage limit is the moisture content at which a further reduction in moisture content will not cause a decrease in volume of the soil mass.

TABLE 3.2 CLASSIFICATION OF SOILS FOR AIRPORT PAVEMENT CONSTRUCTION
(Courtesy: Federal Aviation Agency)

		Soil group description	Material retained on sieve No. 10 in percent[a]	Mechanical analysis			Liquid limit	Plasticity index
				Material finer than sieve No. 10 in percent				
				Coarse sand, passing No. 10, retained on No. 60	Fine sand, passing No. 60, retained on No. 270	Combined silt and clay, passing No. 270		
Granular	E-1	Well-graded coarse granular soils, not subject to detrimental frost heave, satisfies requirements for base course	0–45	≥40	≤60	≤15	≤25	≤6
	E-2	Less coarse sand than E-1, more silt and clay, unstable when poorly drained, subject to limited frost heave	0–45	≥15	≤85	≤25	≤25	≤6
	E-3	Include fine sandy soils of inferior grading in a cohesionless form or sand-clay with fair-to-good quality of binder, less stable than E-2 under adverse drainage and frost conditions	0–45	—	—	≤25	≤25	≤6
	E-4		0–45	—	—	≤35	≤35	≤10
	E-5	Poorly graded granular soils, contains 35–45% silt and clay combined	0–45	—	—	<45	<40	≤15

	with very low plasticity, stable when dry, lose stability when wet, greatly subject to detrimental frost heave	0–55	—	—	≥45	<40	≤10
E-7	Includes clay loams, silty clays, clays and some sandy clays, stiff and dense at proper moisture content, detrimental frost heave but less severe than E-6	0–55	—	—	≥45	≤50	10–30
E-8	Similar to E-7 but more compressible, subject to expansion and shrinkage, lower stability under adverse moisture conditions	0–55	—	—	≥45	≤60	15–40
E-9	Silts and clays containing micaceous and diatomaceous materials, highly elastic, low stability, subject to frost heave	0–55	—	—	≥45	≥40	≤30
E-10	Silty clay and clay soils, hard when dry and very plastic when wet, compressible, subject to frost heave, require careful control of moisture to produce stable fill	0–55	—	—	≥45	≤70	20–50
E-11	Similar to E-10 but have higher liquid limits	0–55	—	—	≥45	≤80	≥30
E-12	Highly plastic clays or very elastic soils containing excessive amounts of mica, diatomic or organic matter, need maximum corrective measures	0–55	—	—	≥45	>80	—
E-13	Encompasses organic swamp soils, very low stability, very low density and very high moisture contents				Muck and peat—field examination		

Fine grained

[a] If percentage of material retained on sieve No. 10 exceeds that shown, the classification may be raised, provided such material is sound and fairly well graded.

3.4.3 The FAA Classification

Another classification for subgrade soils is the Federal Aviation Agency (FAA) soil classification system [3]. This system requires basically the performance of three tests: (1) mechanical analysis of soils; (2) determination of the liquid limit; and (3) determination of the plastic limit.

As can be noticed from Table 3.2, the mechanical analysis provides the information to permit of the separation of granular soils from fine-grained soils. Granular soils are those classified as E-1 through E-5, while fine-grained soils are those classified as E-6 through E-13. The division between granular and fine-grained soils is made upon the requirement that granular soils must have less than 45% of silt and clay combined. Determination of the sand, silt and clay fractions is made on that portion of the sample passing sieve No. 10, because this is considered to be the critical portion with respect to changes in moisture and other climatic influences.

The several groups shown in Table 3.2 are arranged in order of increasing values of liquid limit and plasticity index. The FAA soil classification system is mainly intended for airport pavement construction.

It is possible that a soil may contain certain constituents that will give test results placing it in more than one group. Such overlapping can be avoided by the use of Fig. 3.9 in conjunction with Table 3.2.

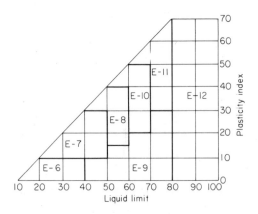

FIG. 3.9. Chart for classification of fine-grained soils. *(Courtesy: Federal Aviation Agency.)*

It should be noticed that in using Table 3.2 and Fig. 3.9, only that portion of the sample passing sieve No. 10 is considered. Naturally, the presence of material retained on the No. 10 sieve should serve to improve the over-all stability of the soil. For this reason, upgrading the soil from

1 to 2 classes is permitted, if percentage of material retained on sieve No. 10 exceeds the values shown in Table 3.2. However, this can only be done if the coarse fraction consists of reasonably sound material which is fairly well graded. Stones or rock fragments scattered through a soil should not be considered of sufficient benefit to warrant upgrading.

REFERENCES AND BIBLIOGRAPHY

1. Aldrich, Harl P., 'Frost Penetration Below Highway and Airfield Pavements', *Highway Research Board Bulletin 135*, 1956.
2. Corps of Engineers, 'Engineering and Design, Pavement Design for Frost Conditions', EM-1110-345-306.
3. Federal Aviation Agency, 'Airport Paving', US Government Printing Office, 1962.
4. Ferrians, O. J. *et al.*, 'Permafrost and Related Engineering Problems in Alaska', *Geological Survey Professional Paper 678*, 1969.
5. *Highway Research Board Proc.*, 'Report of Committee on Classification of Materials for Subgrade and Granular Type Roads', *25*, 1945.
6. *Highway Research Board Special Report 2*, 'Frost Action in Soils', Highway Research Board, Washington, D.C., 1952.
7. *Highway Research Board Special Report 95*, 'Rigid Pavement Design', Highway Research Board, Washington, D.C., 1968.
8. Yoder, E. J., *Principles of Pavement Design*, Wiley, 1959.
9. Motl, C. L. *et al.*, 'Load-Carrying Capacity of Roads as Affected by Frost Action', *Highway Research Board Bulletin 4*, *54* and *96*.
10. Porkhaev, G. V. and Sadovskii, A. V., *Principles of Geocryology Permafrost Studies*, Part 2, Chapter 8, National Research Council of Canada, Technical Translation 1220, Ottawa, 1965.

CHAPTER 4

SUBGRADES, SUB-BASES AND STABILISED SOILS

4.1 INTRODUCTION

The quality and life of a highway or airfield pavement will be greatly affected by the type of subgrade, sub-base and stabilised soil underneath the pavement. For this reason, each of these pavement structure components must be carefully investigated. The decision whether or not to use sub-bases and/or stabilised soils depends upon many factors. The most important of these are the type and quality of the subgrade, the importance of the project and the availability of funds.

4.2 SUBGRADES

4.2.1 Introduction

Desirable properties for the subgrades are: high compressive and shear strength, permanency of strength under all weather and loading conditions, ease of compaction, permanency of compaction, ease of drainage and low susceptibility to volume changes and frost action. Since subgrade soils vary considerably, the interrelationship of texture, density, moisture content and strength of subgrade materials is complex. In addition, behaviour of subgrades under repeated loads is difficult to evaluate and the properties of the subgrade extending underneath the pavement vary along the length and width of a highway or an airfield. For this reason, the principles involved in designing the subgrade should be well understood by the design engineer.

As a first step, a reconnaissance survey of soil conditions followed by an exploration programme should be carried out to investigate the lateral and vertical extent of all the subgrade soil layers underlying the pavement. For this purpose, geological information, aerial photographs and auger borings are needed. Samples of the subgrade soil obtained from the borings should be tested for strength, size, density, natural water content, liquid limit and plasticity limit. The results of these tests will assist the engineer in determining the design values to be used. These tests will also help in finding solutions at locations where the properties of the subgrade soil may be unsatisfactory.

4.2.2 Treatment of Subgrade Soil

If the subgrade material proves to be inadequate as a foundation for the pavement, a special treatment for the subgrade should be followed. This treatment differs from one location to another, depending upon the reasons for inadequacy of the subgrade material, weather conditions and availability of better materials. The cost will generally be a major factor in selecting the method of treatment.

If the design value of the subgrade required for a given situation exceeds, for example, the strength that can be achieved through compacting the existing subgrade soil, then different alternative solutions are open to discussion. If the thickness of the weak subgrade layer is small, i.e. less than 5 ft, then it may be removed and replaced by a better material, if such material is available at a reasonable cost. If the weak layer is thick, the surface grade line should be raised, providing a sufficient depth of cover with good material to protect adequately the low-strength soil.

In situations where it is not feasible to excavate through the entire depth of a weak stratum, it is desirable to permit any accumulated free water to escape—for example, to the side ditches. At the same time, it is important to prevent the silt or clay from pumping up into the voids of relatively coarse granular backfill. For this purpose, a layer of filter sand extending between the side ditches may be used between the fine-grained weak soil and the granular backfill.

In some situations, it may even be possible to increase the strength of a weak subgrade surface layer either mechanically or chemically. The mechanical treatment involves making a sieve analysis of the weak soil and studying its properties. Size and percentage of grains desirable for high strength should be identified for the weak soil which may be lacking in these grains. Through mixing with properly chosen materials, using the right proportions, the strength of the subgrade can be increased significantly.

The chemical treatment involves the use of cementing materials, such as Portland cement, lime or a mixture of lime and flyash. Portland cement can best be used in granular soils, silty soils and lean clays, but it cannot be used in organic materials. Hydrated lime is most efficient when used in granular and lean clays in regions with mild climate, since lime–soil mixtures are generally susceptible to freezing and thawing action. Lime increases strength by pozzolanic action, which is the formation of cementatious silicates and aluminates. The addition to the lime of flyash, a by-product of blast furnaces, to stabilise the subgrade soil, speeds the pozzolanic action. However, the quantity of flyash required for adequate stabilisation is relatively high, which restricts its use to areas where large quantities of flyash are available at low cost.

The chemical treatment of the subgrades is a costly process, especially if the weak subgrade layer needing strengthening is deep and extends over a long distance. Also, the use of chemicals in strengthening weak subgrade soils has its disadvantages. Unless the surrounding subgrade is composed of materials with very low compressibility, the chemically stabilised spot may provide a stronger foundation for the pavement than that provided by the remaining subgrade. A high degree of non-uniformity in the bearing characteristics of the subgrade may be harmful to the pavement system because it causes non-uniform deflections.

In cold regions, subgrades containing frost-susceptible soils need a special treatment. Fine sand and silt, particularly silt with a predominance of particle size of 0·02 mm, is nearly ideal with regard to ability and rate of supplying water by capillarity for the formation of ice lenses under normal field conditions of soil freezing. Clay soil, having smaller particle size, is not as critical as silts because water would be supplied to the freezing surface by capillary rise at a very slow rate. Thus, the rate of penetration of the frost line into the soil will be very slow to accommodate the formation of ice lenses. Granular materials with large bore space generally do not have the problem of frost heave, because the supply of water by capillarity in this case is almost negligible.

Treatment of frost in susceptible subgrades differs from one location to another. One method of treatment is to lower the groundwater table below the frost-susceptible layer through the use of a suitable ground drainage system. Another method that has been successfully used in some situations is the use of a bituminous underseal to prevent or retard capillary rise. Silt lends itself to bituminous treatment because it is easily worked and is fairly stable when confined. A layer of the frost susceptible material a few inches in thickness can be mixed in place with bitumen and then properly spread and compacted before backfill material or sub-base is placed. This treatment also prevents pumping of the silt up into coarser sub-base layers.

4.2.3 Grading the Road or Airfield Bed

The purpose of grading is to provide a clear and stable foundation, to specified cross-sections upon which the pavement structure is to be placed. The level of the grade line is generally governed by alignment and gradient as well as by soil characteristics and drainage conditions.

Any topsoil which is removed during grading should be stockpiled for later use on finished cut-and-fill slopes. This will help planting the finished slopes to improve the appearance of the transportation facility and to prevent erosion.

4.2.4 Subgrade Compaction

Compacting the subgrade soil generally increases its shear strength, reduces its compressibility and reduces its permeability and absorption of water. These improvements are a result of the reduction of the voids of the compacted material and the increase in density.

For these reasons, proper compaction of subgrades for highways and airports is essential. Compaction of the subgrade to high densities and to considerable depths is particularly important if high volumes of heavy traffic are expected. Depth of compaction for subgrades generally ranges between 6 and 12 in. However, for subgrades serving as foundations for areas of channelised traffic on heavily loaded airfields, the depth of compaction may be as high as 5 ft.

Since the density of the soil depends upon the moisture content, it is necessary to carry out laboratory tests on the soil using various moisture contents. The moisture content giving the maximum density is referred to as the optimum moisture content. A moisture content slightly higher than the optimum value obtained from the laboratory tests is generally used.

The moisture content is defined as the weight of water present in a soil expressed as a percentage of the weight of dry soil. Thus, if a soil sample of a test specimen weighs 0·400 lb, and after oven drying to a constant weight it weighs 0·320 lb, the moisture content is $[(0·400 - 0·320)/0·320] \times 100$, or 25%.

Figure 4.1 shows the typical relationship generally obtained between moisture content and dry density for different soils when the standard Proctor density test is used.

As shown in Fig. 4.1, when the moisture content is increased, there occurs a certain amount of lubrication of the soil particles by the water, and it is possible to compact the soil to a greater density with a given amount of compactive effect. A point is then reached where further increase in moisture content will give less density under the same conditions of compaction. This is because the soil will have reached a state of near-saturation and further addition of water will act as an incompressible

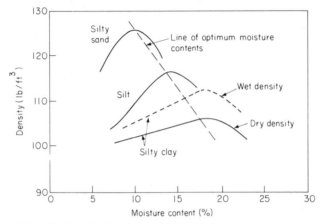

FIG. 4.1. Relationship between density and moisture content.

space filler to hold the soil grains apart. In addition, cohesive soils become plastic with the increase in moisture content and tend to flow around the compacting foot, thus reducing the effectiveness of the applied compactive effort. The moisture content at which maximum dry density is attained is the optimum moisture content.

The standard laboratory compaction test devised by R. R. Proctor is assumed to give soil densities in the laboratory as compatible as possible with those obtained in the field when the same moisture content is used during the compaction of the subgrade. This test is utilised by most high-way departments.

In this test, the soil is compacted in three equal layers in a mould 4 in in diameter and 1/30 ft^3 capacity. Each layer is tamped 25 blows with a 5$\frac{1}{2}$ lb hammer dropped from a height of 12 in. The soil is then levelled with the top of the mould, and the wet density, γ_w, determined by dividing the weight of the soil in the cylinder by the volume. The process is repeated for various moisture contents and the dry density γ_D, determined for each case using equation (4.1):

$$\gamma_D = \frac{\gamma_w}{1+w} \tag{4.1}$$

The optimum moisture content is not a constant value, but rather is a function of soil type and compactive effort. Granular materials have higher densities and lower optimum moisture contents for a given compactive effort than fine-grained materials. Also, the density–moisture content curves are steep for granular materials but relatively flat for clay and silty clay soils. For this reason, moisture control during compaction in the field is more critical for granular materials than for clay soils.

The density of the majority of soils increases as the compactive effort increases, but optimum moisture content decreases. Due to this fact and because of the availability of heavier and more effective compaction equipment for construction, a modified Proctor test is generally used for airfields. In the modified test, the soil is compacted in five layers of equal thickness. Each layer is tamped 25 blows with a 10 lb hammer dropped from a height of 18 in. This results in a compactive effort per unit volume which is about $4\frac{1}{2}$ times that of the standard test.

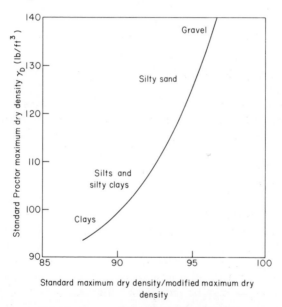

FIG. 4.2. Relationship between standard and modified maximum dry density.

A schematic representation of the relationship between the maximum densities obtained from the standard and modified Proctor tests is shown in Fig. 4.2.

The compaction of clay-like soils can be done most efficiently with sheep's foot rollers, keeping the compaction pressures around 300 psi. Pneumatic, grid-type, smooth-wheeled or sheep's foot rollers can be used in compacting silty soils. Granular materials are compacted efficiently by vibratory equipment.

Some soils can be compacted economically only to relatively low densities. In such a case, although a thicker pavement will be required because of the lower design strength value chosen for the subgrade, the total construction costs may be cheaper at the end.

On the other hand, if a pavement is to be constructed on a water-bearing layer of clay or silt that cannot be drained horizontally, it is important to reduce the moisture content in the soil to the optimum before compaction. By doing this, the strength of subgrade soil will be increased and the stability of the pavement structure maintained. To accomplish this, vertical sand drains can be constructed by drilling holes 15–20 in diameter through the compressible layer to firm soil, and filling the holes with sand. According to the permeability of the soil, the holes are spaced uniformly 8–30 ft centre to centre over the entire area to be drained. A thick layer of porous sand or gravel 18–30 in thickness should connect the tops of the vertical sand drains near the ground surface to carry any excess water outside the pavement edge.

Another problem which is often met is the transition from cut to fill. In this case the transition zone should be about 3–4 ft in depth and should extend gradually in each direction longitudinally for a distance of 30–40 ft. Full transition should also be provided on a transverse section, such as on sidehill fills. In this way, uneven stresses which may result in pavement break-up can be minimised.

4.2.5 Undesirable Types of Subgrades and Methods of Treatment

Subgrade soils which contain large quantities of mica or organic material are elastic and their resilience is high. These materials are subject to rebound upon removal of load and this may cause fatigue failure. Subgrade soils of this type, which are categorised by the HRB classification as A-5 or A-7, should be avoided if possible. If a subgrade of this type must be used, the problem may be reduced by compacting the soil to densities approaching the modified AASHO values.

Soils of high organic content should never be used for subgrades, because they can be dangerous to the stability of the pavement system.

High-volume-change soils, which shrink when water is removed from them, should be kept wet to combat volume change due to drying out. This can be accomplished by encasing the subgrade with plastic sheets or bituminous membranes or by proper utilisation of cover over the subgrade.

Swelling soils should be compacted at a moisture content which is near or slightly higher than the optimum amount. The weight of the pavement should also be great enough to combat the swelling pressures of the subgrades.

Water-bearing strata which will feed water into the pavement structure should be intercepted some distance away from the roadway section. Ditches should be constructed to a sufficient depth to ensure that free water in the ditch will always be below the base or sub-base course level.

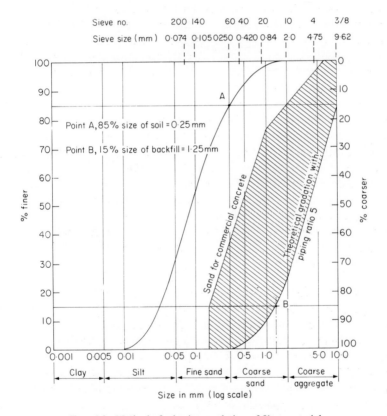

FIG. 4.3. Method of selecting gradation of filter material.

In the case of frost-susceptible soil present in the subgrade, it is necessary to lower the groundwater table by installation of side drains considerably deeper than the water table. The use of proper filter material around the drain is important. Large, open-graded backfill material generally becomes clogged because of infiltration of the surrounding soil, while sand can function as a good filter around drains. The usual control in choosing the gradation of the filter material is that the 15% size of the filter material should not exceed five times the 85% size of the soil being drained. This can be written in the following form:

$$\text{piping ratio} = \frac{15\% \text{ size of filter material}}{85\% \text{ size of soil}} \le 5 \qquad (4.2)$$

In addition, the grain-size curve for filter material should be approximately parallel to that for the soil, as shown in Fig. 4.3, so that most of the finer particles of soil will be effectively prevented from washing into the filter material. If it is not economically feasible to satisfy this condition, then this theoretical curve for the gradation of the filter material should represent the coarse side of the gradation range of particle size. The fine side of the gradation range is generally governed by a compromise between the physical and economical feasibility of modifying the grading and the fact that material with a coarser grading will, in general, perform more effectively in delivering water to the drain pipe.

4.2.6 Subgrade Classification

For each soil group, there are corresponding subgrade classes based on the performance of that soil group as a subgrade for flexible or rigid pavements, under different conditions of drainage and frost. Each subgrade class is determined from the results of soil tests, soil survey and a study of topographical and climatological data. The subgrade classes and their relationship to the soil groups, according to the FAA classification, are shown in Table 4.1 [2]. This classification is mainly for the design of airfield pavements. The prefixes F and R indicate subgrade classes for flexible and rigid pavements, respectively.

For flexible pavements, subgrades classed as F_a furnish adequate subgrade support without the addition of sub-base materials. The E-1 group of soils falls in the F_a subgrade class under all conditions of drainage and frost, while E-2 soils are classed as F_a subgrades only if the drainage is good. The other subgrade classes are designated F_1 through F_{10}, with the larger numbers indicating decreasing value of the soil as a subgrade material.

In the case of rigid pavements, R_a subgrades require no sub-base course. Soils down to the E-5 class may be classified as an R_a subgrade for suitable climatological conditions.

Poor drainage is defined as a condition where the subgrade may be rendered unstable due to: (1) inadequate drainage caused by the character of the soil profile, (2) topographic features such as a flat terrain slightly higher than sea level, (3) capillary rise from a high water table, or (4) any other cause that may result in instability or produce saturation of the subgrade.

TABLE 4.1 FAA CLASSIFICATION OF SUBGRADES
(Courtesy: Federal Aviation Agency)

Soil group	Subgrade class			
	Good drainage		*Poor drainage*	
	No frost	*Severe frost*	*No frost*	*Severe frost*
E-1	F_a or R_a	F_a or R_a	F_a or R_a	F_a or R_a
E-2	F_a or R_a	F_a or R_a	F_1 or R_a	F_2 or R_a
E-3	F_1 or R_a	F_1 or R_a	F_2 or R_a	F_2 or R_a
E-4	F_1 or R_a	F_1 or R_a	F_2 or R_b	F_3 or R_b
E-5	F_1 or R_a	F_2 or R_b	F_3 or R_b	F_4 or R_b
E-6	F_2 or R_b	F_3 or R_b	F_4 or R_b	F_5 or R_c
E-7	F_3 or R_b	F_4 or R_b	F_5 or R_b	F_6 or R_c
E-8	F_4 or R_b	F_5 or R_c	F_6 or R_c	F_7 or R_d
E-9	F_5 or R_c	F_6 or R_c	F_7 or R_c	F_8 or R_d
E-10	F_5 or R_c	F_6 or R_c	F_7 or R_c	F_8 or R_d
E-11	F_6 or R_d	F_7 or R_d	F_8 or R_d	F_9 or R_c
E-12	F_7 or R_d	F_8 or R_c	F_9 or R_c	F_{10} or R_c
E-13	Not suitable for subgrade			

Good drainage can be defined as a condition where: (1) the internal drainage characteristics will not allow of accumulation of water which develops spongy areas in the subgrade, (2) the water table is at such a level that the soil will not become water-logged either by percolation from above or capillarity from below, and (3) the topography is such that surface water will be removed rapidly.

A severe frost condition is assumed to exist if the depth of frost penetration for the particular site is greater than the anticipated thickness of surfacing, base and sub-base as determined for the case of no frost, and the drainage condition as defined in Table 4.1. Otherwise the condition of no frost prevails.

4.2.7 Determination of Subgrade Strength
In the design of flexible and rigid pavements, certain design values are required to represent the subgrade strength. The most important of these values are the California Bearing Ratio (CBR), the Bearing Value and the

Hveem's Resistance value (R) for flexible pavements, and the modulus of subgrade reaction (K) for rigid pavements. The test procedures used in determining these values are explained in Appendix C [15].

4.2.8 Test for Potential Frost Heaving Subgrades

The apparatus shown in Fig. 4.4 can be used to test the susceptibility of the subgrade soil to frost heave. Specimens isolated on their sides are

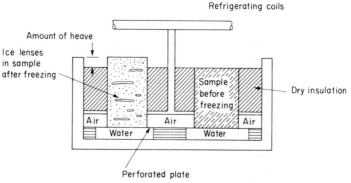

FIG. 4.4. Apparatus for determining soil heave.

subjected to a temperature gradient applied slowly from the top downward with a water resource available at the bottom of the specimen. A water reservoir placed underneath the specimen will allow it to take up water by capillary action. The increase in the height of the specimen will show the amount of heave.

4.3 SUB-BASES

The sub-base is the layer provided between the subgrade and the pavement. Major highways carrying heavy traffic and important airports usually include a sub-base as a part of the pavement structure, unless the subgrade is already of a very high quality.

In Germany, for example, new technical specifications and codes of practice for the construction of concrete pavements were issued in the spring of 1972 by the Ministry of Transport [12]. According to these specifications, roads carrying on the average more than 5000 vehicles per double lane per day, or more than 500 trucks of over 11 kips (5 t) useful load, should have a supporting sub-base course under the concrete pavement. This layer may be of compacted gravel, crushed rock, asphalt course, soil–cement or lean concrete.

4.3.1 Sub-base Materials and their Function

The sub-base may consist of one or more layers properly compacted. Availability of good borrow material will usually dictate what is to be used for sub-base construction. The sub-base should have greater stability and bearing power, better capability to drain accumulated water and less susceptibility to volumetric changes and frost action than the subgrade. The use of granular or stabilised materials for the sub-base has been common in much pavement construction. The sub-base may serve one or more of the following purposes: (1) to provide uniform support, (2) to increase the supporting capacity above that provided by the subgrade soil, (3) to minimise the detrimental effects produced by volume changes in the subgrade, (4) to minimise or eliminate the detrimental effects of frost action by using well-drained sandy and gravelly sub-base layers, and (5) to prevent pumping.

If sub-base grading is such that underlying fine-grained subgrade soils are prevented from penetrating any appreciable distance upward into this layer, then the sub-base can also serve as a filter course.

A sub-base thickness greater than 6 in may be used to replace frost-susceptible subgrade soils (these consist of materials possessing sufficient permeability and a relatively high degree of capillarity, such as silts and varved clays), minimise the effect of volume change, provide increased sub-base pavement strength and prevent pumping.

Pumping action is defined as ejection of water and subgrade soil through joints and cracks and along the edges of pavements; it is caused by downward slab movement actuated by the passage of heavy axle loads over the pavement after the accumulation of free water on or in the subgrade. Three basic conditions must be present to create pumping: (1) frequent heavy loads, (2) fine-grained soils that will go into suspension with water, and (3) free water under the pavement, usually accumulating from infiltration of water through the pavement cracks and joints and along the edges of the pavement.

If any of the foregoing factors is absent, pumping will not occur. Breaking of pavements will result from continued pumping.

To prevent pumping, surface water should be drained away over the shoulders as quickly as possible. Also, the sub-base should be well-graded from coarse to fine. If the sub-base consists of open-graded materials, the No. 200 material should be kept to a minimum with a maximum of 10%, while up to 15% may be permissible in dense-graded materials and up to 10% in sandy materials.

A 3–12 in granular sub-base layer may be required for combating a swelling subgrade, so that the weight of this layer plus the weight of the pavement will hold down the subgrade and absorb most of the expansion.

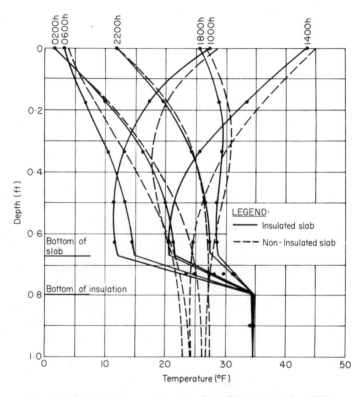

FIG. 4.5. Slab tautochrones showing effect of insulation. (From [8].)

Where economical, the finished sub-base, for a major highway, should have a modulus of subgrade reaction, K, of not less than 200 psi/in for cumulative traffic of less than 0·5 million standard axle and not less than 300 psi/in for cumulative traffic in excess of this figure. For airfields of major airports, the K value should generally be in excess of 400 psi/in. When frost action is anticipated with an E-6 subgrade soil (which is the worst from the standpoint of frost susceptibility), a total thickness of sub-base and pavement equal to the depth of frost penetration may be needed. In such a situation, it may be more economical to reduce the depth of the sub-base and either lower the level of the groundwater table or use an insulation layer (Styrofoam or similar). During the wintertime, such a layer may raise the temperature below the pavement above the freezing temperature, as shown in Fig. 4.5. In this way the frost action can be eliminated or significantly reduced.

4.4 STABILISED SOILS

The basic ingredients for a stabilised soil are reasonably well-graded granular material and a cementing agent or binder. The most elementary binder is clay. These ingredients are frequently found in natural sand–clay mixtures, or they can be built up by mixing sand with a soil having an overabundance of clay, or by mixing clay with a sandy soil.

4.4.1 Sand–Clay Stabilised Soils

Because clay has the ability to hold moisture in thin films on individual particle surfaces, thus producing powerful molecular attraction, it possesses high adhesive and cohesive properties. As the clay's moisture content increases, the adsorbed films of water become thick enough to provide lubrication between particles and the clay mass becomes plastic. If more water is added, the clay will become liquid and will lose its adhesive and cohesive properties.

The amount of clay required for stabilising sand or coarse material should not be excessive; otherwise it will disperse the coarser particles, with a resulting loss of interlock and friction between the sand particles.

If the stabilised sand–clay layer is designed to serve as a sub-base for

TABLE 4.2 GRADING FOR STABILIZED SAND–CLAY SOILS AS RECOMMENDED BY AASHO

| | | Grading type | |
| | *1* | *2* | *3* |
Rainfall conditions in the region	*Heavy*	*Moderate*	*Rare*
Sand portion:			
Passing sieve No. 10	100	100	100
Passing sieve No. 40	40–80	40–80	40–80
Passing sieve No. 60	30–70	40–55	55–70
Passing sieve No. 270	10–40	20–35	30–50
Silt portion, 0·05–0·005 mm	3–20	0–15	10–20
Clay portion, finer than 0·005 mm	7–20	9–18	15–25

the pavement, then it is desirable that from 10 to 25% of the material be coarse aggregate retained on sieve No. 10, but smaller than $1\frac{1}{2}$ in. The gradings of the portion passing sieve No. 10, as recommended by AASHO. are shown in Table 4.2.

In addition, the soil–mortar material passing sieve No. 40 should have a liquid limit less than 30 and a plasticity index less than 10.

If natural material with these requirements is available, it is spread over the roadbed to the desired depth and then compacted with a sheep's foot roller or similar, that produces compaction from bottom to top. It should be pointed out that soil generally compacts to about two-thirds of its loose depth. The depth of the compacted layer may be 4–9 in thick, according to the situation, and the moisture content should be kept near optimum to attain maximum density.

4.4.2 Cement Stabilised Soils

In some situations, especially where swelling subgrades are present, chemical stabilisation of the subgrade may be required so that a sufficient weight of pavement structure can be obtained to hold down the subgrade and absorb most of the expansion. Stabilisation of soils using cement has shown superiority in recent years over mechanically stabilised soils. The AASHO road tests showed that cement stabilised materials perform in an excellent way so long as they are not overstressed.

Most soils can be stabilised with cement, except highly organic soils such as agricultural top soils and peats. Also, heavy cohesive soils are difficult to break up sufficiently to mix in the cement. For this reason, it is important to make a soil survey before stabilisation to ensure that the soil will respond to treatment.

Clays with a liquid limit up to 40% can generally be stabilised. Granular soils with the following range of gradations can be stabilised with cement:

Maximum size	3 in	
Passing sieve No. $\frac{3}{16}$ in	more than 50%	
Passing sieve No. 36	more than 15%	by weight
Passing sieve No. 200	less than 50%	
Finer than 0·002 mm	less than 30%	

Tests have shown that the highest strength of cement stabilised soils is obtained at approximately the optimum moisture content which provides maximum density during compaction. This amount of moisture is generally enough for the hydration of the cement. To determine the proportion of cement required, tests should be done with different cement contents to determine which is most satisfactory for strength with economy. A compressive strength of 400 psi after 7 days from mixing is usually accepted as the desirable minimum. The cement content affects not only the strength but also the durability. The pavement foundation must be able to withstand the effects of water and frost in addition to the applied loads. Soils such as silt and chalk, which exhibit large susceptibility to frost heave in their natural state, can be satisfactorily stabilised and rendered immune to frost action by mixing with cement.

The single-pass method can be used in constructing these cement stabilised soils when the proper equipment is available. The whole operation of construction is completed in one pass of the equipment. This includes the soil pulverisation, grading to the required levels and cement spreading and mixing, as well as application and mixing of water, compacting and finishing. The depth to which the soil is pulverised should be such as to give the required finished thickness after compaction.

No expansion or contraction joints are normally provided in cement stabilised soils, and the construction joints are simple vertical butt joints. Curing of these stabilised soils is essential to reduce the size of shrinkage cracks.

4.4.3 Advantages of Cement Stabilisation
Natural soils, industrial wastes and poor granular materials can very often be converted into good materials by the addition of cement. The stabilisation procedure allows of the use of materials that are generally unstable, in their natural state, as base and sub-base layers for roads, airfields and other pavements.

Cement stabilised materials prevent disruption by frost and provide a uniform working platform for subsequent construction. Their use as base and sub-base layers, instead of granular materials, helps to conserve high-grade aggregates for the construction of the pavement itself. In addition, stabilisation of low-cost local materials may result in a big saving in pavement construction costs.

4.4.4 Types of Soil–Cement Mixtures
Soil–cement, which is a mixture of pulverised soil and measured amounts of Portland cement and water compacted to high density, is used primarily as a base course for roads, streets, secondary airports, shoulders, and parking areas. It can also be used as a sub-base for rigid and flexible pavements.

There are three general types of soil–cement mixture.

1. Compacted soil–cement. This type contains sufficient cement both to withstand standard laboratory freeze–thaw and wet–dry tests and to meet weight loss criteria. It also contains enough moisture for maximum compaction. Since this type of cement stabilised soil is the most commonly used, it is generally referred to simply as soil–cement.

2. Cement modified soil. This type is an unhardened or semihardened mixture of soil and cement. By adding relatively small quantities of Portland cement and moisture to a soil, the chemical and physical properties of that soil are changed. This results in a reduction in the soil's plasticity and volume-change capacity, and causes an increase in its bearing capacity.

The use of cement in producing a cement modified soil can be applied both to silt clay soils and to granular soils.

3. Plastic soil–cement. This type of soil–cement is a hardened mixture of soil and cement that contains, at the time of placing, sufficient water to produce a consistency similar to that of plastering mortar. It differs from compacted soil–cement in the amount of water used.

Granular layers can also be strengthened by adding cement to them to obtain lean concrete. This type of concrete consists of clean, well-graded aggregate, mixed with cement in proportion of 15:1 to 20:1 by weight. It differs from the standard concrete in that the amount of cement used is much less, accordingly a weaker product is obtained.

Lean concrete is generally used in situations where the upper layer of a granular, frost-resisting sub-base has to be strengthened by a cement binder.

4.4.5 Materials and Engineering Properties of Soil–Cement

Most soils can be hardened with Portland cement. On the basis of gradation, soils for soil–cement can be divided into three main groups.

1. Sandy and gravelly soils with about 10–35% silt and clay combined. These have the most favourable characteristics and generally require the least amount of cement for adequate hardening.

2. Sandy soils that are deficient in fines, such as some beach, glacial and wind-blown sands. They make good soil–cement if higher amounts of cement are used. Construction equipment may have difficulty in obtaining traction due to the poor gradation of these sands. However, traction can be vastly improved by keeping the sand wet and by using track-type equipment.

Since these soils are likely to be 'tender', they require special procedures during compaction and finishing to obtain a smooth dense surface.

3. Silty and clayey soils. Soils of this group, which can be pulverised, will provide satisfactory soil–cement. However, difficulties are sometimes experienced in pulverising these soils, especially if the clay content is high. Another useful system for classifying soils is the pedological system, which considers the soil profile. A soil profile is a vertical cross-section of the earth's surface, exposing the different soil horizons or layers. The typical soil profile, as applied to civil engineering, consists of three layers or horizons. The lower horizon, designated parent material or C horizon, consists of the original unweathered soil. The top and middle layers, designated as A and B horizons, constitute the weathered layers. The A horizon experiences maximum illuviation or leaching as a result of the relatively high degree of weathering. Normally, the organic matter content is rather high in the A horizon because of local vegetation. The B horizon is the layer where illuviation or deposition of the material leached from the A horizon occurs. As such, the B horizon normally contains a much higher clay content and displays higher plasticity than the A horizon; however, the organic matter content of the B horizon is generally much lower. The C horizon, consisting of relatively unaltered parent material, does not reflect the influence of the various chemical and physical weathering forces of the environment.

Each soil horizon is generally of a different texture, structure and colour. Soils formed from similar parent materials and under similar conditions

TABLE 4.3 NORMAL RANGE OF CEMENT REQUIREMENTS
FOR B AND C HORIZON SOILS[a]
(Courtesy: Portland Cement Association)

A A S H O soil group	*Percent by vol.*	*Percent by wt.*
A-1-a	5·7	3·5
A-1-b	7·9	5·8
A-2-4		
A-2-5		
A-2-6	7·10	5·9
A-2-7		
A-3	8·12	7·11
A-4	8·12	7·12
A-5	8·12	8·13
A-6	10·14	9·15
A-7	10·14	10·16

[a] A horizon soils (topsoils) may contain organic or other material detrimental to cement reaction and may require higher cement factors. For dark grey to grey A horizon soils, increase the above cement contents 4 percentage points; for black A horizon soils, 6 percentage points.

of climate, topography, drainage and vegetation are alike with similar profiles. These soils have been identified, by soil series, by the U S Department of Agriculture's Soil Conservation Service.

Studies have indicated that soils of the same soil series and horizon and of similar texture, wherever they may be found, will require the same amount of cement. Table 4.3 shows the usual range of cement requirements of AASHO soils groups for B and C horizon soils, while Table 4.4 provides a guide for average cement requirements of miscellaneous materials. Almost all types of cement used in ordinary concrete can function properly in soil–cement.

TABLE 4.4 AVERAGE CEMENT REQUIREMENTS OF
MISCELLANEOUS MATERIALS
(Courtesy: Portland Cement Association)

Material	Percent by vol.	Percent by wt.
Caliche	8	7
Chat	8	7
Chert	9	8
Cinders	8	8
Limestone screenings	7	5
Marl	11	11
Red dog	9	8
Scoria containing plus No. 4 material	12	11
Scoria (minus No. 4 material only)	8	7
Shale or disintegrated shale	11	10
Shell soils	8	7
Slag (air-cooled)	9	7
Slag (water-cooled)	10	12

Standardised tests are generally used to determine the quantities of Portland cement and water to be added and the density to which the mixture must be compacted. The water serves two purposes: the first is to help obtain maximum density by lubricating the soil grains, while the second is that water is necessary for cement hydration. This is the process of hardening and binding the soil into a solid mass.

Soil–cement is compacted to a high density during construction. As the cement hydrates, the mixture hardens to produce a structural slab-like material which consolidates no further under traffic; it neither ruts nor shoves during spring thaws.

Depending on soil type the 7 day compressive strength of saturated soil–cement specimens with the minimum cement content required to meet soil–cement criteria is generally higher than 300 psi. As shown in Fig. 4.6, the 28 day flexural strength is approximately 20% of the compressive strength. The modulus of elasticity after 28 days is about 1 million psi. These strength properties increase significantly with any increase in curing time.

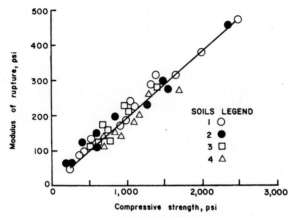

FIG. 4.6. Relationship between modulus of rupture and compressive strength of soil–cement. *(Courtesy: Portland Cement Association.)*

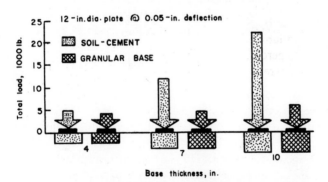

FIG. 4.7. Load-carrying capacity of soil–cement and granular base. *(Courtesy: Portland Cement Association.)*

Because soil–cement acts as a structural material, it has relatively high load-carrying capacity. Figure 4.7 shows that soil–cement can support up to three times greater loads than granular bases of the same thickness, when the thickness of the layer exceeds 7 in.

4.4.6 Design Procedure for Soil–Cement Pavements

The design procedure described hereafter is for a soil–cement pavement that will act as a base for a bituminous surface layer. The factors considered to determine the pavement thickness are: (1) strength of the subgrade, (2) design period for the pavement, (3) volume and distribution of traffic axle loads, (4) thickness of soil–cement base course, and (5) thickness of bituminous surface layer.

Research done by the Portland Cement Assocation (PCA) in the USA has demonstrated that the strength of the pavement can be more accurately assessed by the degree of bending than by deflection measurements alone. For this reason, radius of curvature rather than deflection has been used as a main factor in developing the design formulations. Also, fatigue studies have revealed that, for a given design, the number of load repetitions to failure is related to the radius of curvature of bending, and that the effect of soil type is significant. This has necessitated the division of soils into two broad textural types: granular and fine-grained soils. The two types may be differentiated by the AASHO soil classification. Groups A-1, A-3, A-2-4 and A-2-5 belong to granular soils, while A-2-6, A-2-7, A-4, A-5, A-6 and A-7 belong to fine-grained soils.

In developing charts for the thickness design procedure of soil–cement pavements, the Westergaard modulus of subgrade reaction, K, has been used as a measure of subgrade support. Also, an arbitrary design period of 20 years has been selected for use with this procedure. The designer may, however, select a different value for the design period, proportioning the total volume of traffic accordingly.

Since the weights and volumes of axle loads expected during the design period are major factors in determining the design thickness, the traffic analysis used in this procedure involves: (1) determination of the average daily traffic in both directions and the percentage of trucks, (2) projection of the traffic to a future design period, (3) determination of the probable axle-load distribution, and (4) computation of the fatigue factor.

The average daily traffic (ADT) in both directions and the percentage of trucks can be obtained directly from a traffic survey of the project or from data available from other projects carrying similar traffic. In this procedure, the percentage of trucks includes all panels, pickups, and other two-axle, four-tyre single-unit commercial vehicles, in addition to trucks with three or more axles. Rates of traffic growth and the corresponding 20 year projection factors, as given in Table 4.5, are used to estimate design ADT and the number of trucks that will use the pavement during the design period.

If the total number of traffic lanes in two directions is 2, the number of trucks in design lane is 50% of that in both directions. However, for 4-lane and 6-or-more-lane highways, this number becomes 45 to 40%, respectively, of the total number in both directions.

TABLE 4.5 YEARLY RATES OF TRAFFIC GROWTH AND CORRESPONDING PROJECTION FACTORS

Yearly rate of traffic growth (%)	Projection factor for 20 year design period
1	1·1
1½	1·1
2	1·2
2½	1·3
3	1·4
3½	1·5
4	1·5
4½	1·6
5	1·7
5½	1·8
6	1·9

To express the fatigue effects of the numbers and weights of truck-axle loads, it is necessary to compute the numbers of single and tandem axles of various weights expected during the design period. From axle-load data the number of single and tandem axles per 1000 vehicles can be determined for each load category. The expected numbers of the different axle loads can then be computed by multiplying these values by the number of trucks in the design lane during the design period and dividing by 1000.

In a research report prepared by the American Portland Cement Association (PCA), the number of load repetitions allowed on a soil–cement pavement of a certain thickness has been given in the form:

$$N = \left[\frac{(1·77K)^{A_1}}{C/f(h)} \right]^{A_2} \left(\frac{\sqrt{a}}{P} \right)^{A_2} \tag{4.3}$$

where N = allowable number of load repetitions;

K = modulus of subgrade reaction (pci);

A_1 = an exponent equal to 0·3 for granular soil–cements and to 0·315 for fine-grained soil–cements;

A_2 = an exponent equal to 40·0 for granular soil–cements and to 20·0 for fine-grained soil–cements;

C = a constant equal to 10·4 for granular soil–cements and 10·0 for fine-grained soil–cements;

$$f(h) = \frac{(2 \cdot 1h - 1)^2}{h^{1 \cdot 5}};$$

h = pavement thickness (in);
a = radius of load contact area (in); and
P = wheel load (kips).

The total fatigue consumption of the expected number of axle loads, n_i, of various magnitudes, P_i, has been expressed as:

$$\text{fatigue consumption} = \sum \frac{n_i}{N_i} = \sum \frac{n_i}{\left[\dfrac{(1 \cdot 77K)^{A_1}}{C/f(h)}\right]^{A_2} \left(\dfrac{\sqrt{a_i}}{P_i}\right)^{A_2}}$$

$$= \left[\frac{C}{f(h)(1 \cdot 77K)^{A_1}}\right]^{A_2} \cdot \sum \left(\frac{P_i}{\sqrt{a_i}}\right)^{A_2} \cdot n_i \qquad (4.4)$$

If the summation in eqn. (4.4) is divided by an arbitrarily selected value $(P_g/\sqrt{a_g})^{A_2}$ and the term outside the summation is multiplied by the same value, then:

$$\text{fatigue consumption} = \left[\frac{C}{f(h)(1 \cdot 77K)^{A_1}}\right]^{A_2} \left(\frac{P_g}{\sqrt{a_g}}\right)^{A_2} \cdot$$

$$\sum \left(\frac{P_i/\sqrt{a_i}}{P_g/\sqrt{a_g}}\right)^{A_2} \cdot n_i \qquad (4.5)$$

In this computation, an 18 kip single-axle load or a 9 kip dual-wheel load, P_g, has been chosen to determine the relative fatigue consumption of different axle-load magnitudes. The fatigue consumption coefficients, F_i, given in Table 4.6, represent the values of:

$$F_i = \left(\frac{P_i/\sqrt{a_i}}{P_g/\sqrt{a_g}}\right)^{A_2} \qquad (4.6)$$

The fatigue factor, T, represents the summation:

$$T = \sum F_i n_i \qquad (4.7)$$

If the fatigue consumption is set equal to 100%, the fatigue consumption computation procedure can be expressed as:

$$\left(\frac{C}{f(h)(1 \cdot 77K)^{A_1}} \cdot \frac{P_g}{\sqrt{a_g}}\right)^{A_2} \cdot T = 1 \qquad (4.8)$$

Because the values of A_1, A_2 and C are different for granular and fine-grained soil–cements, separate fatigue consumption coefficients are indicated in Table 4.6.

TABLE 4.6 FATIGUE CONSUMPTION COEFFICIENTS[a]
(Courtesy: Portland Cement Association)

Axle load (kips)	Granular soil–cement	Fine-grained soil–cement
Single axles:		
30	12 500 000	3 530
28	1 270 000	1 130
26	113 000	337
24	8 650	93
22	544	23·3
20	27	5·2
18	1·000 0	1·000 0
16	0·025 0	0·160 0
14	0·000 4	0·020 0
12	—	0·001 8
Tandem axles:		
50	12 500 000	3 530
48	3 210 000	1 790
46	792 000	890
44	186 000	431
42	41 400	203
40	8 650	93
38	1 690	41·1
36	305	17·5
34	50·4	7·1
32	7·5	2·74
30	1·000 0	1·000 0
28	0·120 0	0·341 0
26	0·012 0	0·107 0
24	0·001 0	0·031 0
22	—	0·008 1
20	—	0·001 8

[a] These coefficients express the relative fatigue consumption of different axle-load magnitudes.

In many cases axle-load distribution data are not available for the light-traffic category of pavements, used in residential streets and secondary roads. In the absence of this information, the values listed in Table 4.7 may be used to represent the fatigue requirements for soil–cement pavement design.

TABLE 4.7 REPRESENTATIVE FATIGUE FACTORS FOR
LIGHT-TRAFFIC PAVEMENTS
(Courtesy: Portland Cement Association)

Facility	ADT	Total trucks[a] (%) (approx.)	Heavy trucks[b] (%) (approx.)	Fatigue factor[c]
Purely residential streets	300–700	8	3	5–12
Residential collector streets	700–4 000	8	3	12–20
Secondary roads	Up to 2 000+	14–20	5–8	12–30

[a] All commercial vehicles, including two-axle, four-tyre vehicles.

[b] Excludes panels, pickups, and other two-axle, four-tyre vehicles that are seldom heavy enough to affect design thickness.

[c] These particular ranges of values for the Fatigue Factor are based on the following characteristics of street and secondary road traffic: (1) one-half the indicated number of heavy axle loads, one direction; (2) axle-load distributions varying from 12 000 to 20 000 lb on individual axles; (3) weighted averages of axle loads varying between 13 000 and 16 000 lb on individual axles.

4.4.7 Charts for Design of Soil–Cement Pavements

In this design procedure, the total fatigue consumption effect of the volumes and weights of single- and tandem-axle loadings, for a given design problem, is expressed by the fatigue factor, T, alone. The fatigue consumption coefficients, shown in Table 4.6, are multiplied by the numbers (in thousands) of axles in each weight group, and then summed to give the fatigue factor.

Figure 4.8 can be used to determine the thickness of soil–cement base courses for granular soil–cement, while Fig. 4.9 is applicable for the case of fine-grained soil–cement. The curves in both figures will give the thickness of the soil–cement layer if the values of the modulus of subgrade reaction, K, and the fatigue factor, T, are known. The thickness is usually read to the next highest half-inch.

Many factors should be considered in determining the thickness of the bituminous surface required on top of the soil–cement base course. Some of the important factors are: the type of surfacing, the volume and composition of traffic, climatic conditions, availability of materials and local practices. Under favourable conditions indicated by previous local experience, Table 4.8 can be used to determine the recommended and minimum thicknesses of the bituminous surface layer. This table is based on experience covering a wide range of the variables mentioned before.

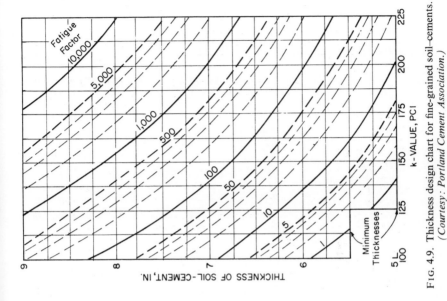

FIG. 4.9. Thickness design chart for fine-grained soil-cements.
(Courtesy: Portland Cement Association.)

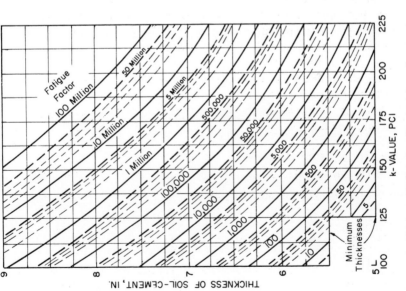

FIG. 4.8. Thickness design chart for granular soil-cements.
(Courtesy: Portland Cement Association.)

It should be stressed that this discussion of pavements which consist of a soil–cement base and a bituminous surface course, is only relevant to secondary roads and airfields with light traffic. However, soil–cement

TABLE 4.8 BITUMINOUS SURFACE THICKNESSES
(Courtesy: Portland Cement Association)

Soil–cement thickness (in)	Recommended bituminous surface thickness (in)	Minimum bituminous surface thickness (in)	
		Non-frost area	Frost area
5–6	$\frac{3}{4}$–$1\frac{1}{2}$	SBST[a]	DBST[a]
7	$1\frac{1}{2}$–2	DBST	1[b]
8	$1\frac{1}{2}$–$2\frac{1}{2}$	1	$1\frac{1}{2}$
9	2–3	2	2

[a] SBST, single bituminous surface treatment; DBST, double bituminous surface treatment.
[b] Where snowploughs are used, a minimum of $1\frac{1}{2}$ in is recommended.

sub-bases of 5–8 in thickness can be used underneath flexible or rigid pavements for major highways and airfields. The design of the rigid or flexible pavement itself follows the procedures explained in the next chapters.

Examples

1. Determine the thickness of a pavement consisting of a granular soil–cement base and bituminous surface course if the following data are given:

Local two-lane road with axle-load distribution as given in column (2) of Table 4.9.

Current average daily traffic (ADT)	= 1046
Yearly rate of traffic growth	= 4%
Truck traffic of all kinds	= 16% of ADT
Design period	= 20 years
Modulus of subgrade reaction	= 125 pci

Solution.

Projection factor for 20 year design period (from Table 4.5) $\quad = 1·5$

Design ADT $= 1·5 \times 1046 = 1569$

Truck traffic $= 0.16 \times 1569 = 251/day$
Each direction $= 50/100 \times 251 = 126/day$
Truck traffic, each direction for the design period
$$= 126 \times 365 \times 20 = 919\ 800$$

Number of axle loads expected during the design period is shown in column (3) of Table 4.9, and is obtained by multiplying 919 800 trucks by the axle loads per 1000 trucks given in column (2) of the same table. The

TABLE 4.9 TYPICAL COMPUTATIONS FOR DETERMINING
AXLE-LOAD DISTRIBUTION

Axle load group (kips) (1)	Axles per 1 000 trucks (2)	Axle loads in design period[a] (3)
Single axles:		
20–22	9·8	9 000
18–20	13·2	12 100
16–18	10·8	9 900
14–16	36·7	33 800
12–14	36·8	33 800
10–12	78·9	72 600
Tandem axles:		
40–42	5·9	5 400
38–40	2·9	2 700
36–38	8·3	7 600
34–36	7·8	7 200
32–34	13·1	12 000
30–32	4·9	4 500
28–30	4·4	4 000
26–28	4·4	4 000
24–26	4·4	4 000
22–24	8·1	7 500
20–22	8·2	7 500
18–20	8·2	7 500

[a] Products of 919 800 trucks times column (2) divided by 1 000.

fatigue factors can be obtained by dividing the number of axle loads in the design period by 1000 and multiplying by the fatigue consumption coefficients given in Table 4.6 for the different axle-load groups. Table 4.10 shows the computation of the fatigue factor, which is 268 000 for this example.

From Fig. 4.8, for granular soil–cements, the ʼrequired thickness of soil–cement base corresponding to a K value of 125 pci and a T value of 268 000, is 7·7 in. When rounded to the next highest half-inch the required thickness of the soil–cement base should be 8·0 in. Table 4.9 gives the corresponding bituminous surface thickness as 2 in. If severe weather conditions prevail, then a thickness of $2\frac{1}{2}$ in is recommended.

TABLE 4.10 TYPICAL COMPUTATIONS FOR FATIGUE FACTOR

Axle load (kips) (1)	Axle loads in design period[a] (thousands) (2)	Fatigue consumption coefficient[b] (3)	Fatigue effects[c] (4)
Single axles:			
22	9·1	544	4 900
20	12·1	27	327
18	9·9	1	10
16	33·8	0·025	1
Tandem axles:			
42	5·4	41 400	223 600
40	2·7	8 650	23 400
38	7·6	1 690	12 800
36	7·2	305	2 200
34	12·0	50·4	600
32	4·5	7·5	34
30	4·0	1·0	4
		Total	267 876
		Fatigue factor	268 000

[a] Number from Table 4.9 column (3), divided by 1 000.
[b] From Table 4.6 for granular soil–cement.
[c] Products of columns (2) and (3).

2. Determine the thicknesses of the soil–cement and bituminous layers of a pavement if the following data are given:

Residential street where no axle-load data is available
　　Fine-grained soil–cement
　　Modulus of subgrade reaction, K　　　　　　　　　　= 100 pci
　　Current average daily traffic (ADT)　　　　　　　　= 600
　　Yearly rate of traffic growth　　　　　　　　　　　= 1%

Solution.

Projection factor for 20 year design period (from Table 4.5)	$= 1{\cdot}1$
Design $ADT = 1{\cdot}1 \times 600$	$= 660$
Fatigue factor estimated from Table 4.7	$= 12$
Thickness of soil–cement base (from Fig. 4.9)	$= 7{\cdot}0$ in
Thickness of bituminous surface layer (from Table 4.8)	$= 1{\cdot}5$ in
Recommended thickness of bituminous layer in case of severe weather conditions	$= 2{\cdot}0$ in

4.5 CEMENT BOUND THERMAL INSULATION LAYERS

In regions where the subgrade is expected to freeze underneath the pavement in Winter, the usual practice is to use thick layers of granular, well-drained sub-bases. These granular materials are becoming difficult to obtain in many areas. An alternative is for an insulating base layer of styropor to be placed directly on the subgrade [1].

The aggregate for the styropor is brought to the site in the form of compact polystyrol granules and is expanded with superheated steam to 50 times its initial volume. In this way styropor balls of $0{\cdot}04{-}0{\cdot}12$ in (1–3 mm) diameter with a bulk weight of 17–25 lb/yd^3 (10–15 kg/m^3) are obtained. A truck with trailer can carry about 1700 yd^3 (1300 m^3).

In general, the mix consists of 25 lb/yd^3 (15 kg/m^3) styropor, 660 lb/yd^3 (390 kg/m^3) cement and 205 lb/yd^3 (120 kg/m^3) fine sand. A water/cement ratio of $0{\cdot}40$ is generally used for the mix. The 7 day compressive strength of the mix should not be less than 300 psi (20 kg/cm^2). The specific weight is generally between 1000 and 1200 lb/yd^3 (600–700 kg/m^3). For light freezing conditions, a styropor layer 6 in (15 cm) in thickness above the subgrade is generally sufficient. However, in locations where the freezing depth of the soil is large and where heavy loads are expected, a layer of 6 in lean concrete may be placed on top of the styropor layer. Another alternative in such a location is to put the styropor layer on top of a granular sub-base of 6–12 in thickness.

REFERENCES AND BIBLIOGRAPHY

1. Austrian Research Institute for Cement Factories, 'Experiences with Cement-Bound Thermal Insulation Layer', *2nd European Symposium on Concrete Roads, Bern, 1973.*

2. Federal Aviation Agency, 'Airport Paving', US Government Printing Office, 1962.
3. Hennes, R. G. and Ekse, M., *Fundamentals of Transportation Engineering*, Second Edition, McGraw-Hill, 1969.
4. *Highway Research Record*, No. 128, 'Frost, Physical Properties and Stabilization', Highway Research Board, Washington, D.C., 1966.
5. *Highway Research Record*, No. 198, 'Stabilized Soils: Mix Design and Properties', Highway Research Board, Washington, D.C., 1967.
6. *Highway Research Record*, No. 315, 'Soil Stabilization: Multiple Aspects', Highway Research Board, Washington, D.C., 1970.
7. Yoder, E. J., *Principles of Pavement Design*, Wiley, 1959.
8. LeMoal, G. A., 'Thermal Behaviour of Rigid Pavement Slabs', M.Sc. Thesis, Department of Civil Engineering, The University of Calgary, 1971.
9. Portland Cement Association, *Soil–Cement Construction Handbook*, 1962.
10. Portland Cement Association, *Soil–Cement Laboratory Handbook*, 1971.
11. Portland Cement Association, 'Thickness Design for Soil–Cement Pavements', *Engineering Bulletin*, 1970.
12. Schuster, F. O., 'The Behaviour of Concrete Pavements on Stabilized Bases', *2nd European Symposium on Concrete Roads, Bern, 1973*.
13. Sharp, D. Raymond, *Concrete in Highway Engineering*, Pergamon Press, 1970.
14. Siedek, P., 'Sub-bases for Concrete Pavements', *2nd European Symposium on Concrete Roads, Bern, 1973*.
15. The Asphalt Institute, 'Soil Manual for Design of Asphalt Pavement Structures', Manual Series No. 10, Second Edition, April 1963.

PART II

FLEXIBLE PAVEMENTS

CHAPTER 5

CHARACTERISATION OF MATERIALS FOR FLEXIBLE PAVEMENTS

5.1 INTRODUCTION

Materials characterisation can be defined as the selection of constitutive equations to adequately model the response of paving materials to the loading and environmental conditions which they will be subjected to as components of a pavement. This characterisation can take the form of determining the elastic constants and failure criteria so that elastic theories may be used to compute critical stresses and strains. The acceptability of these may be assessed in terms of the anticipated life of the pavement.

In addition to the stress–strain characteristics of the materials used in the various layers of the pavement structure, information is required on the likely mode of failure of the various materials, under repeated loads. This information can be used to set design criteria, in the form of maximum allowable stresses or strains to be used in the design.

5.2 FLEXIBLE PAVEMENT MATERIALS

The materials generally used in flexible pavement construction fall essentially into four categories [18]: bituminous bound, cement bound, unbound aggregate, and cohesive subgrade soil. The unbound aggregate category may include non-cohesive soils. These materials are in fact inhomogeneous, anisotropic, non-elastic and non-linear; some of their properties are time-dependent and affected by environmental changes such as temperature and moisture content. For this reason, it is necessary to be selective in the use of material and pavement designs, with due consideration being given to their performance under the prevailing climatic conditions.

Accurate computation of stresses and strains in pavement structures consisting of different layers and materials whose behaviour is usually influenced greatly by time, temperature, moisture, etc., is an extremely complex task. Although the use of modern techniques such as the finite element method, in dealing with the non-linear response of pavement materials, is useful as a research tool, it still lacks practical applicability. The most promising design approach still appears to be that based on the use of linear elastic theory. In this approach, the non-linear characteristics of the material can be handled by an iterative process, whereby a different set of values for the elastic constants is used for each stress level. Other changes in material properties with time, temperature and moisture may be dealt with in a similar manner.

One of the major problems in relating material characteristics to actual pavement design is that pavement performance under traffic does not depend solely on the characteristics of the materials in the individual layers, but rather on the interaction of the various layers. For this reason it appears that the value of those laboratory tests concerned with failure of materials is mainly of a qualitative nature. These tests can be used to compare the performance of different materials and to obtain an understanding of the factors affecting performance. Design criteria will have to be established from relationships between the relevant strains, deflections or stresses, and actual performance on roads with traffic.

5.3 BITUMINOUS BOUND MATERIALS

5.3.1 Determination of Stiffness Modulus for Asphalt Mixes

Bituminous materials are viscoelastic. Their stress–strain characteristics are dependent on both temperature and time of loading.

Van der Poel defined the stiffness, S, of bituminous bound materials as the relationship between stress and strain expressed as a function of loading time, t, and temperature, T.

$$S_{t,\,T} = \frac{\sigma}{\varepsilon} = \frac{\text{stress}}{\text{total strain}} \tag{5.1}$$

The variation of this stiffness, or stiffness modulus, with time and temperature is shown in Fig. 5.1 for a typical mix. Under usual traffic conditions, the stiffness of an asphalt mix may vary from about 3×10^6 psi at low temperatures and fast speeds to about 7×10^4 psi at high temperatures and creep speeds. This indicates that accurate values of mix stiffness can only be obtained if the measurements are made under the

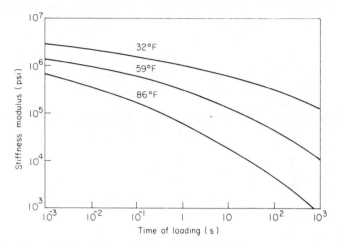

FIG. 5.1. Effect of loading time and temperature on the stiffness of a typical asphalt mix for a base course. (From [18].)

appropriate conditions of time, temperature and stress. The stiffness of asphalt mixes also depends on the mix variables, as well as on environmental factors. Examples of these mix variables are aggregate type and grading, bitumen type and content, degree of mix compaction and percentage of air voids. Also, recent work has indicated that the stiffness is stress-dependent, i.e. the material is non-linear. However, this effect is small compared with that of temperature.

Since pavements serve traffic which is generally in motion, it is preferable to determine the stiffness of the asphalt mixes from dynamic-type tests, utilising sinusoidal or pulse loading. For this purpose, either field tests or tests on samples of the mix can be used.

5.3.2 Field Tests on Flexible Pavements
In the case of field tests, equipment consisting of a special trailer carrying an inverted electromagnetic vibrator can be utilised [23]. The vibrator is supported by a spring on a 1 ton cylinder which makes contact with the pavement surface through a 12 in diameter steel plate. The applied sinusoidal vertical force is measured by a calibrated geophone fixed to

the vibrator, while the resultant velocity is measured by a similar cali-brated geophone, located centrally near the top of the 1 ton mass. The quantities measured are: (1) the vertical oscillating force, F, applied to the system; (2) the resultant vertical velocity, v, or the displacement, y; (3) the phase angle by which v loads on F; and (4) the frequency, f, of the applied load, usually between 14 and 200 c/s. The sinusoidal stress in the pavement due to the vibrating load can be expressed by [2]:

$$\sigma = \sigma_0 \sin \omega t \tag{5.2}$$

where σ_0 is the maximum stress, ω is the angular frequency ($\omega = 2\pi f$) and t is the time.

The resulting strain, ε, is of the same frequency as the stress, but out of phase with it by an angle ϕ; it is expressed as follows:

$$\varepsilon = \varepsilon_0 \sin (\omega t - \phi) \tag{5.3}$$

The complex modulus E^*, which is the ratio between the stress and corresponding strain, can be obtained from eqns. (5.2) and (5.3) as follows:

$$E^* = \frac{\sigma}{\varepsilon} \tag{5.4}$$

and the dynamic stiffness, S, becomes the ratio of amplitude of stress to that of strain:

$$S = \frac{\sigma_0}{\varepsilon_0} \tag{5.5}$$

5.3.3 Laboratory Tests on Asphalt Concrete Specimens

The dynamic stiffness and Poisson's ratio for compacted asphalt concrete specimens can also be obtained by using non-destructive dynamic tech-niques such as ultrasonic waves [22]. This technique, which has been used by acoustic engineers in studies of homogeneous materials such as metals, plastics and glass, can also be applied to asphalt paving materials. By measuring the propagation velocity of high-frequency sound waves through the material, various material constants, such as the dynamic stiffness, can be determined. The test procedure is dynamic, and corresponds to the type of loading that occurs on the actual structure.

The electronic equipment generally used to generate and detect ultra-sonic waves consists of a pulse generator, source and receiver piezoelectric ceramic transducers, and an oscilloscope. The set of transducers used in determining the dynamic stiffness generates primarily compressional waves. The resonant frequency for the compressional ceramic discs may be of the order of 300 kc/s.

The compacted asphalt concrete specimens usually employed are of 4 in diameter and 2 in high. The internal specimen temperatures can be monitored by implanting thermistors in the material. When choosing the locations of these thermistors, it is important to make sure that they do not interfere with the transmission of the sound pulse.

Due to the inherent non-homogeneity of asphalt concrete, longer wavelengths of relatively large amplitude are generally needed to overcome the attenuation tendencies of the material. Because of the high attenuation and scatter in the material, it is recommended that identical source and receiver transducers be used put directly opposite to each other on the top and bottom faces of the test specimen.

A direct transmission technique using a cathode ray oscilloscope can be utilised to measure the time lapse between the actuation of the wave source and the detection of the generated wave at the receiver. Once the longitudinal and transverse wave velocities are determined, the dynamic stiffness and Poisson's ratio can be calculated as follows:

$$S = \left[3 - \frac{1}{\left(\dfrac{V_c}{V_s}\right)^2 - 1} \right] pV_s^2 \cdot \frac{1}{144} \qquad (5.6)$$

$$v = \left[\frac{1 - \frac{1}{2}\left(\dfrac{V_c}{V_s}\right)^2}{1 - \left(\dfrac{V_c}{V_s}\right)^2} \right] \qquad (5.7)$$

where S = dynamic stiffness (psi);
v = Poisson's ratio;
V_c = longitudinal wave velocity (ft/s);
V_s = transverse wave velocity (ft/s); and

$$p = \text{mass density of the material} = \frac{\text{unit wt. (lb/ft}^3)}{32 \cdot 2}.$$

A brief study of some asphalt concrete mixes, using this technique, has resulted in the following observations:

1. Both the dynamic stiffness and the shear modulus of the asphalt concrete mix decreased as the temperature increased.
2. The maximum dynamic stiffness occurred at an 'optimum' asphalt content for wave transmission of 6%.

3. The dynamic Poisson's ratio increased directly with the increase in asphalt content.
4. Poisson's ratio, which is generally taken as 0·35–0·40 for design purposes, increased rapidly towards the theoretical maximum of 0·5 as the temperature exceeded 100°F.
5. While the amount of voids in the specimen had little influence on the rate of wave transmission through the specimens at low temperatures, the influence was more pronounced when the temperature exceeded 80°F.

5.3.4 Effect of Temperature and Load Frequency on Stiffness

The design, in many cases, requires the conversion of measured stiffnesses to other temperatures and frequencies. In such situations, the dynamic

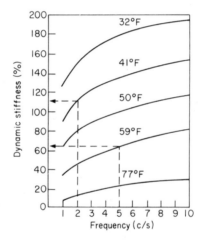

FIG. 5.2. Nomograph for conversion of the dynamic stiffness for asphalt concrete mixes.

stiffness can be estimated approximately by using the nomograph shown in Fig. 5.2, which has been developed by Guericke and Weinert [6].

Example
The dynamic modulus, *S*, of an asphalt concrete mix has been measured at 41°F and 2 c/s; its value is found to be 800 000 psi. It is required to determine the value of *S* for the same mix but at a temperature of 59°F and a frequency of 5 c/s.

Solution. Using Fig. 5.2:

$$S_{59°\,F,\,5\,c/s} = 800\,000 \times \frac{65}{112} = 464\,300 \text{ psi}$$

The relationship between the time of loading of a vehicle in motion and the frequency can be expressed as follows:

$$t = \frac{1}{f} = \frac{2\pi}{\omega} \qquad (5.8)$$

where t = duration of load (s);

ω = angular frequency; and

f = frequency (c/s).

The time of loading of a vehicle moving at a speed of 30 mph is of the order of 0·1 s.

5.3.5 Dynamic Stiffness of Bituminous Materials

If direct measurement of the dynamic stiffness of an asphalt mix is not possible, then an estimate of the stiffness may be made if the stiffness of the bituminous binder is known.

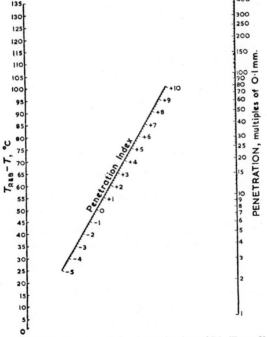

FIG. 5.3. Nomograph for determination of PI. (From [24].)

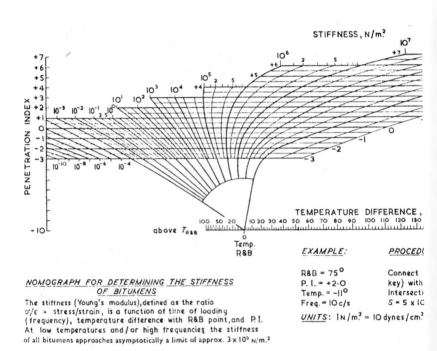

STIFFNESS, N/m²

PENETRATION INDEX

TEMPERATURE DIFFERENCE,

above $T_{R\&B}$

Temp.
R&B

<u>NOMOGRAPH FOR DETERMINING THE STIFFNESS</u>
<u>OF BITUMENS</u>

The stiffness (Young's modulus), defined as the ratio
σ/ϵ = stress/strain, is a function of time of loading
(frequency), temperature difference with R&B point, and P.I.
At low temperatures and/or high frequencies the stiffness
of all bitumens approaches asymptotically a limit of approx. 3×10^9 N/m.²

<u>EXAMPLE:</u>   <u>PROCEDU</u>

R&B = 75°   Connect
P. I. = +2·0   key) with
Temp. = −11°   Intersecti
Freq. = 10 c/s $S = 5 \times 10$

<u>UNITS</u>: I N/m.² = IO dynes/cm.²

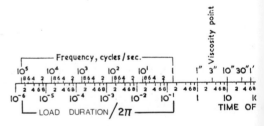

Frequency, cycles/sec.

Viscosity point

LOAD   DURATION $/2\pi$

TIME OF

F I G. 5.4. Nomograph for determining the stiffness of bitumens. (From [24].)

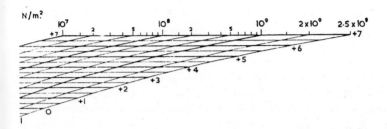

N/m.²

DIFFERENCE , °C
100 110 120 130 140 150 160 170 180 190 200   below $T_{R&B}$

<u>*PROCEDURE TO OBTAIN STIFFNESS MODULUS*</u>

Connect 10 c/s - point on scale A (see key) with 86°(75 + 11) - point on scale B. Intersection with +2·0 P.I. line gives
S = 5 x 10⁸ N/m.²

= 10 dynes / cm.² = ₁1·02 x 10⁻⁵ kg./cm.² = 1·45 x 10⁻⁴ lb./sq.in.

<u>*PROCEDURE TO OBTAIN VISCOSITY*</u>

Similar to stiffness procedure, only connect viscosity point on lower scale with temperature point. Figure reading gives viscosity in N sec./m.²
1 N sec./m.² = 10 poises

KEY:

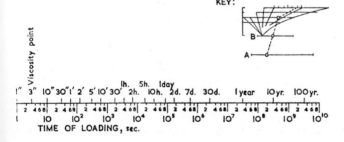

Viscosity point

1″ 3″ 10″30″1′ 2′ 5′10′30′ 2h. 10h. 2d. 7d. 30d. 1 year 10yr. 100yr.
1h. 5h. 1day

1   10   10²   10³   10⁴   10⁵   10⁶   10⁷   10⁸   10⁹   10¹⁰
TIME OF LOADING, sec.

FIG. 5.4—*contd.*

A nomograph developed by Van der Poel [24], which is based on the results of both static and dynamic tests, allows the determination of the stiffness of the bituminous binder at any particular temperature and time of loading. This stiffness is analogous to an elastic modulus at short loading times and low temperatures. Even though the results of static and dynamic measurements in this nomograph are plotted on the same graph, there will never be any doubt as to whether the plot refers to a static or dynamic test. This is because dynamic tests can only be used for times smaller than 1 s, static tests only for longer times.

In plotting the nomograph, a penetration index (PI) was used. This index indicates the temperature susceptibility of the penetration of bitumen and can be determined by carrying out both a penetration test and a softening point, ring-and-ball test on a sample of the bitumen. In the latter test, a ring of given dimensions filled with bitumen is loaded with a steel ball (3·5 g) and the whole system is heated in a bath. The temperature at which the bitumen reaches a certain deformation is reported as the ring-and-ball softening point ($T_{R\&B}$). The nomograph shown in Fig. 5.3 can be used to determine the value of the penetration index for the standard types of bitumen, once the penetration $T_{R\&B}$ and T are obtained. T refers to the temperature in °C at which the penetration test is carried out.

By using the nomograph shown in Fig. 5.4, it is possible to determine the stiffness, S, of bitumen as a function of stress, time (or frequency) and temperature. The nomograph consists of three parallel lines on which suitable scales for t, $(T_{R\&B} - T)$ and S are marked. These lines are located in such a way that a set of values for the variables obeying the formula used in developing the nomograph is on a straight line crossing the three scale lines. The units used are: time, t (s); frequency, f (c/s); temperature difference $(T - T_{R\&B})$ (°C); and stiffness, S (N/m²).

The nomograph can also be used to determine the viscosity of bitumen at a certain temperature. The procedure for this is to connect the 3 s viscosity point on the lower scale with the temperature difference $(T - T_{R\&B})$, on the temperature scale. Intersection with the corresponding penetration index line permits of direct reading of the viscosity in N s/m², on the upper scale by moving from the point of intersection parallel to the adjacent curve. However, it should be stressed that this procedure is allowed only when the behaviour in the range considered is truly viscous, i.e. only when the stiffness value lies in the left-hand lower part of the nomograph.

5.3.6 Calculation of Stiffness for the Asphalt Concrete Mix

Once the stiffness of the bitumen at a specific time of loading and temperature has been established, the stiffness of the asphalt mix containing this bitumen can be calculated from [16]:

$$S_{\text{mix}} = \left[1 + \frac{2\cdot5}{n} \cdot \frac{C_v}{1 - C_v} \right]^n \cdot S_{\text{bit}} \tag{5.9}$$

where S_{mix} = stiffness of the asphalt concrete mixture (kg/cm^2);

S_{bit} = stiffness of the bituminous binder at the desired temperature and time of loading, determined from the nomograph but transferred to kg/cm^2;

C_v = volume concentration of aggregate in the mixture

$$= \frac{\text{volume of aggregate}}{\text{volume of aggregate and bituminous binder}}; \text{ and}$$

$$n = 0\cdot83 \log \frac{400\,000}{S_{\text{bit}}}.$$

Van der Poel's method is valid for mixtures with C_v values between 0·7 and 0·9 and air void contents of about 3%. Van Draat and Sommer [25] have suggested that air voids of greater magnitude can be considered by using a corrected volume concentration of aggregate, C_v', such that:

$$C_v' = \frac{C_v}{1 + H} \tag{5.10}$$

where H = the difference between the actual air void content and 3%, expressed as a decimal.

For certain asphalts, such as waxy or blown asphalts, Van der Poel's nomograph cannot be applied directly in determining their stiffness. Heukelom's Bitumen Test Data Chart [8, 9] must first be used to obtain a corrected softening point for calculating the Penetration Index.

5.3.7 Fatigue Strengths of Asphalt Mixes

Fatigue can be defined as the phenomenon of fracture under repeated or fluctuating stress having a maximum value generally less than the tensile strength of the material. This tensile strength may be of the order of 300–500 psi, as derived from split-cylinder tensile tests on typical asphalt concrete mixes. The dynamic stiffness plays a predominant role in determining fatigue behaviour. Also, several studies have indicated that the maximum principal tensile strain, at the bottom of the bituminous bound

layer, is a major determinant of fatigue crack initiation. Fatigue life can
be estimated from the following relationship:

$$N_f = K \left(\frac{1}{\varepsilon_m} \right)^n \tag{5.11}$$

where N_f = number of load applications to initiate a fatigue crack;
ε_m = maximum value of applied tensile strain; and
n and K = factors depending upon mix composition.

Figure 5.5 shows the strain–life relationships for some typical wearing
and base course mixes and indicates the considerable difference in fatigue
performance. These results are from laboratory tests which apply continuous cycles of loading of particular magnitudes. It is difficult to relate
them with absolute fatigue lives for design purposes since, in practice, the
pavement is subjected to successive load pulses of varying magnitudes
with varying time intervals between pulses.

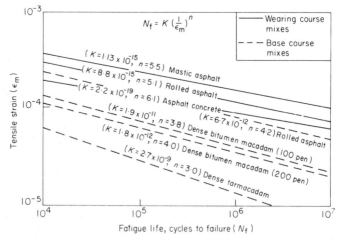

FIG. 5.5. Typical fatigue lines for wearing and base course mixes. (From [18].)

Results of recent work in this field indicate that for maximum fatigue
life the amount of filler and bitumen in a mix should be such that a
condition of maximum stiffness associated with minimum voids is produced.

Also, it appears from these results that for mixes with similar binder
contents the aggregate type and bitumen grade have a small effect on the
strain–life relationship, under normal weather conditions.

Further, for a given mineral aggregate with a given gradation and constant volumetric composition of mineral aggregate, bitumen and voids, the strain per volume percent binder (ε/V_b) depends only upon the stiffness of the binder. However, the strain per volume percent may increase with increasing maximum size of aggregate.

The influence of loading time, temperature and penetration grade of bitumen on the initial strain corresponding to a given number of cycles to failure, $\varepsilon(N)$, depends for a given mix upon the effect which these variables have on the stiffness of the bitumen [13]. $\varepsilon(N)$ increases with increasing binder content and with increasing maximum size of aggregate. On the other hand, the value of $\varepsilon(N)$ decreases sharply when the filler is reduced below a certain limit.

The above discussion on fatigue is based on the results of tests mainly concerned with cracking phenomena. However, fatigue can also manifest itself as a progressive reduction in stiffness. This decrease in stiffness can lead to overstressing of other components of the pavement structure with consequent increase in deformation.

5.3.8 Low-Temperature Cracking in Flexible Pavements

Low-temperature transverse cracking of bituminous pavements is a type of pavement distress prevalent in Canada and the northern parts of the USA, Europe and Asia. Deterioration in pavement performance can result from these cracks, in the form of spalling, heaving or settling at the cracks.

Comparisons between laboratory and field results have revealed that there is a good correlation between the laboratory-predicted fracture temperatures of the binder and mix and the temperature of initial cracking of the asphaltic pavement in the field. The tendency of an asphaltic pavement to crack can therefore be predicted, for practical purposes, by a knowledge of the binder stiffness modulus at low temperatures and long loading time. Conversely, the stiffness parameter of the binder or mix may be used as a pavement design criterion to alleviate the transverse cracking problem [1, 12, 21].

Studies by Hills and Brien have indicated that asphalt concrete can be expected to crack if accumulated thermal stresses exceed the tensile breaking strength of the compacted mix [11].

The thermal stresses can be calculated approximately for a long, completely restrained strip by using the following equation, suggested by Hills and Brien:

$$\sigma_x(T) = \alpha \frac{T_f}{T_0} \int S(r, T)\,\mathrm{d}T \simeq \alpha \sum_{T_0}^{T_f} S(\Delta T) \cdot \Delta T \qquad (5.12)$$

where $\sigma_x(T)$ = unit tensile stress in the longitudinal direction;

α = average coefficient of thermal contraction over the temperature range, and may be approximately taken equal to $1 \cdot 5 \times 10^{-5}/°F$ in the temperature range $-20°F$ to $+30°F$;

T_0 and T_f = initial and final temperatures, respectively, of the total temperature drop, $T_0 - T_f$;

$S(r, T)$ = stiffness modulus, which varies with temperature and rate of loading;

$S(\Delta T)$ = stiffness modulus, determined at the midpoint of discrete temperature intervals over the range T_0 to T_f, and using a loading time which corresponds to the time intervals for the ΔT change; and

ΔT = discrete temperature interval.

A plot of the thermal stresses and tensile strength versus temperature, for a certain concrete mix, would provide a reasonable prediction for the fracture temperature. Figure 5.6 is an example in which this procedure has been applied to some of the Ste. Anne Test Road data in Canada.

Heukelom has shown that the tensile properties of an asphalt mix are related to the tensile properties of the binder used in that mix, by a mix factor. Accordingly, it should be possible to predict the susceptibility of an asphalt pavement to low-temperature cracking from a knowledge of binder properties alone.

Further, since cracking tendency is a function of tensile strength and thermally induced stress, which, in turn, are functions of binder stiffness, it should be possible to predict cracking tendency solely from binder stiffness.

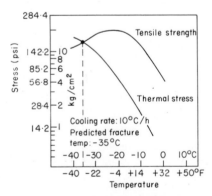

FIG. 5.6. Prediction of fracture temperature.
(Courtesy: Roads and Transportation Association of Canada.)

A recent study has indicated that an asphaltic pavement will not crack because of thermally induced effects unless the pavement thermal temperature becomes lower than that at which the binder stiffness modulus at half-hour loading equals 36 000 psi (2500 kg/cm^2).

The temperature at which initial cracking may be expected will increase with the age of the pavement, because of stiffening of the binder. This increase in cracking tendency with pavement age can, however, be predicted from a knowledge of the ageing characteristics of the binder involved during the service life of the pavement.

It appears from the above discussion that, in selecting materials for mix design, bituminous binders may be chosen on the basis of their stiffness characteristics. They should yield mixes with predicted minimum non-fracture temperatures lower than the pavement surface temperatures likely to be encountered in service. In this respect, low temperature susceptibility and soft grades are beneficial binder properties. Also, the asphalt source (supplier) appears to be a dominant variable associated with low-temperature cracking.

Because the bituminous component of the pavement has the greatest influence on low-temperature cracking, other components have received little attention. Yet it is known that the nature of the subgrade can in certain situations affect crack initiation, cracking frequency and performance of cracked sections. Penetration of water through the cracks can cause appreciable swelling of certain clay subgrades, with resulting bumps in the pavement. On sandy subgrades, severe dips can occur, resulting in a very rough riding pavement. Also, field investigations have indicated that when low-temperature cracking occurs, the frequency is much higher over sand than clay subgrades.

5.3.9 Nomograph for Predicting Cracking Index

The most recent, though only partly developed, technique estimates cracking frequency as a function of several variables [7]. This technique is based on field data from the Provinces of Ontario and Manitoba in Canada. The models tested are in the form:

$$I = f(s, t, a, d, m) \pm E \qquad (5.13)$$

where I = cracking index, defined as the sum of the cracks across the full road width, plus 50% of the half-width cracks, per 500 ft of 2-lane roadway;

s = stiffness modulus of the original asphalt cement (kg/cm^2), according to McLeod's method [17];

t = thickness (in) of the bituminous layer;
a = pavement age (years);
d = subgrade soil type;
m = winter design temperature (°C); and
E = standard error of estimate of I.

The nomograph shown in Fig. 5.7 illustrates one of the models developed with a level of correlation γ^2 in excess of 0·8 and an error E value of about 6.

$$\text{MODEL}: \quad 10^I = 2\cdot4970 \times 10^{30} \cdot s^{(6\cdot79660 - 0\cdot874031 + 1\cdot33884a)} \cdot (7\cdot0539 \times 10^{-3})^d \cdot (3\cdot1928 \times 10^{13})^m \cdot d^{0\cdot60263s}$$

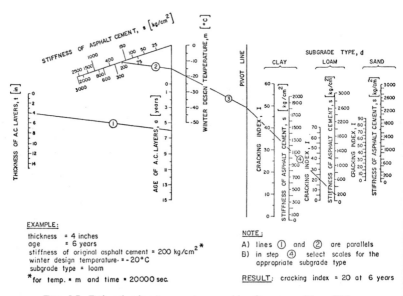

EXAMPLE:
thickness   = 4 inches
age   = 6 years
stiffness of original asphalt cement = 200 kg/cm² *
winter design temperature· = -20°C
subgrade type = loam
*for temp.· m and time · 20000 sec.

NOTE:
A) lines ① and ② are parallels
B) in step ④ select scales for the appropriate subgrade type

RESULT: cracking index = 20 at 6 years

FIG. 5.7. Estimating low-temperature cracking frequency. (From [7].)

5.3.10 Synthetic Rubber Additives in Asphalt Mixtures

A study was recently made [4] to determine the effect of synthetic rubber additives on the properties of asphalt mixtures. The amount of synthetic rubber solids in the rubber asphalt compound was 3% by weight of asphalt. Mixtures containing synthetic rubber were laid at a much higher temperature (47°F higher) than those without rubber. The roller was able to roll the pavement at this hotter temperature with no 'pick up'.

The average effect of the rubber solids was to increase the stability by about 7%. The flow of the asphalt mix containing rubber showed a marked levelling off at 5·5% asphalt content and greater. The average density was less than that of the mixtures without rubber additives. In addition, the mixes with rubber asphalt compound showed an average void content of 0·6% greater than that of the mixes without rubber. No significant difference in viscosity appeared among the mixtures, with or without rubber additives, at initial laying and up to 30 min after the paver-pass. However, the rubber asphalt mixtures appeared to be slightly stiffer because of the rubber present.

Another similar study [5] has indicated that rubber additives do not have a significant effect on creep compliances, ultimate compressive strengths and strains, and tensile strengths of the asphalt mixtures at the initial period.

5.4 CEMENT BOUND MATERIALS

5.4.1 Performance of Cement Bound Materials

The performance of cement bound layers in flexible pavements depends largely on the relationship of their stiffness to their strength. This is particularly true regarding the ability of these layers to resist cracking.

Cement is sometimes utilised as a binder for base and sub-base layers, namely, lean concrete, cement bound granular material and soil–cement. The stronger forms of these cement bound materials such as lean concrete can be very stiff compared with the material of the other layers both above and below it. The dynamic modulus of elasticity of lean concrete may vary between 1×10^6 psi and 5×10^6 psi, depending on factors such as type and amount of cement, water content and aggregate type and grading. Due to this high stiffness, a lean concrete layer will be subjected to considerable tensile stresses, due to both traffic and temperature, which may exceed its flexural strength. The fatigue aspect is also relevant here and a reduction of approximately 30% in strength can be expected under repeated loads.

These considerations, together with the shrinkage cracks which are generally present, may indicate that the characteristics of lean concrete could approach those of an unbound layer, if the cracking is extensive. If reliable information is not available it is suggested that values of 7×10^4 psi and 0·25 for the dynamic modulus and Poisson's ratio, respectively, would be appropriate for design calculations.

The uncracked modulus of elasticity of soil–cement will probably be in the range of $1.5 \times 10^5 - 10 \times 10^5$ psi. However, soil–cement layers become extensively cracked under heavy traffic. For this reason, they should be treated as unbound granular materials having a modulus related to that of the underlying subgrade and a Poisson's ratio of 0.3.

5.5 SOILS AND UNBOUND AGGREGATES

5.5.1 Characterisation of Sub-base and Unbound Base Materials
Repeated-load triaxial tests can be used to determine the elastic constants of soils and granular materials. Two factors are of particular interest when characterising these materials for design purposes. The first factor is the resilient behaviour which may be simply stated as the relationship between applied stress and recoverable strain. The second factor is the relationship between permanent strain and the number of load applications. Failure of the material usually depends on the second factor.

The resilient characteristics of pavement materials are normally specified in terms of the modulus of elasticity and Poisson's ratio. Both cohesive and non-cohesive soils and aggregates exhibit non-linear stress–strain characteristics. The resilient modulus of elasticity is a function of both shear and normal stresses. For cohesive soils the shear stress predominates, while for granular materials the modulus is largely defined by the normal stress level.

When a thick asphalt layer is used, the influence of the non-linear behaviour of soils and aggregates is too small to be significant, and the linear elastic theory can be used for pavement analysis. Simplified models incorporating linear relationships have been suggested, e.g. as in Fig. 5.8 for cohesive soils and granular materials.

The value of Poisson's ratio for cohesive soils may be taken as between

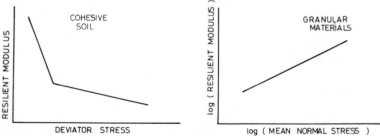

FIG. 5.8. Simplified non-linear models for resilient behaviour of clay and granular materials.
(From [10].)

0·4 and 0·5, with the higher value relating to soft, wet materials. Granular materials exhibit values of Poisson's ratio from 0·25 to 0·4 and, in general, an average value of 0·30 seems appropriate for design purposes. The relationship between permanent deformation and number of loading cycles is generally characterised by a relatively sharp increase in permanent deformation during the early load application, followed by a very small subsequent increase. The relationship between permanent strain and number of stress applications, for a saturated normally consolidated silty clay tested under repeated-load triaxial conditions, is shown in Fig. 5.9. This relationship differs from that for granular materials in that a definite failure point can be established when the rate of permanent deformation shows a sharp increase.

For normally consolidated silty clay tested in undrained conditions, there appears to be a useful relationship between the consolidation pressure, quick shear strength and repeated load shear strength. The repeated-load strength is about 70% of the 'one-shot' strength. Such a relationship is extremely useful, since it relates dynamic repeated-load behaviour to a standard laboratory test.

The mode of failure for granular materials in a pavement system is still a matter of speculation. Initial permanent deformation contributes to the over-all pavement deformation, which may ultimately lead to failure. During this process, however, the material is further compacted and may become stronger.

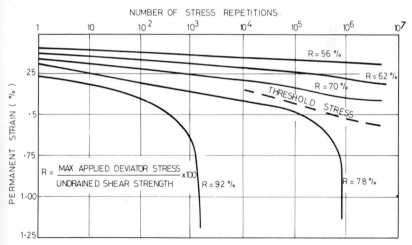

FIG. 5.9. Relationship between permanent strain and number of stress applications for a normally consolidated saturated silty clay. (From [15].)

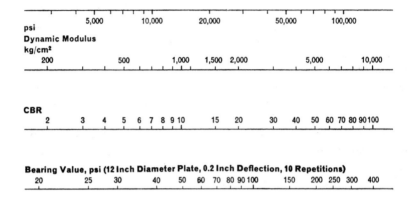

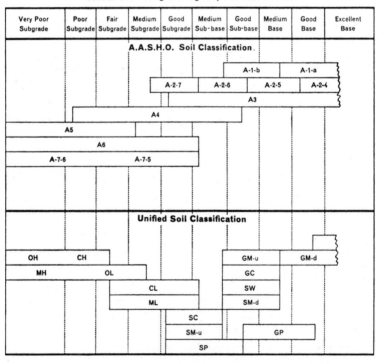

FIG. 5.10. Estimation of dynamic modulus of soils. *(Courtesy: Shell Petroleum Company.)*

5.5.2 Dynamic Modulus of Soils

Some of the methods of design of flexible pavements, like the Shell method, are based on the dynamic elastic modulus of the soil. This modulus is in many situations not easy to measure directly. An approximate guide can be obtained from the results of conventional tests such as the CBR or plate-bearing tests.

For practical purposes the dynamic elastic modulus, E_{dy}, can be taken as equal to 100 CBR (kg/cm^2) or 1500 CBR (psi). The CBR test should preferably be carried out at the likely moisture content of the soil in service under an impermeable surface. This 'equilibrium moisture content' is usually similar to that found at a depth of about 3 ft (0·9 m) in the natural soil. Soils that are liable to be subjected to frost penetration should be tested in a saturated condition.

Figure 5.10 gives conversion factors for determining the dynamic modulus (E_{dy}) of the soil once the CBR or the bearing value is known.

REFERENCES AND BIBLIOGRAPHY

1. Anderson, K. O. and Haas, R. C. G., 'Use of the Stiffness Concept to Characterize Bituminous Materials', *Proc. Can. Tech. Asph. Assoc.*, 1970.
2. Bazin, P. and Saunier, J. B., 'Deformability, Fatigue and Healing Properties of Asphalt Mixes', *Proc. 2nd International Conference on the Structural Design of Asphalt Pavements, University of Michigan, Ann Arbor, August 1967*, 553–570.
3. Burgess, R. A., Kopvillem, O. and Young, F. D., 'Ste Anne Test Road—Flexible Pavement Design to Resist Low Temperature Cracking', *Proc. 3rd International Conference on the Structural Design of Asphalt Pavement, London, September 1972*, 27–40.
4. Darter, M. I., Peterson, D. E., Jones, G. M. and Vokac, R., 'Design, Construction, and Initial Evaluation of Experimental Test Sections of Asphalt Containing Synthetic Rubber', *Highway Research Record*, No. 313, 1970.
5. Fitzgerald, J. E. and Lai, J. S., 'Initial Evaluation of the Effect of Synthetic Rubber Additives on the Thermorheological Properties of Asphalt Mixtures', *Highway Research Record*, No. 313, 1970.
6. Guericke, R. and Weinert, F., 'The Behaviour of Bituminous Mixtures in Laboratory Tests and Under Road Conditions', *Proc. 3rd International Conference on the Structural Design of Asphalt Pavements, London, September 1972*, 233–240.
7. Hajek, J. and Haas, R. C. G., 'Predicting Low-Temperature Cracking Frequency of Asphalt Pavements', *Highway Research Record*, No. 407, 1972, 39–54.
8. Heukelom, W., 'A Bitumen Test Data Chart for Showing the Effect of Temperature on the Mechanical Behaviour of Asphaltic Bitumens', *J. Inst. Petroleum*, November 1969.

9. Heukelom, W. and Klomp, A. J. G., 'Road Design and Dynamic Loading', *Proc. Assoc. Asph. Paving Tech.*, 1964.
10. Hicks, R. G., 'Factors Influencing the Resilient Properties of Granular Materials', Ph.D. Thesis, University of California, 1970.
11. Hills, J. F. and Brien, D., 'The Fracture of Bitumens and Asphalt Mixes by Temperature Induced Stresses', *Proc. Assoc. Asph. Paving Tech.*, **35**, 1966.
12. Kasianchuk, D. A., Terrel, R. L. and Haas, R. C. G., 'A Design System for Minimizing Fatigue, Permanent Deformation and Shrinkage Fracture Distress of Asphalt Pavements', *Proc 3rd International Conference on the Structural Design of Asphalt Pavements, London, September 1972*, 629–655.
13. Kirk, J. M., 'Relations between Mix Design and Fatigue Properties of Asphalt Concrete', *Proc. 3rd International Conference on the Structural Design of Asphalt Pavements, London, September 1972*, 241–247.
14. Kopvillem, O. and Heukelom, W., 'The Effect of Temperature on the Mechanical Behaviour of Some Canadian Asphalts as Shown by a Test Data Chart', *Proc. Can. Tech. Asph. Assoc.*, 1969.
15. Lashine, A. K., 'Some Aspects of the Characteristics of Keuper Marl under Repeated Loading', Thesis, University of Nottingham, 1971.
16. Marek, C. R. and Dempsey, B. J., 'A Model Utilizing Climatic Factors for Determining Stresses and Deflections in Flexible Pavement Systems', *Proc. 3rd International Conference on the Structural Design of Asphalt Pavements, London, September 1972*, 101–114.
17. McLeod, N. W., 'Prepared Discussion on Ste. Anne Test Road Paper', *Proc. Can. Tech. Asph. Assoc.*, 1969.
18. Pell, P. S. and Brown, S. F., 'The Characteristics of Materials for The Design of Flexible Pavement Structures', *Proc. 3rd International Conference on the Structural Design of Asphalt Pavements, London, September 1972*, 326–342.
19. Roads and Transportation Association of Canada, 'Low-Temperature Pavement Cracking Studies in Canada', *Proc. 3rd International Conference on the Structural Design of Asphalt Pavements, London, September 1972*, 581–589.
20. Shell Ltd, 'Shell Design Charts for Flexible Pavements', Fifth Reprint, London, 1972.
21. Shields, B. P., Anderson, K. O. and Dacyszyn, J. M., 'An Investigation of Low-Temperature Cracking of Flexible Pavements', *Proc. Can. Good Roads Assoc.*, 1969.
22. Stephenson, R. W. and Manke, P. G., 'Ultrasonic Moduli of Asphalt Concrete'. *Highway Research Record*, No. 404, 1972, 8–21.
23. Szendrei, M. E. and Freeme, C. R., 'The Computation of Road Deflections under Impulsive Loads from the Results of Vibration Measurements', *Proc. 2nd International Conference on the Structural Design of Asphalt Pavements, University of Michigan, Ann Arbor, August 1967*, 141–149.
24. Van der Poel, C. 'A General System Describing the Viscoelastic Properties of Bitumens and its Relation to Routine Test Data', *J. Applied Chemistry*, May 1954.
25. Van Draat, W. E. F. and Sommer, P., 'Ein Gerat zur Bestimmung der Dynamischen Elastizitätsmoduln von Asphalt', *Strasse und Autobahn*, **35**, 1966.

CHAPTER 6

DESIGN OF FLEXIBLE PAVEMENTS FOR HIGHWAYS

6.1 INTRODUCTION

Flexible pavements are those consisting of a base course of suitable granular material with or without bituminous binder, and a bituminous surface course. They have little beam strength and carry the loads by distributing them and reducing their intensity until they may be carried safely on the subgrade.

Flexible pavements can be divided into two groups, according to the type of base used in the pavement system. These are:

1. Flexible pavements with untreated granular bases.
2. Full-depth asphalt pavement, in which asphalt mixtures are employed for all courses above the subgrade; in other words, an asphalt base is used in this group.

It is strongly recommended that full-depth asphalt pavements be used, since untreated granular bases frequently act as moisture reservoirs which hold water continually in contact with the subgrade, causing a gradual decrease in subgrade strength. In addition, untreated granular bases cannot withstand tensile stresses and thus perform in an inferior manner when compared with asphalt bases.

6.2 FULL-DEPTH ASPHALT

The main advantages of full-depth asphalt pavements are:

1. The asphalt base can resist load tensile stresses, spreading the load over broader areas; thus, less pavement structure thickness is required.

2. Aggregates not suitable for asphalt surface courses may be used in asphalt base courses.

3. With proper construction of the asphalt concrete base, the riding quality of the pavement will improve.

4. Asphalt concrete bases afford an excellent means for stage construction and may be used by haul traffic before the surface course is placed, thus expediting construction and reducing completion time.

5. Full-depth asphalt pavements have no permeable granular layers, which entrap water, reduce subgrade strength and impair performance.

6. Subsurface drainage is normally not required unless the groundwater table is high and needs to be lowered.

7. The need for base drainage under the shoulder is eliminated; this allows of a substantial reduction in the quantity of high quality granular materials.

8. Because they are thinner than flexible pavements with untreated granular bases, less interference with utilities in city street construction is expected.

9. They are unaffected by frost or moisture and provide a retained uniformity in the pavement structure.

Flexible pavements can also be categorised into two groups according to the type of binder used in the wearing course, as follows:

1. The intermediate type of flexible-pavement wearing course, in which liquid bituminous materials are used as the binder.

2. The high type of flexible-pavement wearing course, in which asphalt cements and the heaviest grades of tars are used. Because asphalt cements are semi-solid, much higher mixing temperatures are needed for high-type bituminous mats.

6.3 METHODS OF DESIGN OF FLEXIBLE PAVEMENTS

There are numerous methods of flexible pavement design. In one group of methods of design, elastic and viscoelastic approaches are used to determine the stresses, strains and displacements within the layers of flexible pavements. In another group, empirical and semi-empirical formulae are developed from analysing field test data using the serviceability–performance concept. In a third design group, the pavement structural design is based upon limiting permanent deformation of the subgrade and preventing fatigue fracture of the wearing course, using limiting strain criteria.

Some of the most important methods of design of flexible pavements are discussed in this chapter.

6.4 ASPHALT INSTITUTE DESIGN METHOD

This method was developed from the statistical analysis of data from the AASHO Road Test after incorporating information from the WASHO Road Test, British Test Roads and various other existing design procedures. The revised method which appeared in the 8th edition of the *Asphalt Institute Manual Series 1* [13], August 1970, has incorporated for the first time a new traffic analysis method. The design procedure is as follows:

(a) Determine the initial and future traffic conditions throughout the design period.
(b) Evaluate the subgrade and available construction materials.
(c) Evaluate the environmental factors which may affect pavement behaviour or service.

6.4.1 Traffic Analysis
Definitions of the special terms that will be used in the traffic analysis are:

1. Design lane. The lane on which the greatest number of equivalent 18 000 lb single-axle loads is expected. Normally, this will be either lane of a two-lane roadway or the outside lane of multi-lane highways.

2. Design period. The number of years from the initial application of traffic until the first major resurfacing or overlay is anticipated. This term should not be confused with pavement life. By adding asphalt overlays as required, pavement life may be extended indefinitely, or until geometric considerations or other factors may make the pavement obsolete.

3. Design traffic number (D TN). The average daily number of equivalent 18 000 lb single-axle loads estimated for the design lane during the design period, which is considered 20 years in developing the thickness design charts.

4. Equivalent 18 000 lb single-axle load. The effect on pavement performance of any combination of axle loads of varying magnitude, equated to the number of 18 000 lb single-axle loads required to produce an equivalent effect.

5. Initial daily traffic (I D T). The average daily number of vehicles expected to use the roadway, in both directions, during the first year.

6. Initial traffic number (I TN). The average daily number of equivalent

18 000 lb single-axle load applications expected on the design lane during the first year.

7. *Traffic classification.*

Light: Traffic conditions resulting in a Design Traffic Number (DTN) less than ten (10).

Medium: Traffic conditions resulting in a Design Traffic Number (DTN) between ten (10) and one hundred (100).

Heavy: Traffic conditions resulting in a Design Traffic Number (DTN) above one hundred (100).

To determine the ITN for a new pavement facility, the designer must depend upon traffic studies of similar facilities, and community or regional

TABLE 6.1 DETERMINATION OF TRUCK FACTOR

Axle-load group (1 000 lb)	Load equivalency factor	Axles per day per 1 000 trucks and combinations	Equivalent 18 000 lb single-axle loads per 1 000 trucks and combinations
(1)	(2)	(3)	(4)
Single axles:			
under 8	—	1 135·4	—
8–12	0·11	487·3	53·6
12–16	0·34	282·7	96·1
16–18	0·76	118·6	90·1
18–20	1·31	31·9	41·8
20–22	2·26	2·6	5·9
22–24	3·91	6·5	25·4
24–26	6·74		
		Subtotal	312·9
Tandem axles:			
under 14	—	189·3	—
14–20	0·11	141·6	15·6
20–26	0·27	168·4	45·5
26–30	0·57	99·4	56·7
30–32	0·92	2·6	2·4
32–34	1·25		
34–36	1·70		
36–38	2·33		
38–40	3·15		
40–42	4·36		
42–44	5·88		
44–46	8·15		
		Subtotal	120·2
		Total single plus tandem axles	433·1
		Truck factor	$\dfrac{433\cdot1}{1\,000} = 0\cdot43$

planning studies, to provide the information needed for traffic analysis. Also, traffic growth for comparable facilities will help in determining anticipated traffic growth on the new facility.

Load equivalency factors are then applied to convert single- and tandem-axle loads of given magnitudes to equivalent 18 000 lb single-axle load applications using the chart of Fig. 1.6. The average number of equivalent 18 000 lb single-axle loads per truck, called the truck factor, is used in computing ITN. To determine the truck factor, axle loads are grouped as illustrated in column 1 of Table 6.1. Then Fig. 1.6 is used to determine the equivalency factor for the average axle load of each group, as shown in the second column of the table. The axles per day per 1000 trucks and combinations, given in column 3, can be obtained from traffic counts and loadometer data. Multiplying these values in column 3 by the equivalency factors in column 2 gives the number of equivalent 18 000 lb single-axle loads, shown in column 4.

The load equivalency factors for single-axle loads less than 8000 lb and tandem-axle loads less than 14 000 lb are omitted from the table, although these axle loads are included in the truck count. The effect of these lighter loads on the pavement may be disregarded, except that if the computed truck factor is less than 0·05, a truck factor of 0·05 should be used.

Multiplying the truck factor by the estimated number of daily trucks expected to use the design lane during the first year of service, will give the ITN.

TABLE 6.2 ESTIMATED RANGES IN PERCENT TRUCKS AND AVERAGE GROSS WEIGHT IN THE US[a]
(Courtesy: The Asphalt Institute)

Description of highway or street	*Percent heavy trucks (excluding pick-up and light panel trucks)*	*Average gross weight (1 000 lb)*
City streets (local)	5 or less	15–25
Urban highways:		
Primary	5[b]–15	20–30
Interstate	5–10	35–45
Local rural roads	15 or less	15–25
Interurban highways:		
Primary	5–20	30–40
Interstate	10–25	35–45

[a] Average US conditions only. Other countries and local US conditions, depending on land use and industry, may require special considerations.
[b] Sometimes less.

When sufficient data are not available for an accurate estimate of ITN, the following approximate method can be used:

1. Estimate IDT, which is the average daily number of vehicles expected during the first year following the opening of the finished roadway to traffic.
2. Use Table 6.2 to estimate the percentage of heavy trucks, *A*.
3. Use Table 6.3 to determine the percentage of heavy trucks, *B*, in the design lane. Normally, most trucks operate in the outermost traffic lanes, and may be considered equally divided in both directions.
4. Average daily number of heavy trucks on the design lane

$$= (IDT) \times \frac{A}{100} \times \frac{B}{100} \qquad (6.1)$$

where IDT, *A*, and *B* are as described in steps 1, 2 and 3, above.

TABLE 6.3 PERCENTAGE OF TOTAL TRUCK
TRAFFIC IN DESIGN LANE
(Courtesy: The Asphalt Institute)

Number of traffic lanes (two directions)	Percentage of trucks in design lane
2	50
4	45 (35–48)
6 or more	40 (25–48)

5. Estimate the average gross weight of the heavy trucks using Table 6.2, and determine the legal single-axle load limit established by state or local statutes.
6. Use Fig. 6.1 to determine the ITN as follows:
   (a) Enter the chart with the average gross weight at the proper point on line D.
   (b) Locate the number of heavy trucks, daily average on the design lane, at the proper point on line C.
   (c) Connect the points on lines D and C with a straight line and extend it to line B. Where this line intersects line B, is the pivot point.
   (d) Locate the proper single-axle load limit point on line E.
   (e) Connect the single-axle load limit point on line E with a straight line to the pivot point on line B, and extend it to line A.
   (f) Read the ITN on line A where the extended line E–B intersects it.

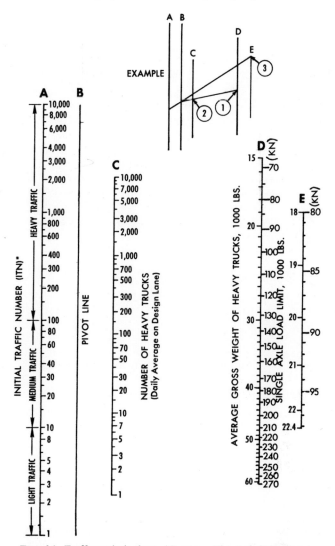

FIG. 6.1. Traffic analysis chart. *(Courtesy: The Asphalt Institute.)*

7. When the resultant ITN is 10 or less, and when a relatively large number of automobiles and light trucks are expected to use the roadway, a correction of ITN is required using Fig. 6.2 as follows:
   (a) Enter Fig. 6.2 on the horizontal scale at a point representing the daily volume of automobile and light trucks in the design lane.

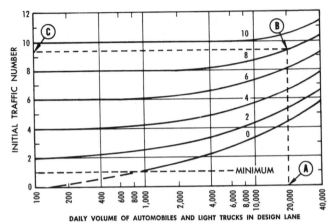

FIG. 6.2. Chart for adjusting Initial Traffic Number (ITN) for daily volume of automobiles and light trucks. *(Courtesy: The Asphalt Institute.)*

(b) Move vertically to the curve representing the ITN based on heavy trucks, determined previously.

(c) Read the corrected ITN on the initial traffic number scale.

Example

Daily volume of automobiles and light trucks in design lane = 20 000 vehicles

ITN (based on heavy trucks) = 8

Enter the chart in Fig. 6.2 at Daily Volume = 20 000 (point A) and move vertically to ITN 8 line (point B)

The corrected ITN is 9·5 (point C)

The Design Traffic Number (DTN) can be obtained from the ITN by using Table 6.4 as follows:

1. Establish the design period. For new construction, the Design Period will normally be 20 years.

2. Estimate the annual growth rate of traffic. Currently, traffic in the USA increases, on average, by about 3–5% per year. In other countries, especially in Europe, the annual growth rate may be more than 5%.

3. For the selected design period and annual growth rate, determine the ITN adjustment factor from Table 6.4.

4. Multiply the ITN by the adjustment factor to obtain DTN_{20} for use in the thickness design charts.

The thickness design charts developed are based on a 20 year design period. For other periods, an adjustment must be made to reflect the fewer, or additional, equivalent 18 000 lb single-axle loads, by multiplying the ITN by the proper factor from Table 6.4. The DTN thus obtained is the average daily number of equivalent 18 000 lb single-axle load applications for the selected design period, adjusted to an equivalent DTN for a 20 year design period.

TABLE 6.4 INITIAL TRAFFIC NUMBER (ITN)
ADJUSTMENT FACTORS
(Courtesy: The Asphalt Institute)

Design period, years (n)	Annual growth rate, percent (r)					
	0	*2*	*4*	*6*	*8*	*10*
1	0·05	0·05	0·05	0·05	0·05	0·05
2	0·10	0·10	0·10	0·10	0·10	0·10
4	0·20	0·21	0·21	0·22	0·22	0·23
6	0·30	0·32	0·33	0·35	0·37	0·39
8	0·40	0·43	0·46	0·50	0·53	0·57
10	0·50	0·55	0·60	0·66	0·72	0·80
12	0·60	0·67	0·75	0·84	0·95	1·07
14	0·70	0·80	0·92	1·05	1·21	1·40
16	0·80	0·93	1·09	1·28	1·52	1·80
18	0·90	1·07	1·28	1·55	1·87	2·28
20	1·00	1·21	1·49	1·84	2·29	2·86
25	1·25	1·60	2·08	2·74	3·66	4·92
30	1·50	2·03	2·80	3·95	5·66	8·22
35	1·75	2·50	3·68	5·57	8·62	13·55

$$\text{Factor} = \frac{(1+r)^n - 1}{20\,r}$$

Note: Truck growth rate, which includes both number and weight of trucks, may increase faster than over-all traffic growth rate, particularly on roads with large volumes of heavy trucks. Growth rates for these roads should be determined from truck weight study data, if possible.

Example

A six-lane interstate interurban highway will have an estimated ADT = 38 000. The annual growth rate of traffic is expected to be 4%. The legal single-axle load limit is 18 000 lb and the expected average gross weight is 40 000 lb. The heavy trucks in the traffic stream are expected to be 11% of the total traffic volume. The heavy trucks in the design lane are estimated to be 40% of the total number of heavy trucks. Find the DTN for a 20 year design period.

1. $IDT = 38\,000$ vehicles per day.
2. Percent heavy trucks in both directions, $A = 11$.
3. Percent heavy trucks in design lane, $B = 40$.
4. Number of heavy trucks in design lane, using eqn. (6.1) $= 38\,000 \times (11/100) \times (40/100) = 1672$.
5. Draw a line to connect the number of 1672 trucks on line C to the average gross weight of 40 000 lb on line D in Fig. 6.1, and project to line B.
6. Single-axle load limit—18 000 lb. Plot on line E, Fig. 6.1.
7. Draw a line through the points on lines B and E, and project to line A.
8. Read off the ITN on line A; the value is 1400.
9. The ITN is more than 10, so no correction for automobiles and light trucks is necessary.
10. Design Period $= 20$ years.
11. Annual Growth Rate $= 4\%$.
12. Initial Traffic Adjustment Factor $= 1.49$ from Table 6.4.
13. $DTN_{20} = 1400 \times 1.49 = 2086$ or, rounding off, 2100.

6.4.2 Thickness Design for Full-Depth Asphalt Pavement
The charts given in Figs. 6.3 and 6.4 are used to determine the total thickness of hot mix asphalt concrete pavement, T_A, required above the subgrade for given traffic and subgrade conditions. The asphalt paving mixtures used should be similar to Asphalt Institute Type IV mixes. Figure 6.3 is used when the subgrade strength is measured by means of the California Bearing Ratio (CBR) Test or Plate-bearing Test, while Fig. 6.4 is used when subgrade strength is measured by the R value, derived from the Stabilometer Test.

Example
Determine the total asphalt pavement structure thickness, T_A, if the CBR of subgrade is 7 and the DTN is 2100.
   Solution.
   1. Locate CBR value of 7 on Scale B in Fig. 6.3.
   2. Locate DTN value of 2100 on Scale C.
   3. Draw a line connecting CBR value and DTN value and extend to Scale A.
   4. Read value of 10·4 in on Scale A and round to the next highest one-half inch, or 10·5 in.

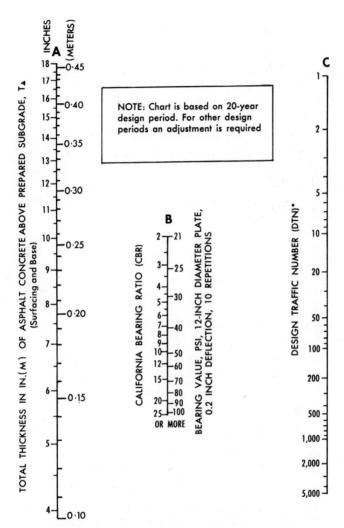

FIG. 6.3. Thickness design chart for asphalt pavement structures using subgrade soil CBR or plate-bearing values. *(Courtesy: The Asphalt Institute.)*

Thus, a total asphalt pavement structure thickness, T_A, of 10·5 in is required for the assumed conditions of CBR and DTN. The recommended minimum thicknesses of total asphalt pavement structure, T_A, are given in Table 6.5.

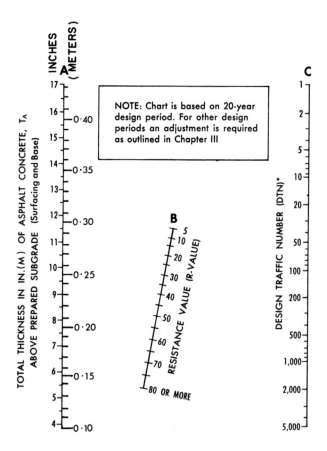

FIG. 6.4. Thickness design chart for asphalt pavement structures using subgrade soil resistance value. *(Courtesy: The Asphalt Institute.)*

TABLE 6.5 RECOMMENDED MINIMUM THICKNESS
OF FULL-DEPTH ASPHALT PAVEMENT
(Courtesy: The Asphalt Institute)

Design traffic number (DTN)	Minimum T_A (in)
Less than 10	4
10–100	5
100–1 000	6
More than 1 000	7

As this thickness design procedure calls for the use of asphalt paving mixtures similar to Asphalt Institute Type IV mixes, there is no need for a different surface course mix for structural strength. However, a different surface course mix may be needed to provide desirable surface properties for skid resistance, hydroplaning prevention, light reflection or other such characteristics. Usually, these surfaces are $\frac{1}{2}$–1 in thick and are added to the total design thickness of the pavement structure.

6.4.3 Asphalt Bases Other Than Asphalt Concrete

Aggregate is often not available at an economical price to produce paving mixtures for the asphalt base similar to Asphalt Institute Type IV mixes. Many deposits of natural sand, for example, have gradings similar to Asphalt Institute Mix Types VI, VII or VIII. These materials, when hot-mixed with asphalt cement, provide economical and durable asphalt bases.

The tentative method for thickness design of asphalt pavement structures using hot-mix sand asphalt bases, is as follows:

1. Determine the required total thickness of asphalt concrete pavement structure, T_A, using Fig. 6.3 or Fig. 6.4.

2. Establish the thickness of pavement surface, T_s, to be built with asphalt concrete. The recommended minimum thickness of pavement surface is:

Light traffic, DTN less than 10	2 in
Medium traffic, DTN 10 to 100	3 in
Heavy traffic, DTN above 100	4 in

3. Subtract the thickness of asphalt concrete pavement surface, T_s (step 2), from the total thickness of asphalt concrete pavement structure, T_A (step 1), to determine the required effective thickness of asphalt concrete base, T_e.

4. Multiply the difference found in step 3 by a factor of 1·3 to obtain the required thickness of hot-mix sand asphalt base, T_b.

Example

Assume a road in the Heavy Traffic classification that will require a total thickness of asphalt concrete pavement structure, T_A, of 11·0 in. Further, assume that a pit-run sand is economically available that will result in an asphalt paving mix meeting the criteria specified above.

(a) According to the recommended minima in step 2, above, the thickness of asphalt concrete pavement surface, T_s, should be not less than 4 in.

(b) Subtract the 4 in thickness of asphalt concrete pavement surface, T_s, from the 11 in total thickness of asphalt concrete pavement structure, T_A. The required thickness of asphalt concrete base, T_e, then, is 7 in.

(c) Multiply the required thickness of asphalt concrete base ($T_e = 7$ in) by a factor of 1·3. The required thickness of hot-mix sand asphalt base, T_b, is 9·1 in, or 9·5 in when rounded to the next highest one-half inch.

If liquid asphalts of the Rapid Curing (RC), Medium Curing (MC) and Slow Curing (SC) types, or emulsified asphalts, are used instead of hot-mix asphalts, then the tentative method for thickness design of asphalt pavement structures is as follows:

1. Determine the required total thickness of asphalt concrete pavement structure, T_A, from Figs. 6.3 or 6.4.

2. Establish the thickness of pavement surface, T_s, to be built with asphalt concrete. The recommended minimum thickness of pavement surface is:

(a) Over base mixes with aggregate gradings similar to Type IV mixes:

Light traffic, DTN less than 10	2 in
Medium traffic, DTN 10–100	3 in
Heavy traffic, DTN above 100	4 in

(b) Over other base mixes:

Light traffic, DTN less than 10	3 in
Medium traffic, DTN 10–100	4 in
Heavy traffic, DTN above 100	5 in

3. Subtract the thickness of asphalt concrete pavement surface, T_s (step 2), from the total thickness of asphalt concrete pavement structure, T_A (step 1), to obtain the required effective thickness of asphalt concrete base, T_e.

4. Multiply the difference found in step 3 by a factor of 1·4 to obtain the required thickness of liquid or emulsified asphalt base, T_b.

Example

A road in the Medium Traffic classification requires a total thickness of asphalt concrete pavement structure, T_A, of 10·0 in. A pit-run aggregate is economically available for preparing a plant-mixed base with emulsified asphalt. Determine the thickness required for the road-mixed emulsified asphalt base.

Solution.

1. According to the recommended minima in step 2, above, the thickness of asphalt concrete pavement surface, T_s, should be not less than 4 in.

2. Subtract the 4 in thickness of asphalt concrete pavement surface, T_s, from the 10 in total thickness of asphalt concrete pavement structure, T_A. The required thickness of asphalt concrete base, T_e, is thus 6 in.

3. Multiply the required thickness of asphalt concrete base ($T_e = 6$ in) by a factor of 1·4. The required thickness of road-mixed emulsified asphalt base, T_b, is 8·4 in or 8·5 in when rounded to the next highest one-half inch.

6.4.4 Thickness Design for Untreated Granular Bases

Conditions may sometimes exist that call for consideration of a pavement section with an untreated granular base replacing a portion of the full-depth asphalt pavement.

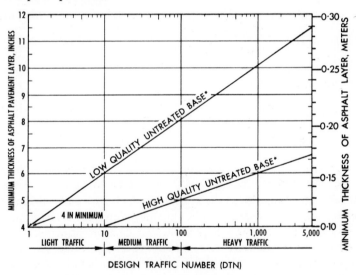

FIG. 6.5. Minimum thickness of asphalt pavement layers over untreated granular bases. *(Courtesy: The Asphalt Institute.)*

The A A S H O Road Test and other pavement studies have demonstrated that untreated granular layers require substantially thicker pavement sections than full-depth asphalt pavements.

Only a limited portion of a full-depth asphalt pavement may normally be converted to untreated granular base. The maximum thickness that may be converted depends on the minimum thickness of asphalt pavement layer required above it to support the anticipated traffic. The chart in Fig. 6.5 may be used as a guide for establishing this minimum thickness. The chart takes into consideration the amount and weight of traffic loadings, expressed as DTN, and the strength properties of untreated granular base.

To determine the maximum thickness of asphalt layer for which granular base may be substituted, subtract the minimum required thickness of asphalt pavement layer (from Fig. 6.5) from the full-depth asphalt pavement thickness, T_A. Lesser thicknesses of asphalt layer may be converted to granular base, if desired.

TABLE 6.6 UNTREATED BASE QUALITY REQUIREMENTS
(Courtesy: The Asphalt Institute)

Test	Test requirements	
	Low quality	*High quality*
CBR, minimum	20	100
or		
R value, minimum	55	80
Liquid limit, maximum	25	25
Plasticity index, maximum	6	NP
Sand equivalent, minimum	25	50
Passing No. 200 sieve, maximum	12	7

Analytical studies of extensive data available from road tests, laboratory experiments and theoretical analyses show that there is no simple, constant factor for converting a given thickness of asphalt layer into a thickness of untreated granular base that will provide equivalent load-supporting capacity. This conversion is a variable that depends principally on the amount of traffic, the magnitude of the wheel loads and the strength properties of the untreated granular base and subgrade. The Asphalt Institute recommends the use of a Substitution Ratio (S_r) for making an approximate thickness conversion from asphalt layer to untreated granular base. Specifically, it is recommended that:

1. 2·0 in of high-quality untreated granular base material be required for each 1·0 in of asphalt layer for which it is substituted. In this case, $S_r = 2·0$.

2. 2·7 in of low-quality untreated base material be required for each 1·0 in of asphalt layer for which it is substituted. In this case, $S_r = 2·7$.

High-quality and low-quality untreated base materials are prescribed in Table 6.6.

Example

Assume these conditions for the design of a highway:

1. A granular base material of high quality is available.
2. DTN = 80.
3. CBR of subgrade = 4.

Using Fig. 6.3, the total thickness of asphalt concrete pavement, T_A, would be 9·5 in. Figure 6.5 indicates that, for these conditions, about 5 in is the least thickness of asphalt pavement layer that should be used. The thickness of asphalt layer that may be converted to untreated granular base is therefore:

$$9·5 \text{ in} - 5·0 \text{ in} = 4·5 \text{ in}$$

A Substitution Ratio, S_r, of 2·0 for high-quality untreated granular base material can be used. Therefore, the thickness of granular base to be substituted for 4·5 in of asphalt layer is:

$$4·5 \text{ in} \times 2·0 = 9·0 \text{ in}$$

Alternative thickness designs are as shown in Fig. 6.6.

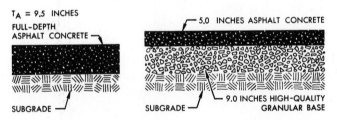

FIG. 6.6. Alternative flexible pavement structure for the example.

6.5 THICKNESS DESIGN FOR SHOULDERS

It is recommended that full-depth asphalt concrete shoulders be used because they contribute high lateral support to the pavement structure. A Design Traffic Number of 1 is recommended for use in shoulder thickness design.

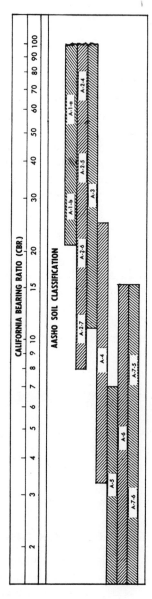

FIG. 6.7. Approximate correlation between C B R and H R B classifications.
(Courtesy: The Asphalt Institute.)

6.6 SUBGRADE STRENGTH VALUES

The thickness design nomographs of Figs. 6.3 and 6.4 call for subgrade strength values determined by California Bearing Ratio (CBR), Resistance (*R*) value or Bearing Value (Plate-Bearing Test).

Evaluation methods employing these mechanical strength tests are by far the most reliable for design purposes and should always be used to evaluate subgrade soils for design of pavements for medium and heavy traffic conditions. However, soil classification systems can sometimes be used to estimate subgrade strength for light traffic conditions if done by an experienced soils engineer. Figure 6.7 is a guide for estimating CBR values from AASHO classification.

In strength evaluation of untreated subgrade materials, the factors which may adversely affect the load supporting properties of the materials must be considered. The three most critical factors are moisture, soil expansion and frost effects. CBR and *R* value methods both take into account the critical effects of strength loss due to saturation and swelling of the soils. But effects of these factors should be estimated for plate-bearing tests made *in situ*.

Frost action can be evaluated on the basis of either frost heave or weakening during the frost melting period. The design method used here takes into account reduced supporting capacity of the subgrade during the frost melt period. It results in a pavement structure that is adequate during the frost melt period but with a load-carrying capacity in excess of that required during other periods of the year. Areas with abrupt changes in subgrade conditions, and localities where soils are highly susceptible to frost heaving and frost boils, should be removed and replaced, or else reworked to unify the upper portion of the subgrade.

The design subgrade strength value, defined as the subgrade strength value equal to or exceeded by 90% of all test values in the section, is determined as shown in the following example.

Example
Given 11 CBR test values (9, 6, 12, 7, 8, 7, 10, 9, 10, 11 and 11) from a roadway section, determine the Design CBR value.
1. CBR = 6, 7, 7, 8, 9, 9, 10, 10, 11, 11 and 12.
2. The percentage of CBR values equal to or greater than each different value is given below.

CBR	Number equal to or greater than	Percent equal to or greater than
6	11	(11/11) 100 = 100
7		
7	10	(10/11) 100 = 90·9
8	8	(8/11) 100 = 72·7
9		
9	7	(7/11) 100 = 63·6
10		
10	5	(5/11) 100 = 45·4
11		
11	3	(3/11) 100 = 27·3
12	1	(1/11) 100 = 9·1

3. The graphical presentation of the above table is shown below.

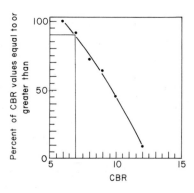

4. The design subgrade strength value, CBR = 7.

If a sample from a test location has a value so low that it indicates a weak area, additional samples should be obtained and tested to determine the extent of the area. Such locations may require local increases in thickness to provide uniform support for the entire length of the section. Test values representing these locations are omitted from the design strength calculations.

6.7 THE AASHO METHOD FOR PAVEMENT DESIGN

This method allows for consideration of economical balance of thickness design for the various pavement system layers. In this method [9] the pavement serviceability performance concept has been used to analyse the results of the AASHO Road Test. Accordingly, the design chart shown in Fig. 6.8 was developed, to relate the structural number SN to

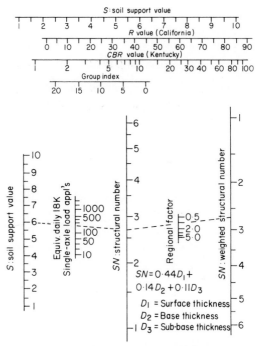

FIG. 6.8. AASHO tentative procedure for design of flexible pavements considering 20 year traffic analysis. (From [9].)

soil support value S and equivalent daily 18 000 lb single-axle load applications. The equivalent number of daily 18 000 lb single-axle load applications is the same as the Design Traffic Number (DTN) mentioned previously in the Asphalt Institute method of design. It can be calculated using the same traffic analysis procedure given in that method. The chart of Fig. 6.8, which is prepared for a terminal serviceability index $p = 2·0$, is based on a 20 year design period. For other design periods, the DTN or the equivalent daily 18 000 lb single-axle load applications should be adjusted in the same way, as mentioned before in the traffic analysis procedure. In the AASHO Road Test, an environmental factor was introduced, and is shown in Fig. 6.8 to reflect the greater damage caused to the roads by traffic under adverse conditions. This factor, referred to as the 'regional factor R' in the chart, is estimated by analysing the duration of certain typical environmental conditions during an annual cycle. A regional factor of 0·5 represents excellent environmental conditions, while a regional factor of 5 indicates very adverse conditions. Suggested regional factors are:

Frozen roadbed	0·20–1·0
Dry soil in the roadbed	0·30–1·5
Saturated roadbed (Spring break-up)	4·0–5·0

Once the structural number SN is determined, from Fig. 6.8, the following equation can be used to determine the thickness of the different layers in the flexible pavement:

$$SN = a_1D_1 + a_2D_2 + a_3D_3 \qquad (6.2)$$

where D_1, D_2 and D_3 = thicknesses of surface, base and sub-base layers respectively; and

a_1, a_2 and a_3 = coefficients whose values depend upon material properties.

The most common values for a_1, a_2 and a_3 are:

$a_1 = 0·44$, and is used if the surface course is a plant-mix asphalt concrete of high stability:

$a_2 = 0·14$, and is used for a base layer consisting of crushed stone;

$a_3 = 0·11$, and is used for a sandy gravel sub-base layer.

However, other values for a_1, a_2 and a_3 can be used, if the surface, base or sub-base courses consist of materials different from the above. These values are:

$$a_1 = \begin{cases} 0·20, \text{ for low-stability, road-mix asphalt} \\ 0·40, \text{ for sand asphalt} \end{cases}$$

$$a_2 = \begin{cases} 0·07, \text{ for sand gravel} \\ \left.\begin{matrix} 0·23 \\ 0·20 \\ 0·15 \end{matrix}\right\} \begin{matrix} \text{for cement treated bases with} \\ \text{compressive strength after} \\ \text{7 days as indicated to the right} \end{matrix} \left\{\begin{matrix} 650 \text{ psi or more} \\ 400–650 \text{ psi} \\ \text{less than 400 psi} \end{matrix}\right. \\ 0·30, \text{ for a bituminous treated coarse-graded base} \\ 0·25, \text{ for a bituminous treated sand asphalt} \\ 0·15–0·30, \text{ for a lime treated base} \end{cases}$$

$a_3 = 0·05–0·10$, for sand or sandy clay sub-bases

The following example illustrates the use of the chart in Fig. 6.8.

Example
Given: CBR = 10.
    Equivalent daily 18 000 lb single-axle load applications = 200.
    Regional factor $R = 2$.
Determine the weighted structural number SN and the thickness of the different layers for the pavement system.
    Solution. The soil support value S corresponding to a CBR of 10 is 6. Using the chart of Fig. 6.8, the structural number SN = 2·7. For a regional factor of 2, the weighted structural number = 3·0.

The structural requirement can be met by a variety of different combinations of D_1, D_2 and D_3, as indicated by the equation in Fig. 6.8. The tentative coefficients for D_1, D_2 and D_3 suggest the relative structural value of the layers of the pavement structure. Some possible methods of determining the pavement thickness are:
1. The structural requirement is assumed to be supplied by D_3, with D_1 and D_2 equal to zero. The thickness requirement would be 3·0 divided by 0·11, or about 27·5 in. Sub-base material is generally gravel or select borrow. The decision to do this would depend on availability (and in-place cost) of a well-graded gravel of excellent quality. Surface treatment (prime and seal) would provide the wearing surface. Then later, as traffic increases and/or as construction funds become available, the layer constructed earlier would provide an excellent base for a paving mat.
2. A 2 in paving mat D_1 is assumed to be combined with a base course D_2 (generally crushed rock or crushed gravel) of minimal thickness, say $3\frac{1}{2}$ in, and a sub-base layer D_3. Then

$$3·0 = (0·44 \times 2) + (0·14 \times 3·5) + 0·11D_3$$

from which $D_3 = 15$ in of sub-base material and the resulting total thickness is 20·5 in.
3. D_1 is assumed to consist of a 2 in asphalt wearing course and the rest is an asphalt concrete base course. The layers combined form a full-depth asphalt pavement. Then, $3·0 = 0·44D_1$; therefore, $D_1 = 6·9$ in. D_1 when rounded gives 7·0 in, from which 2·0 in is wearing course and 5·0 in is asphalt concrete base.

6.8 THE CALIFORNIA BEARING RATIO METHOD

This empirical method for design of flexible pavements is widely used with various modifications by highway and airport organisations in several countries. It is based on the California Bearing Ratio (CBR) test explained in a previous chapter.

Records of performance of existing pavements, accelerated road tests and direct field measurements of soil strength are combined with experience to develop design charts which correlate laboratory CBR values with anticipated field conditions. The daily number of commercial trucks

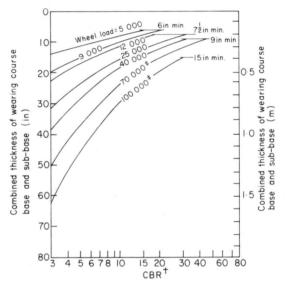

FIG. 6.9. Chart for design of flexible pavements for highways and airfields using the CBR method. † CBR is for compacted and soaked specimen at 0·1 in penetration. ‡ Tyre pressure is 100 psi. *(Courtesy: Corps of Engineers.)*

using the pavement, and the weights, are of significance in modifying thickness requirements. Figure 6.9 shows a typical basic chart that can be used in the design.

Example
Establish a grade line and determine thickness of flexible pavement structure for a highway if the maximum design wheel load is 5000 lb.

Data from a soil survey indicate three definite horizons of soil, as illustrated in Fig. 6.10(a), and sufficient laboratory tests have been made to determine CBR values for the different types of soil indicated.

Solution. Using the design chart of Fig. 6.9 for a soil with a CBR of 5 gives a minimum of 11 in of cover with material of higher bearing strength. Accordingly, the B horizon soil may be used to within 11 in of the top of the pavement or final grade line. It is important that construction procedures ensure optimum density so that the maximum strength of the

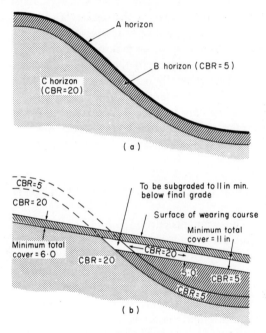

F ɪ ɢ. 6.10. The three soil horizons and CBR values for example. (a) Original soil profile. (b) Final graded profile section.

soil will be realised, and that neither will frost have an adverse effect on the soil nor will annual frost penetrate below 11 in. If the expected number of trucks with wheel loads larger than about 4000 lb exceeds 100 per day, the thickness will usually be increased over the minimum because significant stresses below the design maximum should be given consideration.

Attention should always be given to the transition zone between cut and fill so that weak materials are not left in place directly beneath the pavement. Subgrading and backfilling with more stable material should be carried out, as indicated in Fig. 6.10(b). The soil from horizon C may be used for this purpose. Since a minimum of 6 in of flexible-pavement structure is indicated by the design chart for a 5000 lb wheel load and a CBR value of 20, the soil with a CBR of 20 may be used up to that distance from the surface of the wearing course. Thus, a lift of 5 in of this soil will be utilised in the fill section, where it is to be placed on soil with a CBR of 5.

While the balancing of cut and fill quantities is of economic importance in a highway grading project, it can be seen that establishment of final grade lines should be influenced by the nature of the soils which are to create such a balance. In the above example, for instance, cutting deeper into the C horizon would pay off in making available more of the better material for use on the project, to the extent that overhaul does not become excessive. It is generally better practice to build the subgrade to uniform strength over a considerable length of roadway so that the pavement structure may be built to uniform depth.

The disadvantage of this method is that it does not take into account an accurate traffic analysis similar to that used in the previous methods.

6.9 HVEEM-CARMANY METHOD

Several highway departments currently use this method [6]. The primary soil strength criterion is the resistance value R using the Hveem Stabilometer, as explained previously. The R value is used in conjunction with a traffic index to determine a required thickness of cover, expressed as a gravel equivalent on scale C of the nomograph of Fig. 6.11. The gravel equivalent is the thickness, in inches, of base course and surfacing required for minimal tensile strength materials such as untreated gravel or crushed rock.

A secondary strength criterion is the cohesiometer value C, which measures the relative flexural strength of the paving mat or treated base material. In this procedure the flexural strength of a disc-shaped test specimen is determined in a way similar to the tests for the modulus of rupture of Portland cement concrete. Values of C above 100 serve to reduce the total thickness of cover from that indicated by the gravel equivalent, thus taking advantage of the load-spreading capability of paving or surfacing materials with inherent tensile strength. Continuation of the nomograph of Fig. 6.11 through scale D to scale F gives the adjusted design thickness.

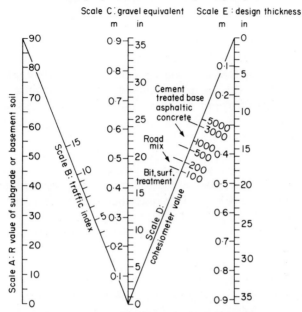

FIG. 6.11. Thickness design chart for base course and/or paving mat using Hveem stabilometer. *(Courtesy: California Division of Highways.)*

The traffic index measures the severity of vehicular traffic, particularly that of the heavier wheel loads of commercial trucks and buses. It can be seen, then, that implications of the design wheel load are inherent in the numerical value of traffic index to be used for a given traffic demand. An estimated correlation of traffic index with design wheel load and the general category of traffic use is given in Table 6.7.

TABLE 6.7 ESTIMATED CORRELATION OF TRAFFIC INDEX WITH DESIGN WHEEL LOAD

Traffic index	Design[a] wheel load (lb)	Traffic severity	General category of traffic use
5	—	Very light	Property access roads and streets
6	5 000	Light	Feeder roads and streets
7	9 000	Medium	County and city neighbourhood arterials
8	12 000	Heavy	Secondary state highways
9	16 000	Very heavy	Primary state highways
10	20 000	Severe	Truck routes and truck lanes
15	30 000	—	Logging roads

[a] One-half of single-axle load or one-half of tandem-axle load.

Another important feature of this method of thickness design is a determination of the swell pressure developed by the compacted soil as it is subjected to soaking in water. In general, the swell pressure varies from zero for predominantly granular material to possibly 2 psi for fine-grained soils. The design procedure requires that the minimum thickness of cover be sufficient to prevent swell from taking place, assuming the cover material to have a unit weight of 130 pcf. On this basis, the required minimum cover thickness t in inches is

$$t = 13 \cdot 3p \tag{6.3}$$

where p is the swell pressure of the soil in psi. The cover thickness so determined will govern if this is greater than the thickness determined from the design chart of Fig. 6.11.

6.10 SHELL METHOD

In this method [11], it is assumed that the flexible pavement structure consists mainly of three layers which behave elastically under the dynamic loads applied by traffic. Accordingly, the elastic theory for layered systems may be used to calculate the distribution of stresses and strains. The three layers are: (1) a bitumen bound layer or layers; (2) a granular unbound layer or layers; (3) the subgrade.

Figure 6.12 illustrates the pavement system and shows the location and nature of the strains which are considered to be critical.

The Shell design method is based on the following criteria for pavement failure under traffic.

(a) Excessive deformation of the surface due to the accumulation of small permanent deformations in the pavement structure. In well-designed pavement structures, these deformations are primarily dependent on the vertical compressive strain in the surface of the subgrade. Large deformations may also lead to cracking of the surface layer.

(b) Cracking of the asphalt layer, which may be induced by a repeated flexing of the asphalt layer under the repeated traffic loads (fatigue). The initiation of such cracks is governed by the horizontal (radial) tensile strain at the underside of the asphalt layer, as shown in Fig. 6.12.

The design procedure is to select the thickness of each of the layers so that the strains developed by traffic at the critical points are acceptable. Design charts have been developed on this basis and cover a wide range of conditions of subgrade and traffic. It is only necessary for the designer to have information on the subgrade and the expected traffic so that he can select the appropriate design chart. The charts are based on the dynamic elastic modulus of the subgrade (E_s), which can be considered equal to 1500 times the CBR, approximately, when no other data are available.

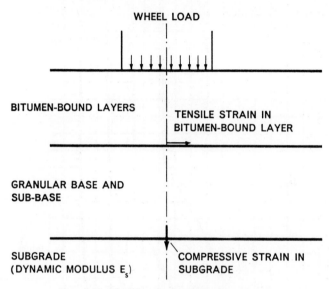

FIG. 6.12. Flexible-pavement three-layer system.

From traffic studies, the total number of axle loads using the design lane during the design period of the pavement should be determined for each load category. These axle loads should then be converted to the equivalent 18 kip standard axle load applications, using Fig. 1.6.

For the equivalent 18 kip axle loads the analysis assumes that the subgrade strains can be determined by considering a 9000 lb wheel load acting over a single circular area of 6 in radius. Tensile strains in the asphalt layer are determined using a circular area with a radius of 4·2 in subjected to a uniform contact pressure of 80 psi.

Two groups of design charts have been developed for selected values of the subgrade modulus, E_s, and the 18 kip load applications. The first group of these charts gives the thickness of the dense asphalt layer in combination with granular layers of varying thickness, for selected values of subgrade modulus. The variables in these charts are the number of 18 kip axle load applications and the minimum CBR or R value of the granular layers. The alternative group of charts can also be used to

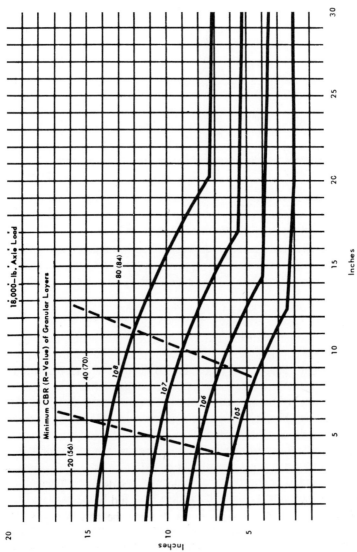

F1G. 6.13. Typical design chart for various 18 kip axle load applications.
(Courtesy: Shell Int. Petroleum Company.)

determine the thicknesses of the dense asphalt and unbound granular layers, for selected values of the 18 kip axle load applications. The only variable in this second group of charts is the subgrade modulus, E_s. Figures 6.13 and 6.14 show typical design charts for the two methods of presentation.

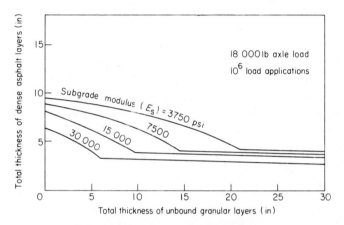

FIG. 6.14. Typical design chart for various subgrade moduli.
(Courtesy: Shell Int. Petroleum Company.)

6.11 DESIGN BASED ON FATIGUE MODE OF DISTRESS

One of the distress mechanisms that can lead to rupture is that of fatigue of asphalt concrete. A working model has been presented by Kasianchuk and co-workers [7, 8] for a subsystem that considers this fatigue mode of distress. The design subsystem is divided into three general sections, termed as follows:

1. Preliminary data acquisition: this involves estimation of traffic and wheel load distribution, a survey of subgrade soils, selection of the most economical materials for construction, determination of the environmental conditions and design of the asphalt concrete mixture to be used.

2. Materials characterisation: this section is concerned with testing the asphalt concrete, subgrade soil and any granular materials that may be used to determine their elastic properties in the range of expected service conditions. It also includes testing the fatigue properties of the asphalt concrete mixture.

3. Analysis and evaluation: in this section, the seasonal variations in the stiffness of the asphalt concrete and the moisture content of the subgrade soil are defined. In addition, the expected response of the asphalt concrete layer to the action of the range of wheel loads and climatic conditions is determined for selected trial design sections. The fatigue life of each trial design under the action of the expected traffic volume is predicted to evaluate the adequacy of the section in providing an acceptable design life.

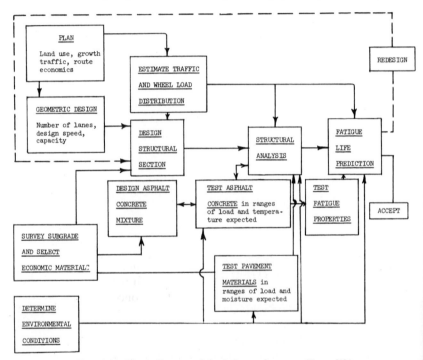

FIG. 6.15. Block diagram of the fatigue subsystem. (From [8].)

It may be noted that there is some similarity between the fatigue subsystem, as indicated in Fig. 6.15, and the classical structural engineering approach. In this approach, a structure is designed, its behaviour under the anticipated service conditions is analysed and its adequacy with respect to some distress criterion is determined. However, the fatigue of asphalt concrete, in flexible pavements, is only one of a number of mechanisms that can lead to distress within the pavement structure. Other subsystems can be developed, following a similar procedure, to analyse such distress modes as distortion, disintegration or low-temperature fracture.

The subgrade soil can be characterised, for example, by subjecting undisturbed samples of the soil to a large number of repetitions of a deviator stress and determining the resilient modulus which corresponds to 25 000 repetitions. The resilient modulus can be calculated from the expression:

$$M_R = \frac{\sigma_d}{\varepsilon_r} \tag{6.4}$$

where M_R = modulus of resilient deformation (psi);

σ_d = repeatedly applied deviator stress (psi); and

ε_r = resilient (recoverable) axial strain (in/in).

The structural analysis phase consists essentially of determining the vertical compressive strain at the subgrade level and the horizontal tensile strain on the underside of the asphalt bound layer. For this purpose the layered elastic theory should be applied. The use of the Chevron computer program developed by Shell can simplify the calculations to a large extent.

In the Shell analysis of pavements designed according to the CBR procedure, the vertical compressive strain at the subgrade level should be kept below about 6.5×10^{-4} in/in. The horizontal tensile strain on the underside of the asphalt bound layer should not exceed 1.5×10^{-4} in/in.

In the evaluation phase, the values of the strains obtained under the expected traffic and environmental conditions are compared with the fatigue life determined for the particular asphalt concrete used. The fatigue life can be predicted by using a compound loading hypothesis, which suggests that fatigue failure occurs when:

$$\sum \frac{n_i}{N_i} = 1 \tag{6.5}$$

where n_i = number of applications at strain level i; and

N_i = number of applications to cause failure in simple loading at strain level i.

The procedure is then to determine the tensile strain in the asphalt bound layer under each wheel load magnitude at the stiffness value representing a certain month of the year. The expected fatigue life under simple loading at that strain level can be determined from the fatigue curve developed for the asphalt concrete. Figure 6.16, for example, shows the relationship between the tensile strain and number of load repetitions to failure for asphalt concrete with an elastic modulus (E_1) equal to 900 000 psi. The cycle ratio for each strain level can be formed using the number of applications, n_i, per month for each axle group, and the number of applications, N_i, to failure for each strain level. Equation (6.5), when used to determine the sum of the cycle ratio per month for consecutive months until this sum reaches unity, will give the fatigue life of the pavement.

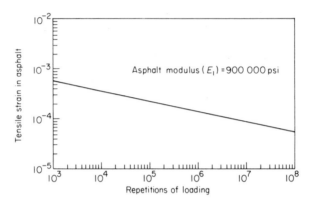

FIG. 6.16. Relationship between tensile strain in asphalt concrete and load application. (From [2].)

REFERENCES AND BIBLIOGRAPHY

1. AASHO Committee on Design, 'AASHO Interim Guide for the Design of Flexible Pavement Structures', American Association of State Highway Officials, 1961.
2. Dormon, G. M., Edwards, J. M. and Kerr, J. E. D., 'The Design of Flexible Pavements', *Transactions of the Engineering Institute of Canada*, Paper No. EIC-64-CIV 5, 1964.
3. Hennes, R. G. and Ekse, M., *Fundamentals of Transportation Engineering*, Second Edition, McGraw-Hill, 1969.
4. *Highway Research Board Bulletin 8*, 'Design of Flexible Pavements Using the Triaxial Compression Test', Washington, D.C., 1947.
5. *Highway Research Record*, No. 90, 'AASHO Road Test Results Applied to Pavement Design in Illinois', Washington, D.C., 1965.
6. Hveem, F. N. and Carmany, R. M., 'The Factors Underlying a Rational Design of Pavements', *Proc. Highway Research Board*, 1948.
7. Kasianchuk, D. A., 'Fatigue Considerations in the Design of Asphalt Concrete Pavements', Ph.D. Thesis, University of California, Berkeley, August 1968.
8. Kasianchuk, D. A., Monismith, C. L. and Garrison, W. A., 'Asphalt Concrete Pavement Design—A Subsystem to Consider the Fatigue Mode of Distress', *Highway Research Record*, No. 291, 1969.

9. Liddle, W. J., 'Application of the AASHO Road Test to the Design of Flexible Pavement Structures', *Proc. 1st International Conference on the Structural Design of Asphalt Pavements, University of Michigan, Ann Arbor, 1962.*

10. Olinger, B. J., 'Modified CBR Flexible Pavements Design', *Highway Research Board Research Report 16-B*, 1954.

11. Shell International Petroleum Company Limited, 'Shell Design Charts for Flexible Pavements', London, 1963.

12. Shook, J. F. and Lepp, T. Y., 'Method for Calculating Equivalent 18-kip Load Applications', *Highway Research Record*, No. 362, 1971.

13. The Asphalt Institute Manual Series No. 1, 'Thickness Design–Full-Depth Asphalt Pavement Structures for Highways and Streets', Revised Eighth Edition, August 1970.

14. Shell Canada Ltd, The 1964 Canadian Supplement to 'Shell 1963 Design Charts for Flexible Pavements', Toronto, 1964.

CHAPTER 7

DESIGN OF FLEXIBLE PAVEMENTS FOR AIRFIELDS

7.1 INTRODUCTION

The base layer of an airfield flexible pavement can consist of untreated granular material, cement stabilised material or bituminous bound material. When asphalt mixtures are used for all courses above the subgrade, a full-depth asphalt pavement is obtained. The advantages of this type of pavement are summarised in Chapter 6 and its use in airfields is recommended, whenever possible, rather than flexible pavements with untreated granular bases.

In this chapter, some of the methods of design of flexible pavements for airfields are discussed. The Asphalt Institute method, explained in subsections 7.5.2 and 7.5.3, is very recent and is the most comprehensive one.

7.2 THE US FEDERAL AVIATION AGENCY METHOD

In this method, the FAA subgrade classification mentioned in Chapter 4 (Table 4.1), has been used in developing charts for determining the required thickness of flexible pavement for airfields [4, 5]. These charts have been based on the following considerations:

1. The gross aircraft weight has been used in plotting the charts, not the equivalent single-wheel load.

2. The required pavement thickness for a specific aircraft at a particular airport can be determined from the readily available information: the gross aircraft weight; the type of main gear undercarriage (single, dual, or dual-tandem); and the FAA subgrade classification of the airport.

3. The main undercarriages of the aircraft are assumed to carry 95% of the total aircraft weight, while the nose gear carries 5% of the total weight.

4. The curves are based on capacity operations for the gross weights indicated.

5. Where a pavement is to be evaluated for limited operations or for less than capacity operations, adjustments can be made as follows. The thickness of the existing pavement is multiplied by 1·1; this increased thickness is then introduced into the appropriate curves; and the load for the limited operations is determined as if the pavement were 10% thicker than it actually is.

6. Critical areas which receive a concentration of aircraft movement, such as aprons, taxiways, turn-arounds and runup areas at runway ends, are taken care of in plotting these charts.

7. Wherever a flexible pavement is to be subjected to concentrated fuel spillage or other solvents, as at aircraft loading positions and maintenance areas, protection should be provided by use of a solvent-resistant seal coat.

8. Full-depth frost protection should be provided for the primary runway(s) at large hub airports. For other paved areas, protection should be provided for depths between 65% and 90% of the total frost penetration.

To convert from equivalent single-wheel load design to that based on gross aircraft weight, some reasonable compromise on the effects of gear dimensions of existing civil aircraft should be made. A study of the wheel spacings for current aircraft has indicated a trend towards increased spacing as the aircraft weight increases. For this reason, a variation in wheel spacings d, S and S_D was used for flexible-design curves in which the narrower spacings were used at the lighter weights and wider spacings were used at the heavier weights. Plots of these dimensions for most multiple-wheel undercarriages indicate that a straight-line variation between 15 000 lb and 50 000 lb with corresponding $d/2$ dimensions of 5 in and 10 in, respectively, would satisfy most of these dual and dual-tandem gears. These plots also indicate that a straight-line variation between 15 000 lb and 100 000 lb for a $2S$ dimension of 35 in and 60 in, respectively, would satisfy most dual-gear aircraft. Additionally, the variation between 50 000 lb and 200 000 lb for $2S_D$ dimensions of 80 in and 130 in, respectively, would satisfy most dual-tandem-gear aircraft. Lines a, b and d in Figs. 7.1 and 7.2 represent the plots corresponding to the spacings $d/2$, $2S$ and $2S_D$, respectively.

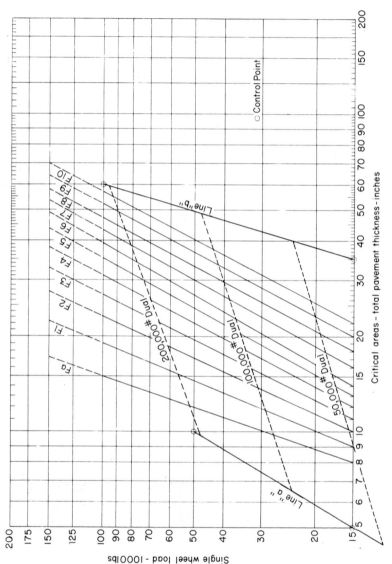

FIG. 7.1. Development of flexible-pavement curves—dual gear. (From [4].)

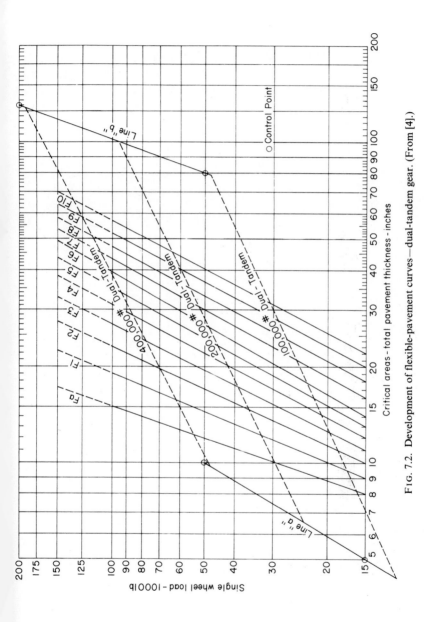

FIG. 7.2. Development of flexible-pavement curves—dual-tandem gear. (From [4].)

Lines representing dual-gear aircraft with gross weights of 50 000, 100 000 and 200 000 lb were plotted on Fig. 7.1. For the 200 000 lb gross weight aircraft, for example, the weight on each dual-wheel main undercarriage is equal to $1/2 \times 95/100 \times 200\,000 = 95\,000$ lb, and on each main wheel is 47 500 lb. As explained previously in Chapter 2, for a pavement of thickness less than or equal to $d/2$, which is 10 in in this case, the equivalent single-wheel load is equal to that of one wheel, which is

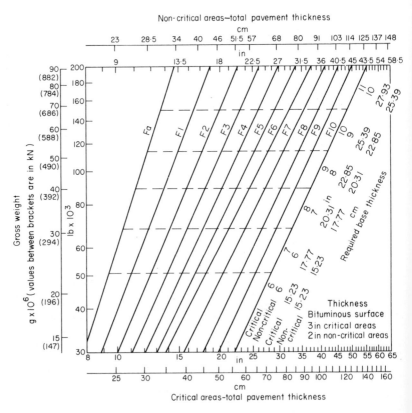

FIG. 7.3. Design curves, flexible-pavement single gear.

47 500 lb. For a pavement thickness greater than or equal to $2S$, which is 60 in in this case, the equivalent single-wheel load is equal to that on two wheels, which is 95 000 lb.

In a similar way, lines representing dual-tandem-gear aircraft with gross weights of 100 000, 200 000 and 400 000 lb were plotted on Fig. 7.2.

For single-wheel-gear aircraft, since the design thickness is determined by the load on one wheel or gear and is independent of depth, it was only necessary to divide the single-wheel load scale by 0·475. This represents the gross aircraft weight requirements for single-gear aircraft and is shown in Fig. 7.3.

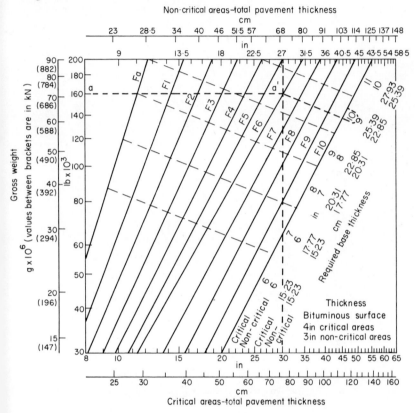

FIG. 7.4. Design curves, flexible-pavement dual gear.

For the case of dual-gear aircraft, a new graph was plotted in Fig. 7.4 with gross aircraft weight on the vertical axis and the total pavement thickness on the horizontal axis, on a log-log scale. Total pavement thickness requirements for each gross aircraft weight, as obtained from Fig. 7.1, were plotted for each subgrade classification. Connection by a straight line of three points for 50 000, 100 000 and 200 000 lb gross weights for each subgrade classification resulted in the reorientation of the subgrade curves. Similarly, Fig. 7.5 for dual-tandem-gear aircraft was developed by using total pavement thickness requirements for each gross aircraft weight as obtained from Fig. 7.2.

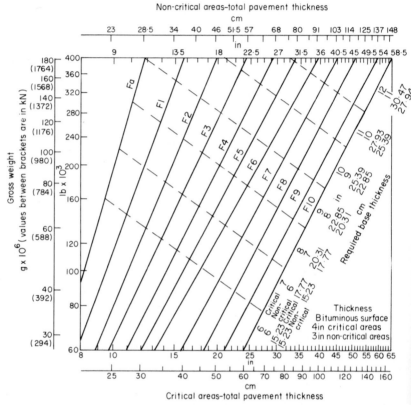

FIG. 7.5. Design curves, flexible-pavement dual-tandem gear.

The dashed lines on Figs. 7.3, 7.4 and 7.5, representing the required non-bituminous base course thickness for critical and non-critical areas, were derived from existing base course thickness requirements. The area between two dashed lines represents those gross weights where the same base course thickness is required. This base course thickness requirement is indicated along the right edge of the F10 subgrade classification line.

The minimum thickness of the bituminous surface course is 3–4 in for critical areas and 2–3 in for non-critical areas. If a bituminous base is recommended, a smaller base thickness can be used.

Example

Determine the thickness of the flexible pavement required for the critical and non-critical areas of an airfield that will receive a dual-gear aircraft of 160 000 lb gross weight. The soil group of the subgrade is E-7 and poor drainage and severe frost conditions exist.

Solution. The subgrade classification for the above-mentioned conditions is F6. Enter Fig. 7.4 on the left at point a, representing 160 000 lb and proceed horizontally until intersecting the F6 subgrade classification line at point a'. From point a', project a vertical line up and down until it intersects the total pavement thickness, giving 27 in at the top and 30 in at the bottom of the graph. These thicknesses determine the non-critical and critical area total pavement thicknesses, respectively. Proceed from point a' to the right, parallel to the sets of dashed lines, until the base thickness scale is intersected. This gives a base thickness of 10 in for the critical area and 9 in for the non-critical area. The required thickness of the asphaltic concrete surface is 4 in for critical areas and 3 in for non-critical areas. Subtracting the thickness of the wearing course and base layer from the total thickness will give the thickness required for the sub-base.

	Critical area	*Non-critical area*
Asphalt concrete wearing course	4 in	3 in
Non-bituminous base	10 in	9 in
Sub-base	16 in	15 in
	30 in	27 in

7.3 DESIGN CHART FOR AIRFIELDS OF SMALL AIRPORTS

Small airports which accommodate personal aircraft and non-scheduled small aircraft for business and industrial activities may not need special consideration of their pavements. For aircraft up to 12 500 lb gross weight, soil stabilisation and bituminous surface treatment may prove adequate. However, for aircraft with larger loads, a proper flexible pavement design should be carried out.

If the FAA subgrade classification system is applied, Fig. 7.6 can be used to determine the thicknesses of the different components of the flexible-pavement structure.

In using this figure, the point of intersection of the horizontal line through the aircraft gross weight with the F curve, representing the subgrade under consideration, will provide the total thickness of the pavement structure. The point of intersection of the same horizontal line with the Fa curve will give the required thickness of base plus surface course. The thickness of the sub-base layer can be obtained by subtracting this value from the total thickness.

Although the minimum thickness of the bituminous bound surface course is 1 in, the use of a $1\frac{1}{2}$–2 in wearing course is recommended.

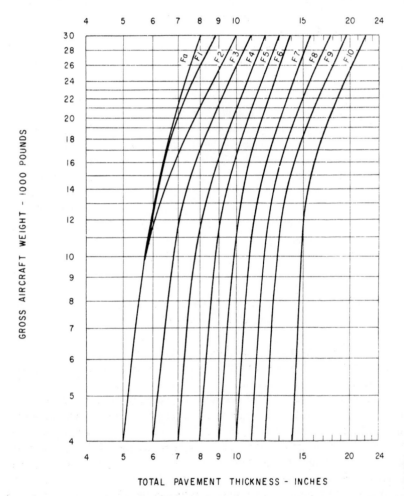

TOTAL PAVEMENT THICKNESS - INCHES

FIG. 7.6. Design curves, flexible pavements—light aircraft. (*Note:* The Fa curve fixes the required base plus surface course thickness. One inch minimum surface thickness assumed for Fa curve.) *(Courtesy: Federal Aviation Agency.)*

7.4 THE UNITED KINGDOM METHOD

7.4.1 The United Kingdom Load Classification Number (LCN) System

An extensive series of loading tests were performed on flexible and rigid pavements of varying thickness and also on a variety of subgrades, to determine a relationship between the value of the failure load and the contact area over which the load was applied. In these tests the failure load was considered to be that load which produced a deflection of 0·1 in.

From these tests the load–contact area relationship given in eqn. (7.1) was established, for contact areas between 200 and 700 in^2:

$$\frac{P_1}{P_2} = \left(\frac{A_1}{A_2}\right)^{0·27} \tag{7.1}$$

where P_1 and P_2 are failure loads on contact areas A_1 and A_2, respectively.

In order to devise a system whereby the capacity of a pavement to carry an aircraft could be expressed as a single number, the idea of a standard load classification curve was introduced. This curve, which is shown in Fig. 7.7, is purely arbitrary and is produced by joining the series of points given in Table 7.1. The values in this table were selected because they were typical of the wheel loads and contact areas of aircraft at the time the LCN system was being developed.

The standard load classification curve of Fig. 7.7 and the relationship in eqn. (7.1) have been combined to produce the chart of Fig. 7.8, which has been drawn in the following manner.

(a) The tyre contact area line is obtained by dividing the load by the tyre pressure.

(b) One point on each LCN line is obtained directly from the standard load classification curve.

(c) Other points on each LCN line have been calculated from the relationship given in eqn. (7.1).

From Figs. 7.7 and 7.8, a clear conception can now be obtained of what the LCN of any pavement represents [7]. Figure 7.7 shows that a pavement having a LCN of, say, 40, is one capable of bearing without failure a load of 40 000 lb on a contact area of 444 in^2 (tyre pressure 90 psi). Furthermore, Fig. 7.8 indicates that the same pavement, if it performs in the manner shown by field test to be typical, will also be capable of bearing any of the infinite number of combinations of load and tyre pressure lying along the LCN 40 line.

The classification of an aerodrome pavement is thus a relatively simple

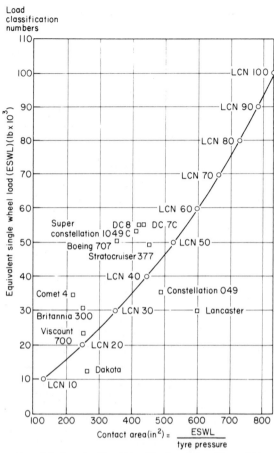

FIG. 7.7. Standard load classification curve. (From [7].)

TABLE 7.1 SELECTED VALUES FOR THE LCN SYSTEM

Wheel loading		*Tyre pressure*		*LCN*
(lb)	*(kg)*	*(psi)*	*(kg/cm²)*	
100 000	45 400	120	8·44	100
90 000	40 800	115	8·09	90
80 000	36 300	110	7·74	80
70 000	31 800	105	7·38	70
60 000	27 200	100	7·03	60
50 000	22 700	95	6·68	50
40 000	18 100	90	6·33	40
30 000	13 600	85	5·98	30
20 000	9 100	80	5·62	20
10 000	4 500	75	5·27	10

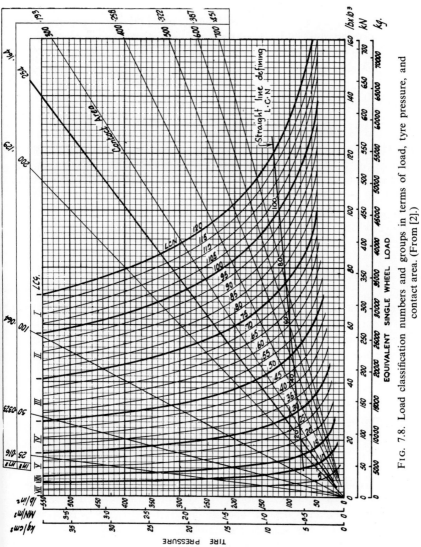

FIG. 7.8. Load classification numbers and groups in terms of load, tyre pressure, and contact area. (From [2].)

matter. It can be done by carrying out a series of loading tests using different sizes of plate to determine the 'failure load contact area' curve for the particular pavement, and then plotting this and the standard load classification curve on one diagram, and reading off the failure load where the two intersect. The LCN of the pavement is then this load in pounds divided by 1000 and is by definition dimensionless.

In many cases it is simpler and more usual to carry out plate-bearing tests using one size of plate only, then to assume the relationship given in eqn. (7.1) to hold good, and read the LCN of the pavement directly from Fig. 7.8. For example, a load of 34 500 lb on an 18 in diameter plate (256 in^2 area) is equivalent to an LCN of 40.

The selection of the LCN to represent any pavement is a matter of statistical analysis and engineering judgement after tests have been carried out, and it is not practicable to use the system to accuracies greater than, say, 10%; for example, a variation in LCN of less than 5 at LCNs of the order of 50 would in practice be disregarded for regular trafficking, as would also a variation of less than 10 at LCNs of the order of 100.

7.4.2 Evaluation of Aircraft

The LCN of an aircraft having one wheel to each undercarriage leg is fixed by the load on the leg and the tyre pressure and can at once be read from the chart in Fig. 7.8. For example, a wheel load of 15 000 lb and tyre pressure of 75 psi is equivalent to an LCN of 15.

For multi-wheeled undercarriages the Equivalent Single-Wheel Load should first be determined. Once this ESWL has been calculated, the LCN of the aircraft is again simply read from Fig. 7.8, and the system operates as if the aircraft had only one wheel per undercarriage leg. For example, an Equivalent Single-Wheel Load of 42 000 lb and tyre pressure of 150 psi is equivalent to an LCN of 50.

7.4.3 Flexible Pavement Design in Terms of LCN Systems

The California Bearing Ratio (CBR) system has been used as the basis of flexible pavement design. The design chart in Fig. 7.9 is obtained by applying the LCN system to the standard US Corps of Engineers CBR curves on the basis that they apply to tyre pressures of 100 psi. The introduction of the LCN system then allows of the use of the curves for any combination of loads and tyre pressures. Repetitions of load are taken into account in developing this chart.

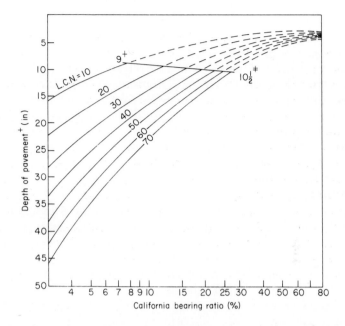

FIG. 7.9. Flexible-pavement design curves for runways. † Total depth of flexible pavement above the subgrade is determined from the CBR for the subgrade; depth of base and wearing courses combined is determined from the CBR of the sub-base; depth of wearing asphalt layer is determined from the CBR of the base. ‡ The line connecting the two values of 9 and $10\frac{1}{2}$ in represents the minimum allowable total thickness of flexible pavement. (From [7].)

7.5 FULL-DEPTH ASPHALT PAVEMENTS

The term 'full-depth' certifies that the pavement is one in which asphalt mixtures are employed for all courses above the subgrade or improved subgrade. A full-depth asphalt pavement is laid directly on the prepared subgrade. Its use in airfield pavements, specially in airports serving air carrier aircraft of greater than 60 000 lb (27 200 kg) gross weight, is more advantageous than using flexible pavements with unbound granular bases.

7.5.1 Theoretical Analysis for the Design
A theoretical procedure has been suggested recently by M. W. Witczak [11] for designing full-depth asphalt concrete pavements for airfields. The design method is based upon using the multi-layered elastic theory and

employing the concept of limiting strains to prevent repetitive permanent deformation and/or shear failure within the subgrade layer, and repetitive load cracking within the asphalt bound layer. The design aircraft considered in the analysis is a DC-8-63F aircraft.

Results developed by Kingham [8] from the AASHO Road Test study on full-depth asphalt concrete pavements have indicated that a limiting vertical compressive strain, ε_c, of 0·001 46 in/in, evaluated at a limiting asphalt concrete modulus of 100 000 psi, is capable of withstanding 1 000 000 strain repetitions. Furthermore, an allowable horizontal tensile strain, ε_t, of 0·000 076 in/in in the asphalt bound layer will allow of 1 000 000 repetitions when evaluated at a critical asphalt concrete modulus of 1 450 000 psi.

The theoretical design procedure involves investigating three important elements which deal with (1) multi-layered stress–strain solutions, (2) materials characterisation, and (3) a failure criteria function expressed in terms of the allowable design stresses or strains.

Shook and Kallas [9], and others, have applied multiple regression techniques to establish a relationship between the dynamic modulus of asphalt concrete mixes, E_1, and temperature for different frequency levels.

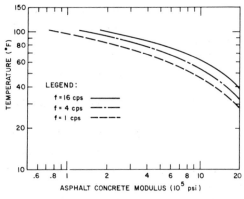

FIG. 7.10. Effect of temperature and frequency upon typical asphalt concrete dynamic modulus. (From [11].)

Average values for the properties of the concrete mix were used in these studies. Figure 7.10 illustrates such a relationship for different frequency levels, while Fig. 7.11 shows this relationship for a frequency of 2 c/s (cps). This frequency was considered to be typical for a dual-tandem gear travelling at a taxiing speed of 10–20 mph. Measured and predicted values for the asphalt concrete modulus indicated good agreement, as shown in Fig. 7.11.

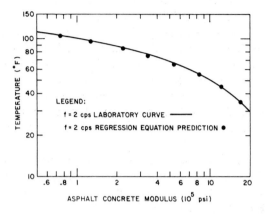

FIG. 7.11. Comparison of measured and predicted asphalt concrete modulus. (From [11].)

Vertical Subgrade Strain

The revised US Corps of Engineers (USACE) design analysis was used to establish flexible airfield pavement thickness requirements for selected aircraft, various CBR values and a number of aircraft passes. The Shell BISTRO computer program for analysing stresses and strains in multi-layered pavement structures was then applied to determine the maximum vertical subgrade strains for a variety of pavement conditions. The following data were used in the theoretical study:

1. Poisson's ratio $v = 0.40$ for asphalt concrete and 0.45 for unbound layers.

2. Dynamic subgrade modulus $E_s = 1500 \times CBR$.

3. The pass per coverage factor for channelised taxiway traffic conditions, established by the Corps of Engineers, was considered valid.

4. One coverage equalled one strain repetition.

5. The modular ratio of the unbound granular layer to the subgrade was variable, and dependent upon the modulus of the subgrade. Approximate ratios used were 2.9, 2.3 and 1.8 for $CBR = 3$, 5 and 10, respectively.

6. The stiffness or dynamic modulus of asphalt concrete mixes is referred to as E_1.

The results of the computer program were used to establish an allowable subgrade strain criterion for each asphalt concrete stiffness. Full-depth thickness requirements at various repetitions to failure were determined

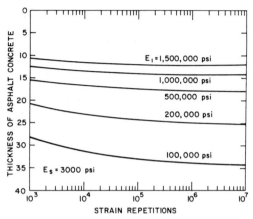

FIG. 7.12. DC-8-63F full-depth thickness–repetition requirements for various assumed asphalt concrete modulus values. (From [11].)

from the vertical subgrade strains for several combinations of E_1 and E_s and for DC-8-63F aircraft. A typical set of curves for a subgrade modulus (E_s) of 3000 psi is shown in Fig. 7.12, while Fig. 7.13 shows similar curves, plotted in a different manner, for E_s equal to 15 000 psi.

The cumulative damage theory was applied to determine the number of strain repetitions to failure (N_f) for a given thickness and subgrade combination, in several locations of the USA and Canada.

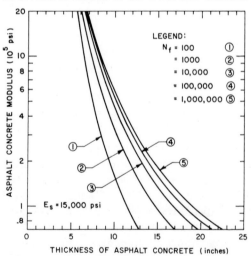

FIG. 7.13. DC-8-63F full-depth thickness–asphalt concrete modulus requirements for various repetition levels. (From [11].)

From this information full-depth thickness requirements were summarised in a tabular form for various subgrade moduli and N_f values for different environments. By defining a standard thickness corresponding to a base N_f equal to one million and an annual average air temperature of approximately 75°F, thickness adjustment factors were calculated; they are shown in Fig. 7.14 as a function of the average yearly air temperature. The thickness reductions within any particular N_f value were found to be nearly equal for any particular temperature. Stated in another way, the percentage reduction in full-depth thickness may be related solely to the average annual air temperature.

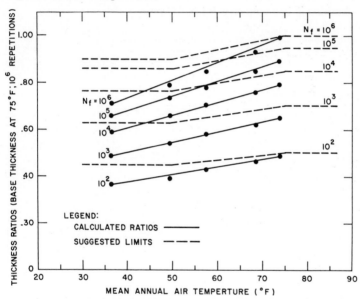

FIG. 7.14. Effect of climatic environment upon percentage thickness requirements for vertical subgrade strain analysis. (From [11].)

To summarise, Figs. 7.12 and 7.13 can be used to determine the full-depth asphalt concrete pavement thickness corresponding to the given E_1 and E_s values, for base values of N_f equal to one million repetitions and an annual average air temperature of 75°F. Figure 7.14 can then be applied to provide the thickness adjustment factor for the mean annual air temperature and strain repetitions applicable to the situation under consideration. Multiplying the pavement thickness by this adjustment factor will give the full-depth asphalt thickness required to satisfy the condition of safe vertical compressive strain, ε_c.

Tensile Strain in the Asphalt Bound Layer

Several investigations have indicated that the design of flexible pavements against repetitive load induced cracking in the bound layer can be related to the maximum horizontal (radial) tensile strain (ε_t) at the bottom of this layer.

Kingham developed tensile strain criteria which involved strain–stiffness relationships defined at two levels of strain repetitions ($N_f =$ 121 000 and $N_f = 1\,120\,000$). From these relationships, a general expression relating ε_t and N_f was derived. Figure 7.15 represents the family of strain–temperature–repetition curves established from this expression. The relationship is linear on a log ε_t–N_f plot.

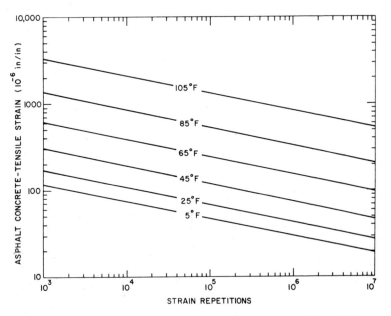

FIG. 7.15. Allowable asphalt concrete tensile strain criteria. (From [11].)

The theoretical tensile strain analysis for various full-depth pavement structures was performed with the Shell BISTRO computer program for the maximum gross weight of the DC-8-63F aircraft. Multiple regression was applied to the computer results for the tensile strains at the bottom of the asphalt concrete layer to establish an expression for the strain ε_t as a function of the thickness and the moduli E_1 and E_s. This in turn allowed of the derivation of a general full-depth thickness design

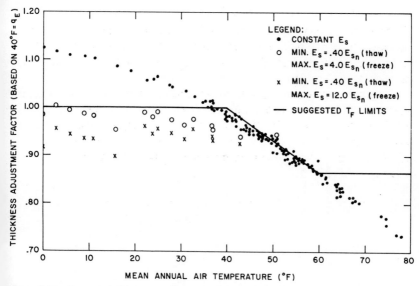

FIG. 7.16. Suggested thickness adjustment factors for asphalt concrete tensile strain analysis. (From [11].)

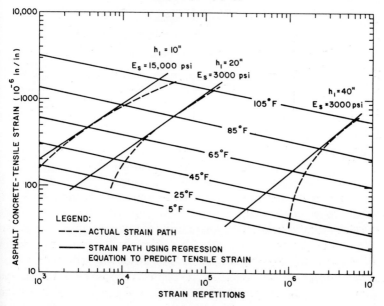

FIG. 7.17. Comparison of actual and predicted effective tensile strain paths for DC-8-63F aircraft. (From [11].)

expression for the DC-8-63F aircraft. This expression takes into account the accumulated damage for a particular combination of monthly temperature and subgrade modulus. An effective mean annual air temperature of 40°F was considered and thickness adjustment factors were established for other temperatures, as shown in Fig. 7.16.

The effective tensile strain path for selected full-depth thicknesses, subgrade moduli and mean annual temperatures is shown in Fig. 7.17 for the DC-8-63F aircraft. Figure 7.18 provides typical design curves for

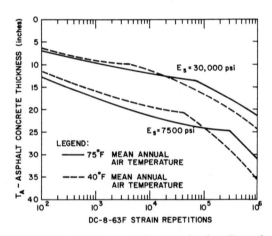

Fig. 7.18. DC-8-63F full-depth thickness requirements showing effects of traffic, subgrade modulus and environment. (From [11].)

the same aircraft, considering two values for the subgrade moduli and also two values for the mean annual temperature. These curves illustrate that for warm climates and weak soils, the failure mode is likely to be by permanent deformation and/or shear occurring in the subgrade, for traffic levels less than about 500 000 passes; this is for a DC-8-63F aircraft moving on a taxiway under channelised traffic conditions. As the climate becomes colder and the subgrade stronger, the likelihood of failure associated with cracking in the asphalt bound layer becomes greater.

7.5.2 The Asphalt Institute Method for Air Carrier Airports

This design method, which has been recently proposed by the Asphalt Institute [10], is the only method that includes a comprehensive analysis of aircraft characteristics and uses aircraft equivalency data. The design principles used in this method are summarised hereafter.

A. Design Analysis

The procedure used in the analysis is similar to what has been suggested by Witczak (see subsection 7.5.1), in that the full-depth asphalt pavement is considered as a multi-layered elastic system. Therefore, the application of a load to the pavement produces two critical elastic strains: a horizontal tensile strain, ε_t, at the bottom of the asphalt concrete layer, and vertical compressive strain, ε_c, at the bottom of the subgrade layer. Each must be examined separately in the design analysis. The location and direction of these strains are shown in Fig. 7.19.

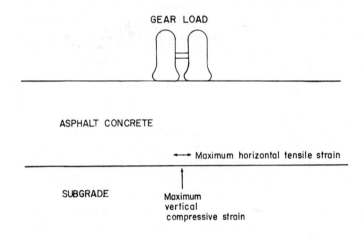

FIG. 7.19. Location and direction of tensile and compressive strains in a full-depth asphalt pavement system. *(Courtesy: The Asphalt Institute.)*

Design criteria, in terms of maximum allowable values for both critical strains, have been established and are used as the basis for selecting the proper thickness, T_A, of full-depth asphalt pavement. The modulus (or relationship between stress and strain) of asphalt concrete is temperature-dependent, so the period of the year causing greatest potential damage to the pavement varies with the temperature condition peculiar to each geographical location. Vertical compressive strain in the subgrade, an indicator of permanent deformation in the subgrade, is greatest when the asphalt concrete modulus is low (that is, the pavement temperature is high). Conversely, when the asphalt concrete has a high modulus value (during cool conditions), horizontal tensile strains at the bottom of the asphalt bound layer are critical, and pavement cracking can occur under repeated load applications. Accordingly, subgrade vertical compressive strain ε_c evaluated at high temperatures, and horizontal tensile strains ε_t within the asphalt concrete layer, evaluated at cool temperatures, constitute the two critical criteria for design of full-depth asphalt. But designing to limit the two types of strain may lead to two different thickness requirements. Therefore, the greater of the thicknesses is selected as the final design thickness.

B. Environment

Temperature has a great influence on the stress–strain distributive characteristics of full-depth asphalt pavements. For this reason, detailed analysis has been carried out to establish relationships between mean annual air temperature and pavement thickness requirements for each design criterion. To satisfy the ε_c criteria, thicknesses are increased as the average annual air temperature gets warmer. On the other hand, thicker pavements are required in cool environments to satisfy the ε_t criteria.

C. Design Location

The choice of airfield location for which the design is to be made is important. Within any airport there is a certain pavement area receiving more traffic than any other area. This area should be selected as the design location for determining the maximum thickness of pavement for the airport. Invariably, the design location for an air-carrier airport will be on a taxiway. Consequently, the critical design location should be considered as the one with the largest number of departures and arrivals, expressed as a percentage of the total operations for the airport.

At airports with multiple runways, there is usually one runway (and its taxiways) which serves more traffic than any of the others. In such cases, it will be more economical to carry out separate pavement thickness designs for each runway/taxiway system. In addition, it is more likely that a taxiway or runway will receive a greater number of stress (or strain) repetitions nearer the central portion of the pavement than at the edges. As thickness requirements are directly related to strain repetitions, substantial savings can be realised by using variable cross-sections.

Variable thickness designs for any desired airfield pavement area (taxiway, runway or apron) may be determined with the aid of thickness percentages. The thicknesses for these areas can be obtained by solving for the full-depth asphalt pavement thickness for the critical (taxiway) location and then using a specific thickness reduction factor for the pavement area in question.

D. Traffic Analysis

One of the most important concepts in this design method is that a given pavement system can withstand a unique number of strain repetitions during its design life. Strain repetitions imposed on the pavement are a function of the type of aircraft, the gear load, the number of aircraft passes and the transverse wander characteristics of the aircraft on the design area. The design thickness is the thickness which will allow the pavement to serve, under the strains imposed by the anticipated traffic, throughout its design period.

Effects on the pavement of strain repetitions of a mixture of different aircraft are considered cumulative. In the traffic analysis procedure included in this design method, the cumulative effects of traffic at different lateral intervals, x, from the taxiway centreline, may be calculated. This is accomplished by equating the number of strain repetitions produced by the aircraft traffic mixture to a number of equivalent strain repetitions produced by an arbitrarily selected 'standard' aircraft. The 'standard' aircraft is a 358 000 lb (162 ton) D C-8-63F at 100% gross weight. Equivalency factors for several major present and proposed jet aircraft types have been developed, and are presented in graphical form in Appendix D for easy use in the traffic analysis. The equivalency factors are functions of the aircraft type, pavement thickness and performance criteria. The effect of wander of the aircraft along the taxiway is directly accounted for in the equivalency factors. It is assumed that all aircraft movements are channelised.

E. Materials Characterisation

All materials under consideration for use in the asphalt pavement structure should be sampled, tested and evaluated for compliance with the quality requirements specified for the job.

One of the best sampling techniques yet devised is the one known as random sampling. With it, the sampling location is selected in such a way that all possible locations in the area to be investigated are equally likely to be chosen. The choice is unbiased because it is made entirely by chance. Details concerning the use of this technique in sampling are given in Appendix E.

Strength tests should be made on subgrade soils, so as to determine the design subgrade value. This value is the basis for the thickness design of the asphalt pavement structure. The design subgrade value used in this thickness design procedure is the modulus of elasticity, E_s, and can be determined using one of the following methods.

1. Direct measurement by means of the resilient modulus (M_r) test. This resilient modulus, which can be substituted directly in the design for the modulus of elasticity, is determined by a repeated load triaxial compression test. The test equipment used to determine M_r is relatively new and may not be readily available.

2. Approximation from the California Bearing Ratio (CBR) test. Thus, E_s (psi) $= 1500 \times CBR$, or E_s (kN/m²) $= 10\,342 \times CBR$.

3. Approximation from the plate-bearing test by using the correlation chart, Fig. 7.20.

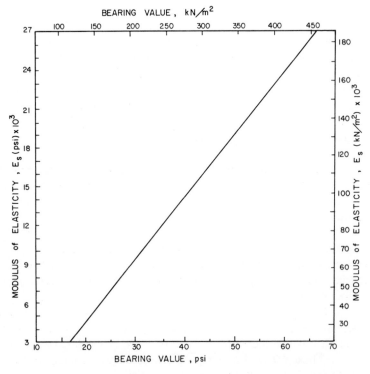

FIG. 7.20. Approximate relationship between plate-bearing value and modulus of elasticity, E_s. (*Courtesy: The Asphalt Institute.*)

4. Approximation from Table 7.2, if the FAA soil classification system is used.

After determining the subgrade strength values, the design subgrade value is fixed. In this method, the design subgrade value is defined as the subgrade modulus of elasticity, E_s, value equal to or exceeded by 85% of all E_s values. The procedure explained in the example given in the preceding chapter (see Section 6.6) can be applied in determining this design value.

TABLE 7.2 APPROXIMATE RELATIONSHIP BETWEEN
SUBGRADE TYPE AND E_s
(Courtesy: The Asphalt Institute)

FAA classification	E_s (psi)	E_s (kN/M²)
F10	5 500	37 900
F9	6 500	44 800
F8	7 700	53 100
F7	8 900	61 400
F6	10 800	74 500
F5	12 600	86 900
F4	14 600	100 700
F3	16 600	114 500
F2	19 900	137 200
F1	22 700	156 500
Fa	31 000	213 700

If a sample from a test location has a value so low that it indicates a weak area, additional samples should be obtained and tested to determine the extent of the area. Such locations may require a local increase in pavement thickness to provide uniform support for the entire length of the section. Test values representing these locations should be omitted from the calculations used to obtain the design strength value.

In strength evaluation of untreated subgrade materials, factors that may adversely affect the load-supporting properties of the materials must be considered. The most critical factors are moisture, volume change and frost. The resilient modulus test and the soaked CBR method take into account the critical effects of strength loss due to degree of saturation and swelling of the soils. Moreover, the Asphalt Institute design method recognises reduced supporting capacity of the subgrade in environments where frost may be a problem.

Because air-carrier airport pavements must support extremely heavy loads, the asphalt aggregate mixtures used must have well-defined mix design criteria. For base courses, dense-graded asphalt concrete mixes with aggregates of $\frac{3}{8}$–1 in nominal size should be used. For surface courses, the nominal size of the aggregates used in the dense-graded asphalt concrete mix should be $\frac{3}{8}$–$\frac{3}{4}$ in. Table 7.3 tabulates suggested criteria for the

TABLE 7.3 SUGGESTED CRITERIA FOR MARSHALL METHOD
AND HVEEM TEST LIMITS FOR DESIGN OF
ASPHALT CONCRETE MIXTURES
(Courtesy: The Asphalt Institute)

| Marshall method mix criteria | Air carrier airports *(more than 60 000 lb (27 200 kg) maximum gross weight)* | | | |
| | Base mix | | Surface mix | |
	minimum	maximum	minimum	maximum
Compaction, number of blows each end of specimen	50		75	
Stability (lb)	1 500	—	1 800	—
(N)	6 700		8 000	
Flow (0·01 in) (0·25 mm)	8	14	8	14
Percent air voids	3[a]	8	3[a]	5
Percent voids in mineral aggregate (VMA)	Table 7.4		Table 7.4	
Hveem method mix criteria				
Stabilometer value	40	—	40	—
Swell (in)	—	0·030	—	0·030
(mm)		(0·076)		(0·076)
Percent air voids	4[a]	—	3[a]	—

Note: All criteria, not stability value alone, must be considered in designing an asphalt paving mixture. For details on mix design see *Mix Design Methods for Asphalt Concrete and Other Hot-Mix Types*, Manual Series No. 2 (MS-2), The Asphalt Institute.

[a] For low-trafficked areas the asphalt content should be increased enough to reduce the percent air voids to near the minimum of the allowable range.

TABLE 7.4 MINIMUM PERCENT VOIDS IN MINERAL AGGREGATE
(Courtesy: The Asphalt Institute)

USA standard sieve designation[a]	Nominal maximum particle size *(in[a])*	*(mm[a])*	Minimum voids in mineral aggregate (%)
No. 16	0·0469	1·18	23·5
No. 8	0·093	2·36	21
No. 4	0·187	4·75	18
$\frac{3}{8}$ in	0·375	9·5	16
$\frac{1}{2}$ in	0·500	12·5	15
$\frac{3}{4}$ in	0·750	19·0	14
1 in	1·0	25·0	13
$1\frac{1}{2}$ in	1·5	37·5	12
2 in	2·0	50	11·5
$2\frac{1}{2}$ in	2·5	63	11

[a] Standard specification for wire cloth sieves for testing purposes, ASTM designation E 11 (AASHO designation M 92).

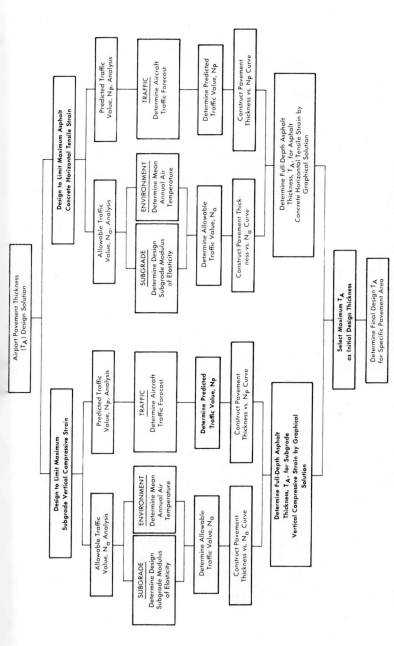

FIG. 7.21. Airport pavement thickness design flow chart. (*Courtesy: The Asphalt Institute.*)

Marshall and Hveem methods of mix design. The suggested minimum percent voids in mineral aggregate (VMA) required in the Marshall method of mix design, so as to ensure sufficient asphalt in the paving mixture, is given in Table 7.4.

7.5.3 The Asphalt Institute Thickness Design Procedure for Air-Carrier Airports

The steps leading to the design thickness are shown schematically in Fig. 7.21. The determination of the design thickness for an airport requires the following input: (1) the subgrade modulus of elasticity; (2) the mean annual air temperature (the 30 year mean is recommended if available, otherwise the mean temperature for the greatest number of years recorded should be used); and (3) a projected forecast of the aircraft traffic mix.

The major steps to be followed for each strain criterion analysed are as follows.

1. Determine the allowable traffic value, N_a, for each strain criterion (ε_c and ε_t) from the design subgrade modulus of elasticity, E_s, and mean annual air temperature, t, for the design location. This yields the number of equivalent DC-8-63F strain repetitions that a full-depth asphalt concrete pavement structure of a given thickness can withstand, for the specific subgrade modulus and environmental combination existing at the design site.

Families of design curves relating full-depth asphalt pavement thickness, T_A, to design subgrade modulus E_s, for various levels of allowable repetitions, have been developed for different temperature values. Two sets of these curves are shown in Figs. 7.22 and 7.23, one of each criterion analysed. Knowledge of the specific design subgrade modulus, E_s, and mean annual air temperature, t, allows a plot to be developed from these curves showing thickness (T_A) requirements versus the allowable strain repetitions, N_a. The plot represents the N_a curve for the specific strain criterion analysed.

To illustrate, assume that a design subgrade modulus, E_s, of 7000 psi (48 000 kN/m^2) and a mean annual air temperature, t, of 60°F (16°C) are the design input variables for a particular airport. The 60°F graph in Fig. 7.22 is used to read off the thickness, T_A, that satisfies the strain ε_c for the various N_a values at an E_s of 7000 psi. These values are plotted on semi-logarithmic graph paper and the curve is drawn as shown in Fig. 7.24. Similarly, the 60°F graph in Fig. 7.23 is used to read off the thickness, T_A, that satisfies the strain, ε_t, for the various N_a values, at an E_s of 7000 psi. These values are then plotted on semi-logarithmic graph paper and the curve is drawn as shown in Fig. 7.25.

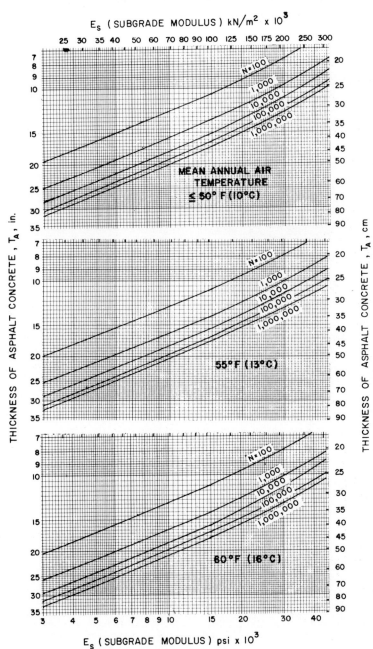

FIG. 7.22. Pavement thickness to limit subgrade vertical compressive strain, ε_c, under DC-8-63F load repetitions for different environments. *(Courtesy: The Asphalt Institute.)*

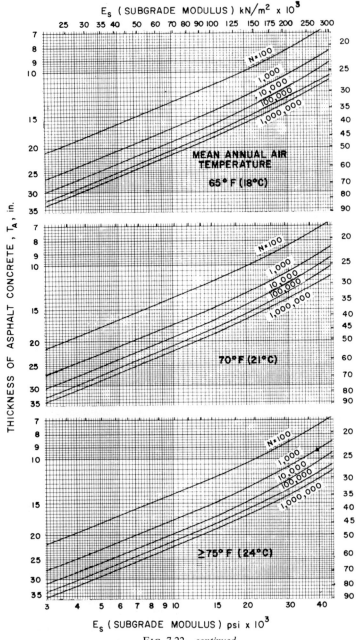

$$F_{IG}.\ 7.22—continued$$

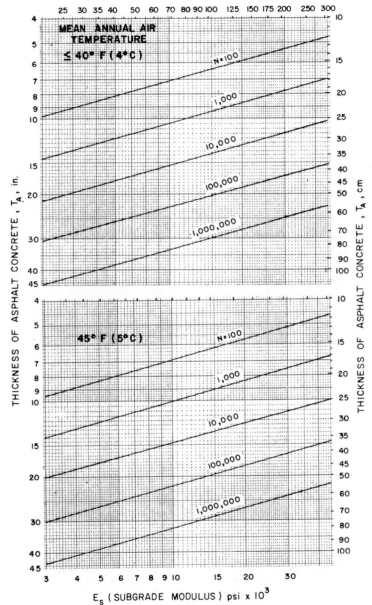

FIG. 7.23. Pavement thickness to limit asphalt concrete horizontal tensile strain, ε_t, under DC-8-63F load repetitions for different environments. *(Courtesy: The Asphalt Institute.)*

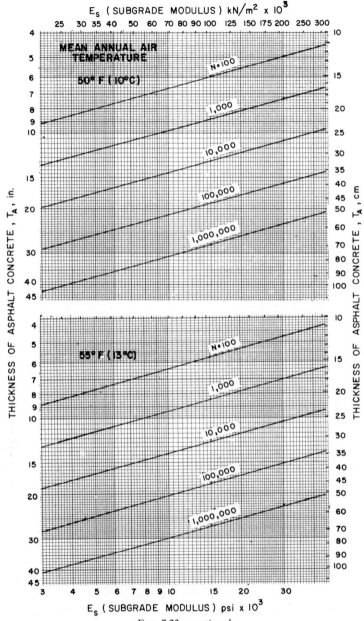

FIG. 7.23—*continued*

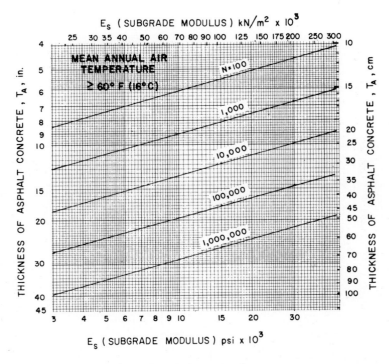

F IG. 7.23—*continued*

2. Determine the predicted traffic value, N_p, from the projected aircraft mix forecast for the pavement's selected design period, and from the aircraft equivalency diagrams given in Appendix D. This yields the number of equivalent D C-8-63F strain repetitions that actually will occur during the selected design period. An aircraft movement is defined as one passage of a single aircraft at the critical design location. Each aircraft is then related to the 'standard' D C-8-63F by use of equivalency factors or the figures of Appendix D. This appendix contains two sets of graphs for various aircraft on which equivalent D C-8-63F strain repetitions can be read off directly for any number of movements from 10 to 10^6; the repetitions can also be read off directly for several distances from centreline, and (depending on the strain criterion) three or four full-depth asphalt concrete pavement thicknesses. Each set of graphs is developed for a specific strain criterion (horizontal tensile, ε_t, or vertical compressive, ε_c), and must therefore be used only to determine thickness requirements for that particular criterion.

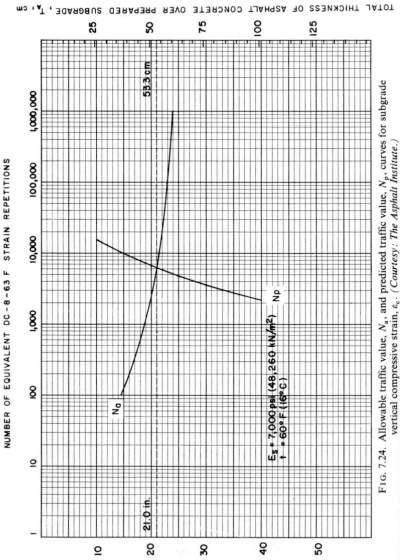

Fig. 7.24. Allowable traffic value, N_a, and predicted traffic value, N_p, curves for subgrade vertical compressive strain, ε_c. (*Courtesy: The Asphalt Institute.*)

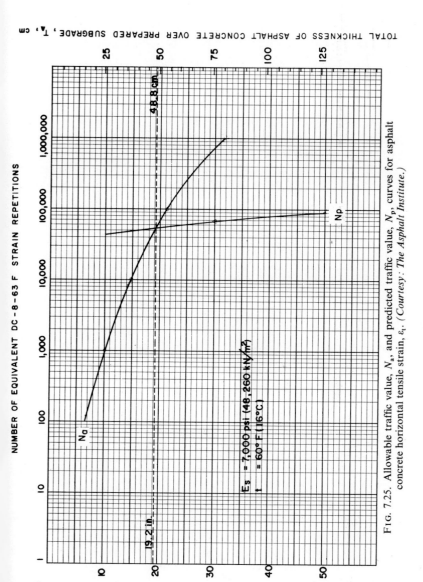

FIG. 7.25. Allowable traffic value, N_a, and predicted traffic value, N_p, curves for asphalt concrete horizontal tensile strain, ε_t. (*Courtesy: The Asphalt Institute.*)

The steps for finding the predicted traffic value, N_p, applicable to traffic analyses for both ε_c and ε_t, are as follows:

(a) Use worksheet No. 1, shown in Fig. 7.26, and record in column A the respective number of movements at the critical design location for each 5 year time period for the anticipated aircraft types.

(b) Sum (horizontally) the movements for each aircraft type to obtain the total cumulative movements expected for the time periods, and record in column B of aircraft traffic worksheet No. 1.

(c) Record on aircraft traffic worksheet No. 2, shown in Fig. 7.27, the appropriate design period, type of strain (ε_c or ε_t), and assumed thickness to be analysed. These thicknesses are 10 in (25 cm), 30 in (76 cm) and 50 in (127 cm) for asphalt concrete horizontal tensile strain; or 10 in (25 cm), 20 in (51 cm), 30 in (76 cm) and 40 in (102 cm) for subgrade vertical compressive strain.

TRAFFIC FORECAST WORKSHEET

Critical Design Location (taxiway) Movements , 5 - year Periods

Aircraft	0 - 5		5 - 10		10 - 15		15 - 20	
	Col. A	Col. B	Col. A	Col. B	Col. A	Col. B	Col. A	Col. B
1. B·707-320 C	7,300	7,300	21,900	29,200	43,800	73,000	73,000	146,000
2.								
3.								
4.								
5.								
6.								
7.								
8.								
9.								
10.								

FIG. 7.26. Aircraft Traffic Worksheet No. 1. Col. A = Estimated number of movements for each aircraft type within the 5 year period indicated. Col. B = Cumulative number of movements for each aircraft type through maximum time period indicated. (*Note:* Record cumulated number of movements for each aircraft type (Col. B) for the selected design period on Aircraft Worksheet No. 2.) (*Courtesy: The Asphalt Institute.*)

(d) Transfer the data obtained in step (b) to the 'movements in design period' column of aircraft traffic worksheet No. 2.

(e) Locate the appropriate aircraft equivalency chart in Appendix D. Be sure that each chart selected corresponds to the thickness values, strain (ε_c or ε_t) and aircraft type recorded on worksheet No. 2.

(f) Using the proper chart, record on worksheet No. 2, for each distance from the centreline, x, the equivalent DC-8-63F strain repetitions deter-

NUMBER OF EQUIVALENT DC-8-63F STRAIN REPETITIONS

TYPE OF STRAIN: ε_c DESIGN PERIOD: **20** YEARS

AIRCRAFT	MOVEMENTS IN DESIGN PERIOD	DISTANCE FROM CENTERLINE				
		5.5 ft (1.7m)	9.5 ft (2.9m)	13.5 ft (4.1m)	17.5 ft (5.3m)	21.5 ft (6.6m)
THICKNESS, T_A = **10** in. (**25** cm)						
1. B-707-320 C	146,000	6,500	15,000	12,000	4,700	—
2.						
3.						
4.						
5.						
6.						
7.						
8.						
9.						
10.						
SUM		6,500	(15,000)	12,000	4,700	—
THICKNESS, T_A = **20** in. (**51** cm)						
1. B-707-320 C	146,000	3,100	6,100	5,500	2,200	—
2.						
3.						
4.						
5.						
6.						
7.						
8.						
9.						
10.						
SUM		3,100	(6,100)	5,500	2,200	—
THICKNESS, T_A = **30** in. (**76** cm)						
1. B-707-320 C	146,000	1,800	3,400	3,000	1,400	—
2.						
3.						
4.						
5.						
6.						
7.						
8.						
9.						
10.						
SUM		1,800	(3,400)	3,000	1,400	—
THICKNESS, T_A = **40** in. (**102** cm)						
1. B-707-320 C	146,000	1,300	2,200	2,000	1,100	—
2.						
3.						
4.						
5.						
6.						
7.						
8.						
9.						
10.						
SUM		1,300	(2,200)	2,000	1,100	—

FIG. 7.27. Aircraft Traffic Worksheet No. 2. Subgrade vertical compressive strain, ε_c. (*Note:* Circle the maximum sum for each thickness. Each circled number and its corresponding thickness is used to plot one point of the actual traffic value curve.)

(Courtesy: The Asphalt Institute.)

mined from the number of movements recorded on the worksheet for the aircraft in question.

(g) Repeat steps (e) and (f) for each aircraft shown on the worksheet.

(h) Determine and record on worksheet No. 2, for each thickness, the sum of the equivalent DC-8-63F strain repetitions at each distance from the centreline.

(i) Circle the maximum sum for each thickness, T_A, as shown in Fig. 7.27.

(j) For each type of strain and design period, plot the maximum values for each assumed thickness, T_A, on the same semi-logarithmic graph paper on which was plotted the N_a curve. This curve represents the solution for the desired predicted traffic value, N_p, for one type of strain (Fig. 7.24).

NUMBER OF EQUIVALENT DC-8-63 F STRAIN REPETITIONS

TYPE OF STRAIN : ε_t　　　　　　　　　DESIGN PERIOD: 20 YEARS

AIRCRAFT	MOVEMENTS IN DESIGN PERIOD	DISTANCE FROM CENTERLINE				
		5.5 ft (1.7m)	9.5 ft (2.9m)	13.5 ft (4.1m)	17.5 ft (5.3m)	21.5 ft (6.6m)
THICKNESS, T_A = 10 in. (25 cm)						
1. B-707-320 C	146,000	17,000	44,000	36,000	11,000	—
2.						
3.						
4.						
5.						
6.						
7.						
8.						
9.						
10.						
SUM		17,000	(44,000)	36,000	11,000	—
THICKNESS, T_A = 30 in. (76 cm)						
1. B-707-320 C	146,000	26,000	68,000	60,000	17,000	—
2.						
3.						
4.						
5.						
6.						
7.						
8.						
9.						
10.						
SUM		26,000	(68,000)	60,000	17,000	—
THICKNESS, T_A = 50 in. (127 cm)						
1. B-707-320 C	146,000	35,000	95,000	78,000	22,500	—
2.						
3.						
4.						
5.						
6.						
7.						
8.						
9.						
10.						
SUM		35,000	(95,000)	78,000	22,500	—
THICKNESS, T_A = in. (cm)						
1.						
2.						
3.						
4.						
5.						
6.						
7.						
8.						
9.						
10.						
SUM						

FIG. 7.28. Aircraft Traffic Worksheet No. 2. Asphalt concrete horizontal tensile strain, ε_t. (*Note:* Circle the maximum sum for each thickness. Each circled number and its corresponding thickness is used to plot one point of the actual traffic value curve.)

(*Courtesy: The Asphalt Institute.*)

(k) Repeat steps (c)–(j) for the other type of strain. (See Figs. 7.25 and 7.28.)

3. Determine the full-depth asphalt concrete pavement thickness, T_A, needed to satisfy the strain criteria for the design input values of subgrade modulus, mean air temperature and traffic mix. This thickness is determined by a simultaneous graphical solution of the plots of the allowable

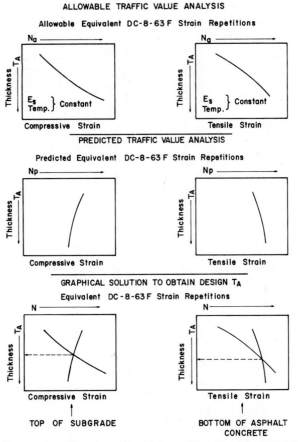

FIG. 7.29. Steps used to determine design thickness. *(Courtesy: The Asphalt Institute.)*

and predicted traffic value results. In Fig. 7.24, for example, the intersection of the N_a and N_p curves yields the full-depth asphalt concrete thickness needed to satisfy the subgrade vertical compressive strain criterion, ε_c (21 in or 53·3 cm). Similarly, the intersection of the N_a and

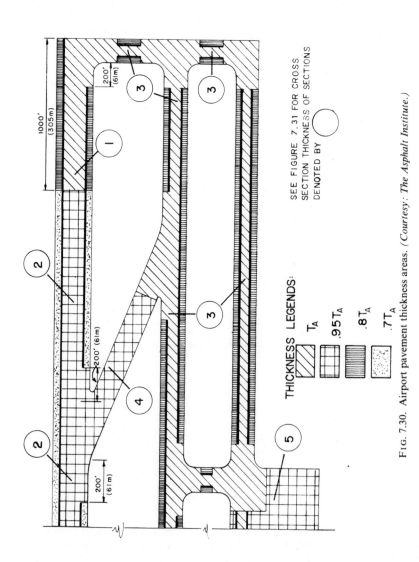

FIG. 7.30. Airport pavement thickness areas. *(Courtesy: The Asphalt Institute.)*

N_p curves in Fig. 7.25 yields the full-depth asphalt concrete thickness required to satisfy the asphalt concrete horizontal tensile strain criterion, ε_t (19·2 in or 48·8 cm). The pavement design thickness is then the greater of the two thicknesses rounded to the nearest $\frac{1}{2}$ in or 1 cm (that is, 21 in or 53 cm).

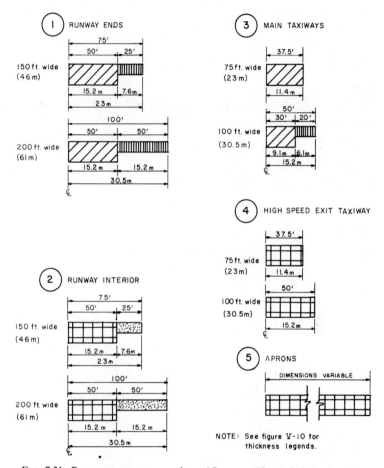

FIG. 7.31. Pavement area cross-sections. *(Courtesy: The Asphalt Institute.)*

A schematic illustration of the steps used to determine the design thickness is shown in Fig. 7.29. In this illustration, the design thickness, T_A, is based on asphalt concrete tensile strain, ε_t, as this gives the greater thickness.

After determining the design thickness, T_A, which represents conditions at the critical design location, reduced thickness designs for the other pavement areas can be obtained with the aid of Figs. 7.30 and 7.31.

For example, using the 21·0 in (53 cm) design thickness for a 20 year design period at the critical design location, a main taxiway 100 ft (30·5 m) wide can have the following variable pavement thicknesses: 21·0 in (53 cm) for a width of 60 ft (18 m)—that is, 30 ft (9·1 m) on either side of the taxiway centreline—and 17·0 in (43 cm) for a width of 20 ft (6·1 m) on each edge of the taxiway.

Example

A full-depth asphalt concrete pavement thickness design is needed for the critical design location, a taxiway 100 ft (30·5 m) wide, of a new airport. This design will be the basis for pavement thickness design for the other areas of the airport. The design period is 20 years. If feasible, variable thickness cross-sections are desired.

The following data are given:

SUBGRADE EVALUATION
Subgrade modulus, E_s (20 test results)

psi	kN/m^2
10 500	72 400
13 500	89 600
10 500	72 400
9 000	62 100
10 500	72 400
12 000	82 700
15 000	103 400
10 500	72 400
12 000	82 700
13 500	89 600
9 000	62 100
13 500	89 600
12 000	82 700
13 500	89 600
10 500	72 400
13 500	89 600
9 000	62 100
10 500	72 400
12 000	82 700
10 500	72 400

MEAN MONTHLY AIR TEMPERATURES

January	41·2°F	(5·1°C)
February	43·3°F	(6·3°C)
March	45·0°F	(7·2°C)
April	62·9°F	(17·2°C)
May	72·1°F	(22·3°C)
June	78·8°F	(26·0°C)
July	84·9°F	(29·4°C)
August	79·3°F	(26·3°C)
September	72·9°F	(22·7°C)
October	62·9°F	(17·2°C)
November	49·0°F	(9·4°C)
December	40·1°F	(4·5°C)

TRAFFIC FORECAST

	Critical design location (taxiway) movements, 5 year periods			
Aircraft	*0–5*	*5–10*	*10–15*	*15–20*
DC-9-41	7 500	11 000	18 000	25 000
B-727-200	18 000	29 000	44 000	51 000
DC-10-10	3 500	7 000	15 000	29 000
DC-8-63F	7 500	9 000	18 000	36 000
Vis-810	7 500	7 500	3 600	1 600

Solution. Allowable traffic value, N_a, *analysis.*

1. Determine the design subgrade modulus of elasticity, E_s. The table below summarises the distribution of subgrade modulus values. In this case, it is not necessary to plot the percentage of each subgrade modulus value to find the 85% level, since it can be read directly from the table.

Subgrade modulus		Number ≥ each different value	Percentage ≥ each different value
(psi)	*(kN/m²)*		
9 000	62 100	20	100
10 500	72 400	17	85
12 000	82 700	10	50
13 500	89 600	6	30
15 000	103 400	1	5

The design subgrade modulus of elasticity, E_s, is 10 500 psi (72 400 kN/m²).

2. The mean annual temperature, t, from the monthly means, is 61°F (16·1°C).

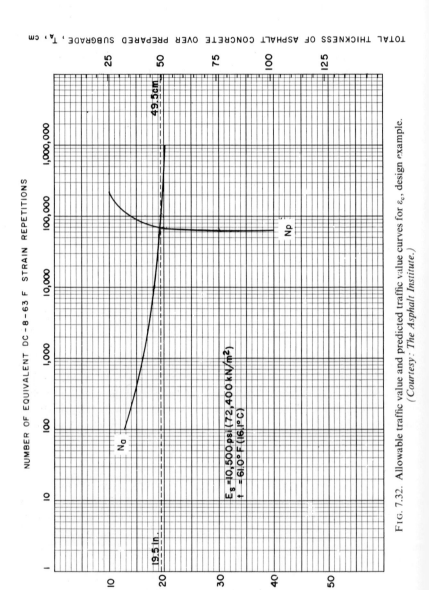

Fɪɢ. 7.32. Allowable traffic value and predicted traffic value curves for ε_c, design example. *(Courtesy: The Asphalt Institute.)*

3. Determine the allowable traffic value, N_a, curve for subgrade vertical compressive strain, ε_c, by using Fig. 7.22 and plotting the thicknesses indicated for each of the strain repetition curves. The full-depth asphalt thicknesses required to satisfy the given strain repetitions are shown in the table below, for $E_s = 10\,500$ psi $(72\,400$ kN/m$^2)$ and $t = 61°$F $(16·1°$C$)$. In addition, Fig. 7.32 illustrates the resulting N_a curve.

| | Number of strain repetitions | | | | |
	100	*1 000*	*10 000*	*100 000*	*1 000 000*
Thickness, T_A	12·8 in	16·2 in	18·3 in	19·6 in	20·4 in
	(32·5 cm)	(41·1 cm)	(46·5 cm)	(49·8 cm)	(51·8 cm)

4. Determine the allowable traffic value, N_a, curve for asphalt concrete horizontal tensile strain, ε_t, by using Fig. 7.23 and plotting the thicknesses indicated for each of the strain repetitions. For $E_s = 10\,500$ psi $(72\,400$ kN/m$^2)$ and $t = 61°$F $(16·1°$C$)$, using the chart for $\geq 60°$F $(16°$C$)$, the full-depth asphalt concrete thicknesses required to satisfy the given strain repetitions are:

| | Number of strain repetitions | | | | |
	100	*1 000*	*10 000*	*100 000*	*1 000 000*
Thickness, T_A	6·0 in	9·1 in	13·1 in	19·1 in	28·3 in
	(15·2 cm)	(23·1 cm)	(33·3 cm)	(48·5 cm)	(71·9 cm)

The resulting N_a curve is illustrated in Fig. 7.33.

Predicted traffic value, N_p*, analysis.*
1. Determine the predicted traffic value, N_p, curve for subgrade vertical compressive strain, ε_c, by using aircraft traffic worksheets 1 and 2. For this purpose, the forecast aircraft movements for all types of aircraft are recorded in column A of worksheet No. 1 (Fig. 7.34), and the cumulative number of movements in column B. As a design period of 20 years is required in the analysis, the data in column B for 15–20 years is transferred to worksheet No. 2.

On worksheet No. 2 (Fig. 7.35), record for each aircraft type the number of equivalent DC-8-63F strain repetitions at various distances from centre-line, derived from the total movements for the 20 year design period.

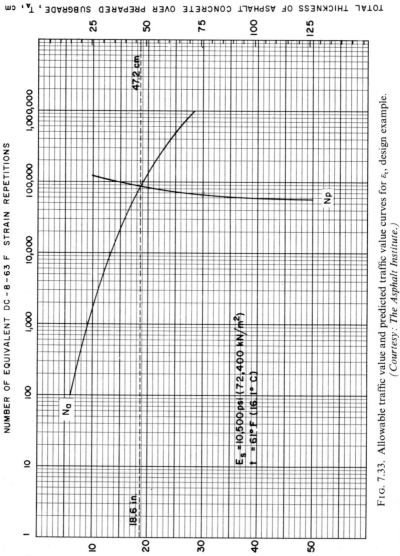

FIG. 7.33. Allowable traffic value and predicted traffic value curves for ε_t, design example. *(Courtesy: The Asphalt Institute.)*

These are taken from the appropriate charts in Appendix D, for each aircraft.

TRAFFIC FORECAST WORKSHEET

Critical Design Location (taxiway) Movements , 5 - year Periods

Aircraft	0 - 5		5 - 10		10 - 15		15 - 20	
	Col. A	Col. B	Col. A	Col. B	Col. A	Col. B	Col. A	Col. B
1. B·727-200	18,000	18,000	29,000	47,000	44,000	91,000	51,000	142,000
2. DC-8-63 F	7,500	7,500	9,000	16,500	18,000	34,500	36,000	70,500
3. DC-9-41	7,500	7,500	11,000	18,500	18,000	36,500	25,000	61,500
4. DC-10-10	3,500	3,500	7,000	10,500	15,000	25,500	29,000	54,500
5. Vis· 810	7,500	7,500	7,500	15,000	3,600	18,600	1,600	20,200
6.								
7.								
8.								
9.								
10.								

FIG. 7.34. Aircraft Traffic Worksheet No. 1. Design example. Col. A = Estimated number of movements for each aircraft within the 5 year period indicated. Col. B = Cumulated number of movements for each aircraft type through maximum time period indicated. *Note:* Record cumulated number of movements for each aircraft type (Col. B) for the selected design period on Aircraft Worksheet No. 2. *(Courtesy: The Asphalt Institute.)*

Sum each 'distance from centreline' column for each thickness, T_A, then circle the maximum sum for each thickness and plot it on the subgrade vertical compressive strain chart (Fig. 7.32).

2. In Fig. 7.32, the intersection of the N_p and N_a curves gives the thickness of asphalt concrete, 19·5 in (49·5 cm), needed to satisfy the subgrade vertical compressive strain criterion.

3. Determine the N_p curve for asphalt concrete horizontal tensile strain, ε_t, by using aircraft traffic sheets 1 and 2 and by following the procedure explained in step 1 (see Figs. 7.33 and 7.36).

4. The intersection of the N_p curve with the N_a curve gives the thickness of asphalt concrete, 18·6 in (47·2 cm), needed to satisfy the asphalt concrete horizontal tensile strain, ε_t, criterion.

Design thickness, T_A. The full-depth asphalt concrete design thickness, T_A, for the proposed taxiway pavement is the greater of the two thicknesses obtained in steps 2 and 4 of the N_p analysis. For this example, the design thickness, T_A, is 19·5 in (50 cm), to the nearest $\frac{1}{2}$ in (nearest cm).

Thickness requirements for various pavement areas. Using Figs. 7.30 and 7.31 as a guide, the following thicknesses can be suggested:

NUMBER OF EQUIVALENT DC-8-63 F STRAIN REPETITIONS

TYPE OF STRAIN : ϵ_c DESIGN PERIOD 20 YEARS

AIRCRAFT	MOVEMENTS IN DESIGN PERIOD	DISTANCE FROM CENTERLINE				
		5.5 ft (1.7m)	9.5 ft (2.9m)	13.5 ft (4.1m)	17.5 ft (5.3m)	21.5 ft (6.6m)
THICKNESS, T_A = 10 in. (25cm)						
1. B-727-200	142,000	9,000	13,000	6,500	1,150	—
2. DC-8-63 F	70,500	32,000	63,000	41,000	6,400	—
3. DC-9-41	61,500	300	340	220	85	—
4. DC-10-10	54,500	—	11,000	110,000	215,000	135,000
5. V1s-810	20,200	28	40	42	31	—
6.						
7.						
8.						
9.						
10.						
SUM		41,328	87,380	157,762	(222,666)	135,000
THICKNESS, T_A = 20 in. (51 cm)						
1. B-727-200	142,000	1,200	1,600	1,000	275	—
2. DC-8-63 F	70,500	32,000	63,000	41,000	6,400	—
3. DC-9-41	61,500	110	120	85	42	—
4. DC-10-10	54,500	—	1,800	10,000	16,000	12,000
5. V1s-810	20,200	21	25	26	22	—
6.						
7.						
8.						
9.						
10.						
SUM		33,331	(66,545)	52,111	22,739	12,000
THICKNESS, T_A = 30 in. (76 cm)						
1. B-727-200	142,000	60	70	54	23	—
2. DC-8-63 F	70,500	32,000	63,000	41,000	6,400	—
3. DC-9-41	61,500	17	18	15	11	—
4. DC-10-10	54,500	—	650	1,800	2,600	2,250
5. V1s-810	20,200	13	12	12	13	—
6.						
7.						
8.						
9.						
10.						
SUM		32,090	(63,750)	42,881	9,047	2,250
THICKNESS, T_A = 40 in. (102 cm)						
1. B-727-200	142,000	1	1	1	1	—
2. DC-8-63 F	70,500	32,000	63,000	41,000	6,400	—
3. DC-9-41	61,500	1	1	1	1	—
4. DC-10-10	54,500	—	250	440	530	470
5. V1s-810	20,200	6	4	4	5	—
6.						
7.						
8.						
9.						
10.						
SUM		32,008	(63,256)	41,446	6,937	470

Fig. 7.35. Aircraft Traffic Worksheet No. 2. Design example, subgrade compressive strain, ε_c. *Note:* Circle the maximum sum for each thickness. Each circled number and its corresponding thickness is used to plot one point of the actual traffic value curve.

(Courtesy: The Asphalt Institute.)

NUMBER OF EQUIVALENT DC-8-63F STRAIN REPETITIONS

TYPE OF STRAIN: ε_t DESIGN PERIOD: 20 YEARS

AIRCRAFT	MOVEMENTS IN DESIGN PERIOD	DISTANCE FROM CENTERLINE				
		5.5 ft (1.7m)	9.5 ft (2.9m)	13.5 ft (4.1m)	17.5 ft (5.3m)	21.5 ft (6.6m)
THICKNESS, T_A = 10 in. (25 cm)						
1. B-727-200	142,000	35,000	75,000	35,000	4,500	—
2. DC-8-63F	70,500	16,500	40,000	29,000	5,600	—
3. DC-9-41	61,500	6,000	7,100	1,700	—	—
4. DC-10-10	54,500	400	5,500	21,000	33,000	21,000
5. VIS-810	20,200	90	410	500	150	—
6.						
7.						
8.						
9.						
10.						
SUM		57,990	(128,010)	87,200	45,250	21,000
THICKNESS, T_A = 30 in. (76 cm)						
1. B-727-200	142,000	9,500	17,500	9,500	1,800	—
2. DC-8-63F	70,500	16,500	40,000	29,000	5,600	—
3. DC-9-41	61,500	1,000	1,100	350	—	—
4. DC-10-10	54,500	450	7,000	25,000	37,500	25,000
5. VIS-810	20,200	8	35	42	14	—
6.						
7.						
8.						
9.						
10.						
SUM		27,458	(65,635)	63,892	44,915	25,000
THICKNESS, T_A = 50 in. (127 cm)						
1. B-727-200	142,000	4,700	8,800	4,700	950	—
2. DC-8-63F	70,500	16,500	40,000	29,000	5,600	—
3. DC-9-41	61,500	350	410	140	—	—
4. DC-10-10	54,500	430	5,100	24,000	35,000	24,000
5. VIS-810	20,200	2	11	13	4	—
6.						
7.						
8.						
9.						
10.						
SUM		21,982	54,321	(57,853)	41,554	24,000
THICKNESS, T_A = in. (cm)						
1.						
2.						
3.						
4.						
5.						
6.						
7.						
8.						
9.						
10.						
SUM						

FIG. 7.36. Aircraft Traffic Worksheet No. 2. Design example, asphalt concrete tensile strain, ε_t. *Note:* Circle the maximum sum for each thickness. Each circled number and its corresponding thickness is used to plot one point of the actual traffic value curve.

(Courtesy: The Asphalt Institute.)

Centre 60 ft (18·3 m) of the design taxiway   19·5 in (50 cm)
Outside 20 ft (6·1 m) on either edge   16·0 in (41 cm)

Thickness of the pavement for the runway and other taxiways can be obtained with the help of Figs. 7.30 and 7.31, if the widths are known.

7.6 CRITICAL RUNWAY AREAS

It is a common procedure in the design to consider 10% of the runway length (but not less than 500 ft) on each end as the critical area, while the central 80% of the runway length is considered as non-critical. This is supported by the fact that during the taking off or landing operations the aircraft is partially airborne due to the speed of the moving aircraft.

However, recent observations at several airports have indicated runway failures beyond the 10% critical length from each end. Different explanations have been given for this phenomenon. Some of these are:

(a) The wing lift is very low throughout the take-off roll until the aircraft rotates to increase the wing angle of attack.

(b) Aircraft in motion are sensitive to runway roughness or undulations and experience accelerations which increase the runway loadings.

Acceleration values which might be expected once per take-off are of the order of 0·3 g.

For the above-mentioned reasons, it is essential to define the boundaries of the critical areas. For large airports, the 10% critical length on each side is most probably not enough, though to date no change has been recommended. Some research programmes are now being conducted by the US Army Corps of Engineers and the Federal Aviation Administration to clarify pavement performance under aircraft landing and take-off.

Transition in pavement section from the critical to the non-critical areas should be made very gradual, since some of the failures take place in this part. Perhaps future recommendations will be for the use of the same thickness for the full length of runway, so as to avoid these problems.

REFERENCES AND BIBLIOGRAPHY

1. American Society of Civil Engineers, 'Development of CBR Flexible Pavement Design Methods for Airfields: Symposium', *Transactions*, **115**, 1950.
2. Department of the Environment, 'Design and Evaluation of Aircraft Pavements', Technical Paper, Directorate of Civil Engineering Development, UK, 1971.

3. Edwards, J. M. and Valkering, C. P., 'Structural Design of Asphalt Pavements for Heavy Aircraft', Shell Construction Service, 1971.
4. Federal Aviation Administration, 'Airport Paving', *Advisory Circular, AC 150/5320-6A*, US Department of Transportation, Reprinted September 15, 1971.
5. Federal Aviation Agency, 'Airport Paving', US Government Printing Office. November 1962.
6. Hennes, R. G. and Ekse, M., *Fundamentals of Transportation Engineering*, Second Edition, McGraw-Hill, 1969.
7. International Civil Aviation Organization, 'Aerodrome Physical Characteristics, Part 2', *Aerodrome Manual*, Second Edition, 1965.
8. Kingham, R. I., 'Failure Criteria Developed from AASHO Road Test Data', *Proc. 3rd International Conference on the Structural Design of Asphalt Pavements. London, September 1972*, 656–669.
9. Shook, J. F. and Kallas, B. F., 'Factors Influencing Dynamic Modulus of Asphalt Concrete', *Proc. Assoc. Asph. Paving Tech.*, 1969.
10. The Asphalt Institute, 'Full-Depth Asphalt Pavements for Air Carrier Airports', Manual Series No. 11 (MS-11), January 1973.
11. Witczak, M. W., 'Design of Full-Depth Asphalt Airfield Pavements', *Proc. 3rd International Conference on the Structural Design of Asphalt Pavements, London, September 1972*, 550–567.

CHAPTER 8

METHODS OF MIX DESIGN, CONSTRUCTION, EVALUATION AND COST ANALYSIS OF FLEXIBLE PAVEMENTS

8.1 INTRODUCTION

Hot-mix asphalt paving mixtures may be produced from a wide range of aggregate combinations. Aside from the amount and grade of asphalt used, the principal characteristics of the mix are determined by the relative amounts of: (1) coarse aggregate retained on sieve No. 8, (2) fine aggregate passing sieve No. 8, and (3) mineral dust passing sieve No. 200.

Accordingly, the aggregate composition may vary from a coarse-textured mix having mostly coarse aggregate to a fine-textured mix having mostly fine aggregate.

The Asphalt Institute classifies hot-mix asphalt paving according to mix type, based on the relative amounts of coarse aggregate, fine aggregate and mineral dust [9]. Figure 8.1 shows the general limits of each of the eight mix types along with the paving mix designation and the maximum size aggregate normally used. The chart is based on the proportions of coarse and fine aggregate in the paving mix and on the proportion of mineral dust, the ranges for which are represented by shaded bands. The area of the classification chart to the left of the shaded band includes the range of mineral dust for each respective mix type, including surface course, binder course (sometimes called intermediate course) and base course mixes. For any mix type, surface course mixes normally contain more mineral dust than either base or binder mixes; and base mixes usually contain the least amount of mineral dust. With a few exceptions, the percentage of mineral dust in base and binder course mixes, for a particular mix type, will normally fall on the left side of the chart, as

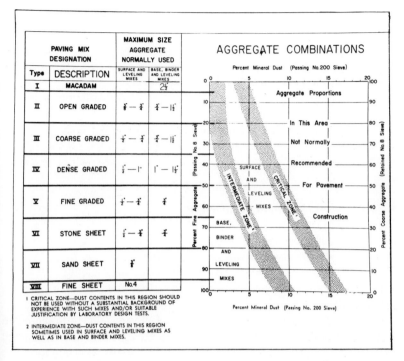

FIG. 8.1. Classification of hot-mix asphalt paving. *(Courtesy: The Asphalt Institute.)*

indicated in Fig. 8.1; while in surface course mixes, it will fall to the right of this.

8.2 ASPHALT PAVING MIX DESIGN

8.2.1 Objectives of Mix Design

The design of asphalt paving mixes is largely a matter of selecting a proportioning material to obtain the desired qualities and properties in the finished construction. The over-all objective for the design of asphalt paving mixes is to determine, through several trial mixes, an economical blend and gradation of aggregates and a corresponding asphalt content that yields a mix having:

1. Sufficient asphalt to ensure a durable pavement by coating thoroughly the aggregate particles, and waterproofing and binding them together under suitable compaction.

2. Sufficient mix stability to satisfy the service requirement and demands of traffic.

3. Sufficient voids in the total compacted mix to provide a space for a slight amount of additional compaction under traffic loading to avoid flushing, bleeding and loss of stability.

4. Sufficient workability to permit an efficient construction operation in placing the paving mix.

There are different possible methods for mix design, such as: (a) the Marshall Method, (b) the Hubbard–Field Method, and (c) the Hveem Method. Only the first method, which is widely known, will be discussed here.

8.2.2 Marshall Method of Mix Design

The method presented here is applicable only to hot-mix asphalt paving mixtures using penetration grades of asphalt cement and containing aggregates with maximum size of 1 in or less. It can be used for both laboratory design and field control of asphalt hot-mix paving.

Before preparing the specimens, it is necessary to check that the materials used and the aggregate gradation meet project specifications.

In the Marshall method, test specimens of $2\frac{1}{2}$ in height $\times$ 4 in diameter, prepared according to a specified procedure, are used. The three principal features of mix design are: (1) bulk specific gravity determination, (2) density–voids analysis, and (3) a stability–flow test of the compacted test specimens. This stability is the maximum load resistance in pounds which the standard test specimen will develop at 140°F when tested, and the flow value is the total movement, in units of $\frac{1}{100}$ in, occurring in the specimen between no load and maximum load during the stability test.

Tests for determining optimum asphalt content should be scheduled on the basis of $\frac{1}{2}\%$ increments of asphalt content, with at least two asphalt contents above and two below 'optimum'. Usually three to five test specimens are prepared for each asphalt content used. Asphalt content may be expressed as a percentage by weight of total mix or as a percentage by weight of dry aggregate.

Preparation of the aggregates should include drying them to constant weight at about 230°F, then separation into the desired size fraction, generally $1-\frac{3}{4}$ in, $\frac{3}{4}-\frac{3}{8}$ in, $\frac{3}{8}$ to No. 4, No. 4 to No. 8, and passing No. 8.

The mixing and compaction temperatures for the asphalt are determined as the temperature to which the asphalt must be heated to produce viscosities of 85 ± 10 s Saybolt Furol and 140 ± 15 s Saybolt Furol, respectively. Mixing the aggregate and asphalt cement should be done as quickly and thoroughly as possible to yield a mixture having a uniform distribution of asphalt throughout. When the entire batch is placed in the mould, 35, 50 or 75 blows should be applied with the compaction hammer, as specified according to the design traffic category, using a free fall of 18 in. The same number of blows should also be applied to the reversed face of the specimen. When the specimen cools sufficiently, it is removed from the mould and placed on a smooth, level surface until ready for testing.

8.2.3 Test Procedure in the Marshall Method

The Marshall testing machine required for testing the 4 in diameter $\times 2\frac{1}{2}$ in height specimens is an electrically powered testing device designed to apply loads through semicircular testing heads at a constant rate of strain of 2 in/min. It is equipped with a calibrated proving ring for determining the applied testing load, a Marshall stability testing head and a Marshall flow meter for determining the amount of strain at the maximum load for test.

In addition, a water bath at least 6 in deep, equipped with a shelf for suspending specimens at least 2 in above the bottom of the bath, and thermostatically controlled to $140°F \pm 1·8°F$, is needed.

The test can be performed as follows:

1. The bulk specified gravity for the freshly compacted cool specimens can be determined from eqn. (8.1) (paraffin-coated specimens):

$$G = \frac{W_a}{V} = \frac{W_a}{W_{pa} - W_{pw} - \left(\dfrac{W_{pa} - W_a}{G_p}\right)} \tag{8.1}$$

where G = bulk specific gravity of specimen;
  W_a = weight of specimen uncoated in air (g);
  W_{pa} = weight of specimen plus paraffin coating in air (g);
  W_{pw} = weight of specimen plus paraffin coating in water (g); and
  G_p = apparent specific gravity of paraffin (0·9 g/cm^3).

2. The stability and flow test can be performed by first adjusting the flow meter to read zero when a 4·00 in diameter metal cylinder is placed in the testing head. Then the specimen is immersed in the water bath at $140°F \pm 1·8°F$ for 30–40 min before test. After removing the test specimen from the water bath and carefully drying its surface, it is placed centrally in the lower testing head; then the upper testing head is fitted into position and the complete assembly is centred in the loading device. The testing load is then applied to the specimen, as shown in Fig. 8.2, at a constant rate of deformation, 2 in/min, until failure occurs.

The total number of pounds required to produce failure of the specimen at 140°F is recorded as the Marshall stability value of the specimen.

The reading of the flow meter, expressed in units of $\frac{1}{100}$ in, is the flow value. For example, if the specimen deforms 0·12 in, the flow value is 12. Care should be taken that the entire procedure for the stability-flow test is completed within 30 s.

3. The density and voids analysis can be made by first determining the average unit weight for each asphalt content, by multiplying the average bulk specific gravity value by 62·4. Then a graphical plot is prepared of unit weight as ordinate versus asphalt content, and a smooth curve giving the best fit for all values is drawn. The unit weight values, read directly from the plotted curve for each asphalt content, can be used to redetermine the bulk specific values by dividing by 62·4.

The maximum theoretical specific gravity of a paving mixture is the specific gravity of the voidless mixture and can be calculated from eqn. (8.2):

$$G_{mm} = \frac{W_{mm}}{V_{mm}} \tag{8.2}$$

where G_{mm} = measured maximum theoretical specific gravity of the void-less mix;

$\quad\quad W_{mm}$ = weight of sample of loose paving mixture in air; and

$\quad\quad V_{mm}$ = volume of voidless mix which is equal to the weight of water replaced by the loose paving mixture of the sample.

It should be noted that the correct value for air voids in a compacted mix and the service volume of the finished pavement, both depend upon the effective asphalt content of a paving mixture, not its total asphalt content. The effective asphalt content is obtained by subtracting the quantity of asphalt lost by absorption into the aggregate particles from the total asphalt content of the mix. Many asphalt pavements that would seem to contain sufficient asphalt, on the basis of their total asphalt binder contents, perform as though the paving mixtures were too lean, because of the portion lost by absorption. To determine the asphalt lost by absorption, it is necessary to determine first the virtual specific gravity of aggregate, which is based on the assumption of no asphalt absorption:

$$G_v = \frac{W_{ag}}{V_{mm} - \dfrac{W_{tac}}{G_{ac}}} \tag{8.3}$$

FIG. 8.2. Marshall stability and flow test. *(Courtesy: The Asphalt Institute.)*

where G_v = virtual specific gravity of the aggregate;

W_{ag} = weight of the aggregate in the sample;

V_{mm} = volume of voidless mixture;

W_{tac} = weight of the total asphalt content in the sample; and

G_{ac} = apparent specific gravity of the asphalt.

The quantity of asphalt binder lost by absorption can then be calculated from eqn. (8.4):

$$A_{ac} = \frac{G_v - G_{ag}}{G_v \cdot G_{ag}} \times 100 \qquad (8.4)$$

where A_{ac} = asphalt lost by absorption into the aggregate particles as percentage by weight of dry aggregate;

G_v = the virtual specific gravity of the aggregate; and

G_{ag} = bulk specific gravity of the aggregate.

Equation (8.5) can be used to determine the effective asphalt content of the paving mixture:

$$P_{eac} = \frac{P_{tac} - \dfrac{A_{ac}}{100}(100 - P_{tac})}{100 - \dfrac{A_{ac}}{100}(100 - P_{tac})} \times 100 \qquad (8.5)$$

where P_{eac} = effective asphalt content as percent by weight of total aggregate plus effective asphalt content;

P_{tac} = total asphalt content as percent by weight of total aggregate plus total asphalt content; and

A_{ac} = asphalt absorption as pounds of asphalt per 100 lb of dry aggregate.

The percent air voids by volume in a compacted paving mixture can be determined from eqn. (8.6):

$$V_v = 100 - P_{gm} \qquad (8.6)$$

where V_v = volume of air voids as percent of bulk volume of compacted mixture; and

$P_{gm} = G_{mb}/G_{mm}$ = measured bulk specific gravity expressed as percent of measured theoretical maximum specific gravity.

8.2.4 Interpretation of Test Data Obtained by the Marshall Method

To determine the optimum asphalt content the following procedure should be followed:

1. Average the flow and stability values for all specimens of a given asphalt content. Values that are obviously in error should not be included.

2. Prepare a separate graphical plot for the following values, as illustrated in Fig. 8.3:

(a) Stability vs. asphalt content
(b) Flow vs. asphalt content
(c) Unit weight of total mix vs. asphalt content
(d) Percent air voids vs. asphalt content
(e) Percent voids in mineral aggregate vs. asphalt content.

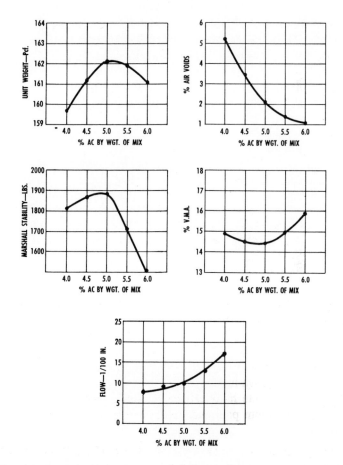

FIG. 8.3. Test property curves for hot-mix design data by the Marshall method. *(Courtesy: The Asphalt Institute.)*

In each graphical plot connect the plotted values with a smooth curve that obtains the 'best-fit' for all values.

The test property curves, plotted as described above, have been found to follow a reasonably consistent pattern for dense-graded asphaltic paving mixes. Trends generally noted are:

1. The stability value increases with increasing asphalt content up to a maximum, after which the stability decreases.

2. The flow value increases with increasing asphalt content.

3. The curve for unit weight of total mix is similar to the stability curve, except that the maximum unit weight normally (but not always) occurs at a slightly higher asphalt content than the maximum stability.

4. The percent of air voids decreases with increasing asphalt content, ultimately approaching a minimum void content.

5. The percent voids in the mineral aggregate generally decreases to a minimum value, then increases with increasing asphalt content.

Using the data curves of Fig. 8.3, asphalt content is determined, which yields the following:

(a) Maximum stability

(b) Maximum unit weight

(c) Median of limits given in Table 8.1 for percent air voids.

The optimum asphalt content of the mix is then the numerical average of the values for the asphalt content determined as noted above.

Example

Assume that data shown in Fig. 8.3 represent laboratory tests on a dense-graded asphalt concrete mix to be used for the Heavy traffic category. The nominal maximum particle size is $\frac{3}{4}$ in. Compute optimum asphalt content.

Solution.

Asphalt content at maximum stability	4·8%
Asphalt content at maximum unit weight	5·1%
Asphalt content providing 4% air voids (median of 3–5% range for surfacing mix, Heavy and Very Heavy traffic category in the design criteria table below)	4·3%
Optimum asphalt content, average	4·7%

Whether or not the asphalt paving mix will be satisfactory at the optimum asphalt content selected is determined as shown in the above example. This determination is made by applying certain limiting criteria to test data for the mix at its optimum asphalt content. The design criteria in Table 8.1 are recommended by The Asphalt Institute.

TABLE 8.1 MARSHALL DESIGN CRITERIA
(Courtesy: The Asphalt Institute)

Traffic category	Heavy		Medium		Light	
No. of compaction blows each end of specimen	75		50		35	
Test property	Min.	Max.	Min.	Max.	Min.	Max.
Stability, all mixtures	750	—	500	—	500	—
Flow, all mixtures	8	16	8	18	8	20
% air voids:						
Surfacing or levelling	3	5	3	5	3	5
Sand or stone sheet	3	5	3	5	3	5
Base	3	8	3	8	3	8
% voids in mineral aggregate:	see Fig.		see Fig.		see Fig.	
Surfacing or levelling	8.4	—	8.4	—	8.4	—
Sand or stone sheet	8.4	—	8.4	—	8.4	—
Base	8.4	—	8.4	—	8.4	—

Note:
1. Laboratory compactive efforts should closely approach the maximum density obtained in the pavement under traffic.
2. The flow value refers to the point where the load begins to decrease.
3. The portion of the asphalt cement lost by absorption into the aggregate particles must be allowed for when calculating percent air voids.
4. Percent voids in the mineral aggregate is to be calculated on the basis of the ASTM bulk specific gravity for the aggregate.
5. All criteria, and not stability value alone, must be considered in designing an asphalt paving mix.

An example of the application of the above criteria may also be illustrated by use of the data shown in Fig. 8.3.

The optimum asphalt content of the mix, as illustrated, is 4.7%. At this asphalt content, properties of the mix are determined from Fig. 8.3 to be as follows:

Stability	1880 lb
Flow	9
Percent air voids	2·8
Percent voids in mineral aggregate	14·4

It will be noted that the stability value exceeds the minimum of 750 lb, that the flow value is within the limiting range of 8–16 and that the percent voids in mineral aggregate exceeds the minimum of 14 (Fig. 8.4). The percent voids in the total mix, however, falls below the lower limit shown in the table of criteria for the design conditions previously cited. Adjustments will normally be made in the mix, to provide a mix design having all test properties within allowable limits.

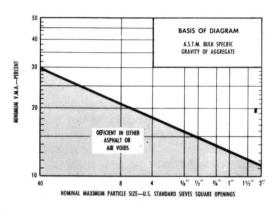

FIG. 8.4. Relationship between minimum voids of mineral aggregate (VMA) and nominal maximum particle size of the aggregate for compacted dense-graded paving mixtures. *(Courtesy: The Asphalt Institute.)*

The selected mix design is usually the most economical one which will also satisfactorily meet all of the established criteria. Mixes with abnormally high values of Marshall stability and abnormally low flow values are often less desirable because pavements of such mixes tend to be more rigid or brittle and may crack under heavy volumes of traffic. This is particularly true where base and subgrade deflections are such as to permit moderate to relatively high deflections of the pavement.

In extreme cases where it is not possible or practical, for economic or other reasons, to meet the requirements of the design criteria, a tolerance of 1% of air voids may be permitted. Under no circumstances, however, shall the allowable flow value be exceeded or shall the stability value be less than that required. It is to be emphasised that these variations should be allowed only under extreme conditions, unless service behaviour with a specific aggregate mixture indicates such a paving mix to be satisfactory.

It should be noted that when the air voids are too low (approaching 1% or less), the pavement is likely to flush or bleed. On the other hand, when the air voids exceed 5%, which is the maximum value permitted for the design of dense-graded paving mixtures, water and air can enter the pavement too easily and cause damage.

8.3 METHODS OF CONSTRUCTION OF BITUMINOUS MATS

8.3.1 Mixed-in-Place Bituminous Mats

The mineral aggregates and bituminous materials are mixed directly on the base on which the mat is to be placed. The compacted thickness of mat is generally between 2 and 3 in. The percentage of voids of compacted crushed aggregates generally used in mixed-in-place bituminous mats is about 15.

The base course must be fully prepared at a considerable time before constructing the mat so that the surface is firm and well established. Before the mixing operations for mat construction take place, the base course receives a prime coat. This is usually done by lightly applying a liquid bituminous material to bind together surface particles and to furnish continuation between the bituminous mat and the base course. The primed base can then be used for the mixing operations, keeping in mind that the base should not be gouged with equipment. Mixing operations are started by uniformly spreading the aggregate that has been deposited along one side of the roadbed, over the entire width of proposed construction. Care should be taken to ensure uniformity of grading and of quantity over the length of the pavement. The aggregate should be dry and the air temperature should preferably be 60°F or above. A pressure distributor is then used to spray the liquid bituminous material at a temperature of 175–225°F over the aggregate in increments of about $\frac{1}{2}$ gal/yd^2 over the full width. Initial mixing after each increment is accomplished by tractor-drawn disk harrows, spring tooth harrows or rotary tillers. When the final increment of bitumen is applied and mixed with the aggregate, the mixture is bladed into a windrow and the rest of the mixing is done thoroughly with blade graders. Once the mixing is complete, as may be indicated by uniformity of texture and colour, the mixed material can be spread over the roadway again, to an accurate section, and compacted by rolling with power rollers. In order to confine material as much as possible, rolling should begin at the edges and proceed towards the centre on the first passes. In some situations, a seal coat is then applied to the surface of the wearing course to provide a waterproof surface and improve visibility and skid resistance. This seal coat is an application of bituminous material followed by a cover of sand or stone chips.

8.3.2 Bituminous Mats Mixed with Travelling Plants

The mineral aggregate is elevated by the machine to a hopper from which it is meted out into the mixing chamber. As the aggregate enters the mixer, it is sprayed with bitumen at a predetermined rate and mechanical mixing continues till all aggregate particles are uniformly coated with bitumen. In this way a continuous flow of materials is provided from the aggregate windrow in front of the travelling plant to a windrow of bituminous mixture behind, which is then spread with blades. A better method is to drop the mixture directly into the hopper of a paver which follows behind, spreading and finishing the mat except for final rolling.

8.3.3 Plant-Mixed Bituminous Mats

Better control of production of bituminous mixtures, resulting in better quality and uniformity of product, can be achieved through the use of stationary and semi-portable plants. A high degree of accuracy in the control of aggregate grading is obtained through the three separate hoppers containing coarse, intermediate and fine aggregates fed through separate control gates into the mixer. Bituminous materials are sprayed on the aggregate as it enters the pugmill and effective mixing is continued till all aggregate particles are coated with bitumen. Afterwards the mixture is hauled to the roadway in trucks and deposited into the hopper of the paver, which spreads it and strikes it off to proper depth with a vibrating screed, covering 10–12 ft at a time. Depending upon the aggregate grading used, the depth of material behind the paver will generally compact about 25%.

When lighter grades of liquid asphalts containing considerable solvent or distillate are used unheated in plant mixes, compaction should be deferred 24–48 h to allow solvents to evaporate. If this is not done, the surface sealing which is sometimes created by rolling may entrap the distillates, thus greatly delaying curing of the mixture. For this reason, the use of heavier grades of bituminous materials applied to hot dry aggregates at the plant is generally preferable.

A seal coat is not immediately necessary on a well-designed and well-constructed bituminous mat.

8.4 STAGE CONSTRUCTION OF ASPHALT PAVEMENTS

Planned stage construction is the construction of roads and streets by applying successive layers of asphalt concrete according to design and a predetermined time schedule.

Two approaches to planned stage construction are described here. In each, the pavement is to be built in two stages. One approach is to use the normal design period of 20 years and then, for stage 1, reduce the design thickness of 1 or 2 in of asphalt concrete. The other is to use a relatively short design period for stage 1 thickness design, say 5 years or less, and then provide for stage 2 construction at the end of that period.

Unforeseen changes in traffic or other conditions may shorten or lengthen the predicted design period for stage 1. With either approach, therefore, performance of the pavement structure should be evaluated every 2 years to determine when stage 2 will actually be needed.

The advantages of planned stage construction include improved pavement performance, more accurate analysis of traffic and, possibly, more effective use of funds.

Improved performance is gained through locating and repairing weak spots that may develop between the first and second stages. Since corrections can be made prior to placing the second stage, a smoother riding pavement and better performance will result.

For the initial pavement design, traffic volume must be estimated, so deferring the final stage makes it possible to secure data on traffic actually using the highway. Thus, corrections to the design may be made for stage 2, either increasing or decreasing the original design, as required.

Both of these advantages should result in a more effective use of funds.

8.4.1 Procedure for Stage Construction by Reducing Full Design Thickness

In the first step of planned stage construction by reducing full design thickness, determine the pavement structure thickness for a 20 year design period using the previously described procedures. Next, reduce the 20 year design thickness by one or more inches of asphalt concrete. This will be the total thickness of asphalt concrete, T_A, for stage 1 construction.

Then, estimate the number of years of service between stage 1 and stage 2 by determining the design period for the stage 1 structural section. This is done by the following steps:

1. Enter the thickness design chart, Fig. 6.3 or 6.4, at the total thickness of asphalt concrete for stage 1, Scale A.

2. Locate the design subgrade strength value on Scale B.

3. Extend a straight line from the point on Scale A through the point on Scale B to the Design Traffic Number (DTN), Scale C.

4. Divide the DTN thus obtained by the Initial Traffic Number (ITN) to determine the ITN Adjustment Factor.

5. Locate the ITN Adjustment Factor in Table 6.4 in the proper Annual Growth Rate column and read, or interpolate, the design period. This is the estimated number of years of service between stages 1 and 2.

Example

A Design Traffic Number (DTN) of 2100 and a design subgrade strength value of CBR 7 result in a 20 year asphalt pavement structure design thickness, T_A, of 10·5 in. 1·5 in of total thickness is withheld from the stage 1 design. Calculate the number of years it must be added (stage 2) to complete the 20 year design thickness. The annual growth rate is assumed to be 4%, which gives an initial traffic number adjustment factor of 1·49 for 20 years.

Solution.

1. Enter Fig. 6.3, Scale A, at 9·0 in (10·5 in − 1·5 in).
2. Locate CBR 7 on Scale B.
3. Extend a straight line from 9·0 in, Scale A, through CBR 7, Scale B, to Scale C.
4. Divide DTN = 450, read on Scale C, by ITN to determine ITN adjustment factor.

$$ITN = \frac{2100}{1·49} = 1400$$

$$ITN \text{ adjustment factor} = \frac{DTN}{ITN} = \frac{450}{1400} = 0·32.$$

5. If the Annual Growth Rate = 4%, then from Table 6.4, the Design Period = 5·8 years by interpolation. Therefore, the estimated length of service between stages 1 and 2 is approximately 6 years.

8.4.2 Planned Stage Construction for a Given Time Period

It is often desirable to construct a pavement for an estimated design period of only a few years, anticipating that the second stage can be more adequately designed as traffic patterns become established. This approach to stage construction design is particularly valuable for city streets or low-volume rural roads where little information on present or future traffic is available.

1. Select the length of time for stage 1, usually less than 5 years.
2. Estimate the Initial Daily Traffic (IDT) and the percent and average gross weight of heavy trucks. Then determine the DTN for a 20 year design period and for stage 1 design period.
3. Determine the design subgrade strength value.
4. Determine the thickness of asphalt pavement structure, T_A, needed for the design conditions for a 20 year design period and also for stage 1 design period.

Example

A section of interurban highway is to be built by planned stage construction. IDT is estimated to be 2500 vehicles, of which 10% are heavy trucks of 35 000 lb average gross weight. The legal single-axle load limit is 18 000 lb. Annual Growth Rate is estimated to be 4% and the ITN is 70. Determine the thickness that should be used in each stage.

Solution.

1. Stage 1 design period selected to be 3 years.
2. $DTN_{20} = 70 \times 1.49 = 104$ and $DTN_3 = 70 \times 0.16 = 11$.
3. Design subgrade strength value, R value $= 45$.
4. Total thickness of asphalt concrete pavement structure, T_A, for 20 year design period $= 8.0$ in. Total thickness of asphalt concrete pavement structure, T_A, for 3 year design period $= 5.5$ in.
5. Stage 2 estimated thickness is then 8.0 in $- 5.5$ in $= 2.5$ in.

8.5 PERFORMANCE EVALUATION OF ASPHALT PAVEMENTS

8.5.1 Methods of Evaluating Pavement Performance

Pavement performance can be defined as the ability to serve traffic safely and comfortably over a period of time. In other words, it is a measure of the adequacy with which the pavement fulfils its design purpose.

The study of pavement performance requires both a functional evaluation in terms of user benefits and a mechanistic evaluation related to the mechanical condition of the pavement. The methods used in performance evaluation of pavement structures generally depend upon the importance of the project and the availability of funds. They may range from simple subjective estimates of general pavement quality and appearance to elaborate procedures involving strength evaluation, determination of riding quality and quantification of distress mechanisms.

A summary of a few well-known methods of pavement performance evaluation is given below.

8.5.2 Pavement Roughness (Distortion)

Three methods of measuring pavement roughness are summarised; these are:

1. Present Performance Rating (P P R) method. In this method a large panel, preferably more than six people, estimates subjectively the riding quality of a pavement section. It is a rapid method of determining road user opinion; its drawback, however, is the inherently subjective nature of the riding quality estimate being made.

2. Car road meter method. In this method, the differential movement between the rear axle and the body of a standard late-model sedan car is summed in $\frac{1}{8}$ in increments. Each Car Road Meter constructed must be calibrated against a present performance rating obtained by a large panel. Calibration should be repeated at least on a seasonal basis.

This is a rapid and economical method of obtaining mass inventory data.

3. Rut depth measurements. Rut depth can be measured by using a minimum 4 ft long straight edge, in accordance with the A A S H O road test procedure.

8.5.3 Pavement Strength

The following two methods can be used in measuring pavement strength:

1. Benkelman beam method. In this method a standard Benkelman Beam is used to determine the rebound deflection due to the design load. Determination of outer wheel path rebounds on 200–400 ft longitudinal spacing is extremely useful in detecting areas of abnormal strength. These Benkelman Beam measurements provide an extremely economical and powerful tool in determining over-all pavement strength and strength variation for highways. Tests done in Canada using the Benkelman Beam have shown that high-performance pavements with ages up to 10 years with average daily traffic per lane of greater than 1000 vehicles exhibit a Benkelman Beam rebound less than 0·035 in.

2. Static plate-bearing tests. The plate-bearing test described in Appendix C can be used to measure the strength of the wearing asphalt layer. The value of the load, which when applied to a circular plate of standard diameter will produce a defined deflection, is considered the failure load. The value of the design load can be compared with the failure load after readjusting the design load for the load intensity used in the test.

However, this method is expensive and time-consuming; it is used more in testing the strength of airfield pavements than in highways.

8.5.4 Fracture Patterns and Disintegration

Table 8.2 shows the diagnostic features of pavement distress. The features can indicate, to a large extent, the degree of performance of the pavement.

TABLE 8.2 MODES AND MECHANISMS OF DISTRESS TYPES

Mode	Manifestation	Mechanism
Fracture	Cracking	Excessive loading Repeated loading (i.e. fatigue) Thermal changes Moisture changes Slippage due to horizontal forces Shrinkage
	Spalling	Excessive loading Repeated loading (i.e. fatigue) Thermal changes Moisture changes
Distortion	Permanent deformation	Excessive loading Time-dependent deformation (e.g. creep) Densification (i.e. compaction) Consolidation Swelling
	Faulting	Excessive loading Densification (i.e. compaction) Consolidation Swelling
Disintegration	Stripping	Adhesion (i.e. loss of bond) Chemical reactivity Abrasion by traffic
	Ravelling and scaling	Adhesion (i.e. loss of bond) Chemical reactivity Abrasion by traffic Degradation of aggregate Durability of binder

8.5.5 Present Serviceability of a Pavement System

Present Serviceability is the ability of a specific type of pavement or section to serve a designated amount and type of traffic in its existing condition. The Present Serviceability Rating was developed at the AASHO road test. Basically it represents the mean subjective rating of a large panel of observers as to the riding quality of a given pavement section at a given time, on a scale ranging from 0 (very poor) to 5 (very good). The Present Serviceability Index (PSI) equation developed for flexible pavements at the AASHO road test is:

$$PSI = 5.03 - 1.91 \log(1 + \overline{SV}) - 0.01 \sqrt{C+P} - 1.38 \, \overline{RD}^2 \qquad (8.7)$$

where PSI = Present Serviceability Index;
      C = amount of cracking (types 2 and 3) (ft^2/1000 ft^2);
      P = amount of patching (ft^2/1000 ft^2);
     $\overline{RD}$ = mean rut depth in wheel path (in); and
     $\overline{SV}$ = mean slope variance in wheel path, which can be measured by a CHLOE Profilometer.

Equation (8.7) applies specifically to the loadings, materials and environment at the AASHO road test.

8.6 COST ANALYSIS

8.6.1 Highway Costs

Most of the methods available for determining the annual cost of highways are similar, in that each considers a highway to be a capital investment of funds. Each method also presents a procedure for determining annual costs which includes many factors in addition to construction costs. R. H. Baldock proposed a realistic procedure which more conclusively describes the total costs involved.

This procedure for evaluation of highways includes all factors affecting the annual cost with respect to a reasonable period of analysis. To avoid obsolescence due to major technological changes affecting transportation, an analysis period of 40 years was selected. The entire investment is amortised during this period, although the highway will almost undoubtedly continue to serve as a portion of the original or some lesser system.

Two methods are proposed for analysis of annual cost. The first involves all costs pertaining to the complete highway and is used to evaluate the whole facility. The second analyses only those costs pertaining to the travelled way or mainline section, including pavement structure, shoulders and, when appropriate, structural drainage features. Only the latter method is needed to evaluate and compare alternative pavement designs to determine the most appropriate design for a specific highway.

The basic factors involved in computing the annual cost per mile of a highway are as follows.

1. First Cost (per mile)
First cost should include construction and right-of-way. Construction costs should be divided between the pavement structure and shoulders, and all other construction expenses. This division makes it easier to compare the annual costs of alternative pavement designs.

2. Maintenance Cost (per mile)
The total per mile maintenance cost should be divided into pavement structure and shoulder maintenance expense, and the expense of all other maintenance. The sum of these represents maintenance in determining total annual cost per mile but only the former is used in comparing alternative designs.

3. Operation Cost (per mile)
Operation costs should include the expense, other than maintaining the capital investment, of providing service to the road user. This includes snow removal, sanding, signs, signals, striping and marking, and similar services. Many states charge some of the above items to maintenance and for determining annual highway costs the separation of these items is not necessary.

4. Administration and Overhead Costs (per mile)
The administration and overhead costs, including field surveys and office design, are considerable and must be charged. It is suggested that they be prorated over the miles on the system, on the basis of first cost of construction.

5. Cost of Resurfacing and Resurfacing Frequency (per mile)
Resurfacing costs are estimated on the basis of past experience. The pavement structural design presented in the Asphalt Institute method of design is based upon resurfacing after 20 years of service life. This period should be used for resurfacing frequency.

6. Salvage Value (per mile)
This procedure amortises the entire investment in a highway over the

analysis period of 40 years. For this reason, the salvage value at the end of the analysis life of the project may be considered as having a zero value and does not enter into the computation.

7. Interest Rate

Money has a definite rental value and interest on the investment in a highway must be charged to permit of economic evaluation of the project. If constructed with borrowed funds, interest payments accrue to the security holder. If, on the other hand, the project is funded from the owners' revenues, interest is in the nature of a fixed charge against the project to compensate for the loss of earning power of the funds 'frozen' therein. In the case of public funds derived from taxes, these funds, if not so captured, could have been invested by the public to yield a safe and reasonable return, and the interest charge therefore represents a cost.

Economists have used interest rates varying from 5 to 10% in studies of highway economics. It is recommended that the interest rate used in determination of the annual highway cost be 6% annually.

TABLE 8.3 SUMMARY OF DISCOUNTING FORMULAE

Situation	Formula	Factor
Future value S of current sum P	$S = PF$	$F = (1+r)^n$, is compound amount factor
Present value P of future sum S	$P = S\left(\dfrac{1}{F}\right)$	$\dfrac{1}{F}$ is present worth factor (PWF)
Cumulative future value of periodic payment R	$S = \dfrac{RF}{\mathrm{CRF}}$	$\dfrac{1}{\mathrm{CRF}}$ is series present worth factor
Present value of periodic payments	$P = \dfrac{R}{\mathrm{CRF}}$	$\mathrm{CRF} = \dfrac{rF}{F-1}$, is capital recovery factor

Four basic formulae can be applied in discounting the value of future costs or in determining the future value of a current amount of investment. These formulae are summarised in Table 8.3 and are based on the assumption that the discount rate is constant over the period being considered in the study.

8. Annual Costs of Travelled Way or Mainline Pavement Only (per mile)

When economic studies are made to determine the most appropriate of the several alternative pavement designs being considered, only the initial construction and maintenance cost of the travelled way, or mainline section, should be used. Right-of-way, Administration and Overhead, Operation and other costs may be disregarded because they apply equally to all alternatives.

Two formulas are presented by Baldock for determining the annual cost of highways. Formula No. 1 includes all costs of building, maintaining, operating and administering the highway. This formula is used to calculate the total annual cost.

Formula No. 1. Formula No. 1 calculates the total annual cost as follows:

$$C = CRF_n \left[A + R_1 \, PWFn_1 + R_2 \, PWFn_1 - \left(1 - \frac{Y}{X}\right)(R_1 \right.$$
$$\left. \text{or } R_2) \, PWF_n \right] + M + O + D \qquad (8.8)$$

where C = the complete annual cost per mile of highway;

CRF = the Capital Recovery Factor = $r(1+r)^n/(1+r)^n - 1$;

PWF = Present Worth Factor, for a single payment = $1/(1+r)^{n_1}$;

r = the interest rate (6%);

n = the analysis period (40 years);

n_1 = the number of years after construction that future work is performed (n_1 will have different values in the same analysis depending upon whether it is used with R_1 or R_2);

A = total construction and right-of-way costs per mile;

R_1 = first resurfacing costs per mile;

R_2 = second resurfacing costs per mile;

Y = number of years between the last resurfacing and the end of analysis period;

X = estimated service life, in years, of last resurfacing;

M = total annual maintenance cost per mile;

O = annual operation cost per mile; and

D = annual administration and overhead cost per mile.

Formula No. 2. Formula No. 2 includes only costs necessary to compare alternate pavement designs. This one calculates the annual cost per mile of the travelled way, or mainline section.

$$C_1 = CRF_n \left[A_1 + R_1 \, PWFn_1 + R_2 \, PWFn_1 - \left(1 - \frac{Y}{X}\right)(R_1 \right.$$
$$\left. \text{or } R_2) \, PWF_n \right] + M_1 \qquad (8.9)$$

where C_1 = annual cost of travelled way, or mainline section, per mile;
        A_1 = initial construction cost of travelled way, or mainline section per mile; and
        M_1 = annual maintenance cost of travelled way, or mainline section per mile.
    All other terms are defined under Formula No. 1.

Example 1. Annual cost of travelled way—complete construction. All asphalt concrete section:

Cost Elements

Analysis period, n		40 years
Interest rate, r		6%
Initial cost, A		$70 710
Resurfacing, R_1		$11 705
Estimated life of resurfacing, X		20 years
Annual maintenance cost, M_1		$190
Time between last resurfacing and end of analysis period, Y		20 years
Capital Recovery Factor, $CRFn$	−40 years	0·066 46
Present Worth Factor, $PWFn$	−40 years	0·097 22
Present Worth Factor, $PWFn_1$	−20 years	0·311 80

$$C_1 = 0{\cdot}066\ 46\ [70\ 710+(11\ 705)\ (0{\cdot}311\ 80)+0-0]+190$$
$$= 0{\cdot}066\ 46\ [70\ 710+3650]+190$$
$$= 0{\cdot}066\ 46\ [74\ 360]+190$$
$$= 4942+190$$
$$= \$5132 \text{ annual cost per mile}$$

Example 2. Annual cost of travelled way—stage construction. All asphalt concrete section:

Cost Elements

Analysis period, n	40 years
Interest rate, r	6%
Initial cost, A_1 ($1\frac{1}{2}$ in A.C. withheld for future)	$61 980
Second stage cost, R_1 (place $1\frac{1}{2}$ in at 5th year)	$8810
Resurfacing, R_2 (place 2 in at 25th year)	$11 705
Estimated life of second stage and resurfacing	20 years
Annual maintenance cost, M_1	$190

Time between last resurfacing and end of
analysis period, Y 15 years
Capital Recovery Factor, $CRFn$ -40 years 0·066 46
Present Worth Factor, $PWFn$ -40 years 0·097 22
Present Worth Factor, $PWFn_1$ $- 5$ years 0·747 26
Present Worth Factor, $PWFn_1$ -25 years 0·233 00

$$C_1 = 0·066\ 46\ [61\ 980 + (8810)(0·747\ 26) + (11\ 705)$$

$$(0·233\ 00) - \left(1 - \frac{15}{20}\right)(11\ 705)(0·097\ 22)] + 190$$

$$= 0·066\ 46\ [61\ 980 + 6583 + 2727 - 284] + 190$$
$$= 0·066\ 46\ [71\ 006] + 190$$
$$= 4719 + 190$$
$$= \$4909\ \text{annual cost per mile}$$

These two examples for the same project but using different methods of construction, show that staged construction is cheaper than complete construction in one stage.

The same procedure can be used to evaluate the costs of alternative projects in which the variables may be: subgrade compaction, thickness of pavement structure, etc.

8.6.2 Airfield Costs
The costs of airfield pavements can be determined by using the same procedure explained above.

8.6.3 Optimal Selection of Flexible Pavement Components
In the AASHO method of design of flexible pavements, a structural number, SN, is used as a means of determining the thickness of the pavement system. This structural number is obtained from the nomograph relating the subgrade strength to the equivalent daily 18 000 lb single-axle load applications; a regional factor, R, is also introduced in the chart to reflect the environmental conditions in the region. This factor should be used to determine the weighted structural number, SN, which, in turn, will define the thickness of the different layers of the pavement structure.

As has been shown before, several different thickness combinations for the sub-base, base and wearing layers can be used to satisfy adequately the structural design of the pavement. However, not all the combinations will provide economical solutions, and only one pavement structure is an optimal selection of the flexible-pavement components for the given design conditions.

The formula for the AASHO method of design of flexible pavements can be written in the general form given in eqn. (8.10):

$$a_1 d_1 + a_2 d_2 + a_3 d_3 + a_4 d_4 \geq SN \qquad (8.10)$$

where d_i = thickness of layer i (in);

$\quad a_i$ = coefficient of relative strength of material of layer i; and

$\quad SN$ = weighted structural number for design.

If the unit cost of the material of the different layers is known, then an objective function which may be optimised for the minimum cost can be formulated as follows [4]:

$$\text{Min. } Z = \left[\frac{c_1 \gamma_1 k_1}{12 \times 2000} \right] \times d_1 + \left[\frac{c_2 \gamma_2 k_2}{12 \times 2000} \right] d_2$$

$$+ \left[\frac{c_3 \gamma_3 k_3}{12 \times 2000} \right] \times d_3 + \left[\frac{c_4 k_4}{12 \times 27} \right] d_4 \qquad (8.11)$$

where Z = total cost of pavement system ($\$/\text{ft}^2$);

$\quad c_i$ = unit cost of material of layer i, in dollars per ton for materials 1, 2 and 3, and dollars per cubic yard for material 4;

$\quad \gamma_i$ = density of material of layer i (lb/ft^3); and

$\quad k_i$ = adjustment factor for increase in width of pavement layers as shown in Fig. 8.5:

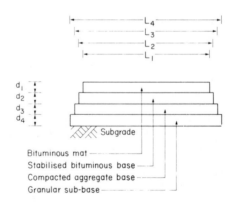

FIG. 8.5. Flexible-pavement structure.

$$k_1 = \frac{L_1}{L_1} = 1; \; k_2 = \frac{L_2}{L_1}; \; k_3 = \frac{L_3}{L_1}; \; k_4 = \frac{L_4}{L_1}$$

In addition to the constraint condition given in eqn. (8.10), there may be other constraints that should be satisfied. These constraints may take the form given in eqns. (8.12)–(8.16):

$$d_1 + d_2 + d_3 + d_4 \geq T_{\text{min.}} \tag{8.12}$$

where $T_{\text{min.}}$ = total minimum thickness of flexible pavement (in) to satisfy certain environmental conditions;

$$b_1' \leq d_1 \leq b_1'' \tag{8.13}$$

$$b_2' \leq d_2 \leq b_2'' \quad \text{or} \quad d_2 = 0 \tag{8.14}$$

$$b_3' \leq d_3 \leq b_3'' \quad \text{or} \quad d_3 = 0 \tag{8.15}$$

$$b_4' \leq d_4 \leq b_4'' \quad \text{or} \quad d_4 = 0 \tag{8.16}$$

where b_1', b_2', b_3' and b_4' are lower limits for thickness (in) that should be exceeded to avoid certain structural, constructional or safety problems.

b_1'', b_2'', b_3'' and b_4'' are higher limits for thicknesses (in) that should not be exceeded for practical or other reasons. Linear or integer programming can be used to determine the thicknesses of the different layers giving the minimum cost for the pavement structure under the given constraint conditions. Computer programs for solving this problem are already available [4].

REFERENCES AND BIBLIOGRAPHY

1. Brokaw, M. P., 'Development of the PCA Road Meter: A Rapid Method for Measuring Slope Variance', *Highway Research Record*, No. 189, 1967.
2. Carey, W. N., Huckins, H. C. and Leathers, R. C., 'Slope Variance as a Measure of Roughness and the Chloe Profilometer', The AASHO Road Test Special Report No. 73, Highway Research Board, 1962.
3. De Neuville, R., Discussion, *ASCE Transportation Journal*, November 1971.
4. Hejal, S. S., Buick, T. R. and Oppenlander, J. C., 'Optimal Selection of Flexible Pavements Components', *ASCE Transportation Engineering Journal*, February 1971.
5. Hennes, R. G. and Ekse, M., *Fundamentals of Transportation Engineering*, Second Edition, McGraw-Hill, 1969.
6. McCullough, B. F., 'Distress Mechanism—General', *Highway Research Board Special Report 126*, 1971.

7. Pavement Design and Evaluation Committee, 'Evaluation of the Car Road Meter: A Device for the Rapid Measurement of Road Roughness', Road and Transportation Association of Canada, Proceedings of Annual Convention, 1971.

8. The Asphalt Institute Manual Series No. 1, 'Thickness Design–Full-Depth Asphalt Pavement Structures for Highways and Streets', Revised Eighth Edition, August 1970.

9. The Asphalt Institute Manual Series No. 2, 'Mix Design Methods for Asphalt Concrete and Other Hot-Mix Types', Second Edition, February 1962.

10. The Asphalt Institute Manual Series No. 3, 'Asphalt Plant Manual', Third Edition, March 1967.

PART III

RIGID PAVEMENTS

CHAPTER 9

THEORIES OF DESIGN OF RIGID PAVEMENTS

9.1 INTRODUCTION

The theories of design of rigid pavements are different from those of flexible pavements. While a flexible-type pavement basically distributes the load gradually to the layers underneath, a rigid pavement acts as a structural element (a plate) resting on an elastic foundation.

In reality, the numerous so-called theories of rigid pavement design all spring from a single theory, the theory of elasticity.

Timoshenko distinguishes three kinds of plate bendings: (1) thin plates with small deflections, (2) thin plates with large deflections, and (3) thick plates.

Since the deflection of pavement slabs is small compared with their thickness, they can belong to category (1).

To develop a theory for the bending of these slabs when they are subjected to vertical loads, the following approximations can be assumed:

1. There is no deformation in the middle plane of the slab (this plane remains neutral during bending).
2. Those planes of the slab initially lying normal to the middle plane of the slab remain normal after bending.
3. The normal stresses in the direction transverse to the plane of the slab can be disregarded.

With these assumptions, all stress components can be expressed in terms of the deflected shape of the slab. The function representing the deflection, w, must satisfy a linear partial differential equation, which together with the boundary conditions completely defines the deflection. The solution of this differential equation gives all the necessary information for calculating the stresses at any point in the plate.

The differential equation describing the deflected surface of pavement slabs subjected to a uniform load, q, applied perpendicular to their surface, is:

$$\frac{\partial^4 w}{\partial x^4} + 2\frac{\partial^4 w}{\partial x^2 \partial y^2} + \frac{\partial^4 w}{\partial y^4} = \frac{q - Kw}{D} \text{ [obtained by LaGrange in 1811]} \qquad (9.1)$$

where K is the modulus of subgrade reaction, D is the flexural rigidity of the slab and w is the deflection of the slab:

$$D = \frac{Eh^3}{12(1 - v^2)}$$

9.2 WESTERGAARD SOLUTION

Westergaard's work [12, 13] was the first serious attempt to find a theoretical solution for rigid pavement design. In the solution which he presented in 1920, he made the following important assumptions:

1. The concrete slab acts as a homogeneous elastic solid in equilibrium.
2. The reaction of the subgrade is solely vertical, and proportional to the deflection of the slab.
3. The reaction of the subgrade per unit area at any given point is equal to a constant, K (modulus of subgrade reaction), multiplied by the deflection at that point.
4. The thickness of the slab is uniform.
5. The load at the interior and at the corner of the slab is distributed uniformly over a circular contact area; for the corner loading, the circumference of this circular area is tangential to the edge of the slab.
6. The load at the interior edge of the slab is distributed uniformly over a semicircular contact area, the diameter of the semicircle being along the edge of the slab.

In his study, Westergaard considered the following cases:

(i) *Maximum tensile stress at the bottom of the slab due to a central load.*

$$f_t = 0.316\,25\,\frac{P_i}{h^2}\left[4\log_{10}\left(\frac{l}{b}\right) + 1.069\,3\right] \qquad (9.2)$$

where l = the radius of relative stiffness = $\sqrt[4]{\dfrac{E_c h^3}{12(1 - v^2)K}}$;

P_i = the load P increased by the impact factor (lb);

h = the slab thickness (in);

a = the radius of the load contact area (in);

b = the radius of equivalent distribution of pressure at the bottom of the slab

$$= \sqrt{1 \cdot 2a^2 + h^2} - 0 \cdot 675\, h; \quad \text{and}$$

f_t = tensile stress due to load (psi).

(ii) *Maximum tensile stress at the bottom of the slab for an interior edge loading in a direction parallel to the edge.*

$$f_t = 0 \cdot 571\,85\, \frac{P'_i}{h^2}\left[4 \log_{10}\left(\frac{l}{b}\right) + 0 \cdot 359\,3 \right] \tag{9.3}$$

where P'_i = the load P' (half the load P) acting on a semicircle, increased by the impact factor (lb).

(iii) *Maximum tensile stress at the top of the slab in a direction parallel to the bisector of the corner angle, for a corner loading.*

$$f_t = \frac{3P_i}{h^2}\left[1 - \left(\frac{a\sqrt{2}}{l}\right)^{0 \cdot 6} \right] \tag{9.4}$$

9.3 PICKETT SOLUTION

In 1951, Pickett developed formulae [8] for the case of a corner load, considering two conditions.

1. Protected corners. Provision is made to transfer at least 20% of the load from one slab corner to the other by some adequate mechanical means or by aggregate interlock. For this case:

$$f_t = \frac{3 \cdot 36 P_i}{h^2}\left(1 - \frac{\sqrt{a/l}}{0 \cdot 925 + 0 \cdot 22 a/l} \right) \tag{9.5}$$

2. Unprotected corners. There is no adequate provision for load transfer and one corner must carry over 80% of the load. For this case

$$f_t = \frac{4 \cdot 2 P_i}{h^2}\left(1 - \frac{\sqrt{a/l}}{0 \cdot 925 + 0 \cdot 22 a/l} \right) \tag{9.6}$$

where a = the radius of tyre imprint for a single-wheel load (in); or, in the case of dual wheels, is the radius of a circle having an area equal to the tyre imprint area of one wheel plus the area of a rectangle, of length equal to the centre-to-centre spacing of the dual wheels, and width equal to the diameter of the tyre imprint area for one wheel.

f_t, l and P_i are as defined before.

These last two formulae consider the effect on the stress due to load of warping at the corner.

9.4 WESTERGAARD FORMULAE FOR LOADS ON ELLIPTICAL AREAS

In 1948, Westergaard developed a new set of formulae for the case of a load acting on an elliptical area, for the various loading positions (Fig. 9.1).

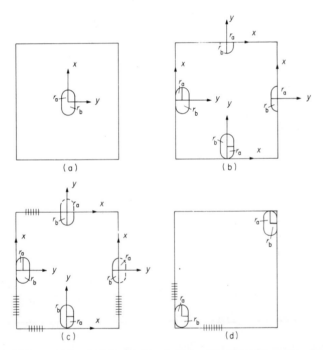

FIG. 9.1. Possible positions of single-wheel load. (a) Load in the interior of the area of the pavement. (b) Load at or next to edge or joint without load transfer. (c) Load at or next to edge or joint with load transfer. (d) Load at the corner of the pavement.

He also considered the effect on the values of stresses of load transfer at the joints. This case of elliptical tyre imprint is mainly applicable to airfield pavements. The new formulae for the different locations of the load are

Case 1. Tensile stresses at the bottom of pavement slab due to a central load.

$$\begin{matrix} f_x \\ f_y \end{matrix} = \frac{P_i}{h^2} \left[0 \cdot 275 \, (1+v) \log_{10} \frac{Eh^3}{K\left(\frac{a+b}{2}\right)^4} \mp 0 \cdot 239 \, (1-v) \frac{a-b}{a+b} \right] \qquad (9.7)$$

where a, b = semi-axes of an ellipse representing the foot print of a tyre; and

x, y = horizontal rectangular co-ordinates, the x axis being in the direction of the semi-axis a.

Case 2. Tensile stresses at the bottom of an edge or a joint that has no capacity for load transfer with load distributed uniformly over the area of an ellipse tangent to the edge or joint.

$$f_e = \frac{3(1+v)P_i}{\pi(3+v)h^2} \left[\log \frac{Eh^3}{100K\left(\frac{a+b}{2}\right)^4} + 1 \cdot 84 - \frac{4}{3}v + (1+v) \frac{a-b}{a+b} + \right.$$

$$\left. + 2(1-v) \frac{ab}{(a+b)^2} + 1 \cdot 18(1+2v) \frac{b}{l} \right] \qquad (9.8)$$

where f_e = maximum tensile edge stress;

l = radius of relative stiffness = $\sqrt[4]{\dfrac{Eh^3}{12(1-v^2)K}}$;

a = the semi-axis parallel to the edge or joint.

The axis x is along the line of the edge or joint.

Case 3. Tensile stresses at the bottom of an edge or a joint that has no capacity for load transfer with load distributed uniformly over the area of one-half of an ellipse.

$$f_e = \frac{2.2(1+v)P_i}{(3+v)h^2} \log_{10} \frac{Eh^3}{100K\left(\frac{a+b}{2}\right)^4} +$$

$$+ \frac{3(1+v)P_i'}{\pi(3+v)h^2} \left[3 \cdot 84 - \frac{4}{3}v - (1-v) \frac{a-b}{a+b} + 0 \cdot 5(1+2v) \frac{b}{l} \right] \qquad (9.9)$$

The load P_i' in this case is half the total load P_i.

Case 4. Tensile stresses at the bottom of a joint that has some capacity for load transfer.

$$f_j = \left(1 - \frac{1}{2}j\right)f_e + \frac{1}{2}jf_e' \tag{9.10}$$

$$f_j' = \frac{1}{2}jf_e + \left(1 - \frac{1}{2}j\right)f_e' \tag{9.11}$$

f_e and f_e' = corresponding stresses that would be produced if the joint had no capacity for load transfer.

Example
$p = 50\,000$ lb $a+b = 30$ in $E = 4 \times 10^6$ psi $K = 200$ psi/in
$h = 10$ in   either a or $b = 20$ in $v = 0.15$

Find the stresses for the four cases of loading.

TABLE 9.1 COMPARISON BETWEEN THE STRESSES
FOR THE EXAMPLE

CASE 1	$f_x = 376.7$ psi in the longitudinal direction $f_y = 444.5$ psi in the transverse direction
CASE 2 *or*	$f_e = 740.8$ psi edge parallel to direction of traffic $f_e = 663.8$ psi joint transverse to direction of traffic
CASE 3	$f_e = 428.0$ psi with half the load on each side of a longitudinal joint
CASE 4 with j = 0.8	$f_j = 444.5$ psi $\Big\}$ edge parallel to the direction of traffic $f_j' = 296.3$ psi

Table 9.1 shows that the maximum stress at a joint with 80% load transfer is equal to the maximum stress at the centre of the slab.

9.5 OTHER METHODS FOR ANALYTICAL ANALYSIS OF RIGID PAVEMENTS

9.5.1 Finite Difference Method

The finite difference method is one of the main numerical methods that can be used in solving such types of problems. In this method the slab surface is divided into a grid, and a general fourth-order difference equation is developed for each nodal point in the grid. The deflection at any specific point is expressed in an individual difference equation for that particular point; then the differential equation

$$\frac{\partial^4 w}{\partial x^4} + 2\frac{\partial^2 w}{\partial x^2 \partial y^2} + \frac{\partial^4 w}{\partial y^4} = \frac{p - Kw}{D} \tag{9.12}$$

is replaced by finite difference equations, one at each node:

$$\frac{D}{\lambda^4}[s]\{w\} = \{p\} - K\{w\} \tag{9.13}$$

where λ is the length of a square mesh of the grid and $[s]$ is the finite difference coefficients matrix (equivalent stiffness matrix).

If the term $K\{w\}$ on the right-hand side of eqn. (9.13) is transferred to the diagonal of the matrix $[s]$, the following equation is obtained:

$$\frac{D}{\lambda^4}\begin{bmatrix} S_{11} + \dfrac{\lambda^4 K}{D} & S_{12} & \cdots \\ S_{21} & S_{22} + \dfrac{\lambda^4 K}{D} & \cdots \\ \vdots & \vdots & \end{bmatrix}\{w\} = \{p\} \tag{9.14}$$

and the problem reduces to solving a large number of simultaneous algebraic equations instead of one complex differential equation. The solution produces the value of the deflection at each individual grid point. Once the deflections are obtained, they are substituted into the appropriate difference expressions to determine moments, shears and reactions at the nodal points in the slab. The difference equations for mesh points near or on the boundaries of the slab have to be modified from the general pattern of finite difference coefficients at interior points, in order to satisfy the boundary conditions. The modification of boundary conditions is one of the difficulties of the finite difference method, and sometimes limits its use.

9.5.2 Linear-Elastic Layered Method
The stresses due to loading in two- or three-layered elastic systems such as a slab or slab-and-base resting on a uniform subgrade were analysed by Burmister, Acum and Fox between 1943 and 1951. The full development of the situation was not feasible until the computer age. It is assumed that the modulus of elasticity, Poisson's ratio and the thickness of each layer

are variables. This analysis permits of determination of the complete state of stress due to static load, at any point in the pavement structure.

9.5.3 Finite Element Method

This method has been used for analysing rigid-pavement slabs on an elastic foundation, considering the subgrade either as an elastic continuum or as a Winkler-type foundation.

The difference in the behaviour of the two types of foundation models is illustrated in Fig. 9.2.

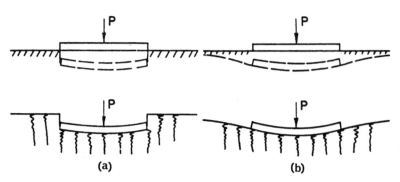

FIG. 9.2. Deflections of a pavement slab considering two different assumptions for the subgrade. (a) Winkler foundation. (b) Elastic continuum.

If a Winkler-type foundation is considered, the subgrade is represented by springs having a constant modulus of reaction K_s. This means that the subgrade reaction per unit area at any point is proportional to the vertical deflection at that point, but independent of the vertical deflection at any other point. If an elastic continuum foundation is considered in the finite element solution, the slab is treated as an assemblage of plate elements, while the foundation is treated as an idealised half-space. The flexibility matrix for the foundation is obtained by determining the deflections at all points for each location of a unit vertical point load.

The stiffness matrix of the foundation is obtained by inverting the flexibility matrix. It is then combined with that of the slab to obtain the complete stiffness matrix of the structure.

When the finite element method is used, the slab is first divided into individual rectangular elements joined at discrete finite numbers of nodal points. The foundation is considered as consisting of a series of rectangular pressure areas whose centres coincide with and remain in contact with the nodal points of the slab. The pressure is assumed to be constant within

each rectangle. The stiffness coefficients at the nodal points of the foundation are determined by inverting the flexibility matrix of the foundation.
This flexibility matrix can be obtained by using the Boussinesq eqn. (9.15)
relating the vertical displacement, of the upper surface of an elastic half-
space, to an applied vertical force:

$$w_{ni} = \frac{F(1-v_s^2)}{\pi E_s d_{ni}} \tag{9.15}$$

where w_{ni} is the deflection at n due to load at i, F is the force on the element
$d\xi d\eta$, v_s and E_s are Poisson's ratio and Young's modulus of the subgrade
soil, respectively, and d_{ni} is the radial distance between points i and n.

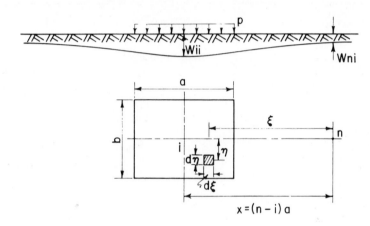

FIG. 9.3. Vertical displacement due to a load uniformly distributed over a rectangular area
on an isotropic semi-infinite solid.

If eqn. (9.5) is integrated for a uniformly loaded rectangular area a times
b, as in Fig. 9.3, then:

$$w_{ni} = 2 \int_{\xi=0}^{\xi=\frac{a}{2}} 2 \int_{\eta=0}^{\eta=\frac{b}{2}} \frac{P_i(1-v_s^2)}{ab\pi E_s} \frac{d\xi d\eta}{\sqrt{\xi^2+\eta^2}}$$

$$= \frac{P_i(1-v_s^2)}{a\pi E_s} \left[B \sinh^{-1} \frac{1}{B} + \sinh^{-1} B - C \sinh^{-1} \frac{1}{C} - \sinh^{-1} C \right]$$

$$= \frac{(1-v_s^2)}{a\pi E_s} f_{ni} P_i \tag{9.16}$$

where $B = [2(n-i)+1]\dfrac{a}{b}$;

$\qquad C = [2(n-i)-1]\dfrac{a}{b}$; and

$\qquad P_i$ = the total load over the rectangular area $a \times b$ at node i.

If $n = i$, then:

$$w_{ii} = \frac{P_i(1-v_s^2)}{a\pi E_s}\left(\left[B\sinh^{-1}\frac{1}{B}+\sinh^{-1}B\right]\times 2\right)$$

$$= \frac{(1-v_s^2)}{a\pi E_s}f_{ii}P_i \tag{9.17}$$

Some values of f_{ii} are given in Table 9.2 for various ratios of b/a.

TABLE 9.2 VALUES OF f_{ii} FOR VARIOUS RATIOS OF b/a

$\dfrac{b}{a}$	$\dfrac{2}{3}$	1	2	3	4	5
f_{ii}	4·265	3·525	2·406	1·867	1·543	1·322

Therefore, for any set of grid points, the deflections can be written as:

$$\{w\} = \frac{1-v_s^2}{a\pi E_s}[f]\{P\} \tag{9.18}$$

where $[f]$ is the flexibility matrix of the foundation, obtained for off-diagonal terms by eqn. (9.16) and for diagonal terms by eqn. (9.17). Inverting, one can write:

$$\{P\} = \frac{\pi E_s a}{1-v_s^2}[K_s]\{w\} \tag{9.19}$$

The stiffness matrix of the whole system is the sum of the slab stiffness and the foundation stiffness. The slab deflects under the combined action of the known applied loads, V, and the unknown reactions:

$$\{V\}-\{P\} = [K_c]\{w\} \tag{9.20}$$

where $[K_c]$ is the stiffness matrix of the concrete slab. The foundation medium deflects under a loading system equal and opposite to the reactions. Therefore:

$$\{V\} = [K_c]\{w\} + [K_s']\{w\}$$
$$= [K_c + K_s']\{w\} \tag{9.21}$$

where $$[K_s'] = \frac{\pi E_s a}{1 - v_s^2}[K_s]$$

Figure 9.4 shows part of a rigid pavement slab analysed by using the finite element method and inserting the actual boundary conditions along

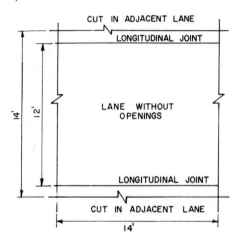

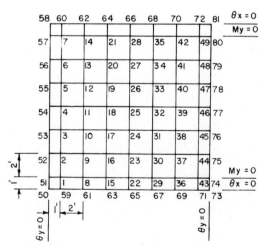

Fɪɢ. 9.4. Finite element mesh and boundary conditions.

the longitudinal joints. Cuts were taken at a sufficient distance from the location of load application; the rotation was assumed zero along these cuts [11].

9.6 DEFLECTIONS IN RIGID PAVEMENTS

Pickett and Ray [10] developed influence charts which can be used in obtaining theoretical deflection of concrete pavement slabs under load.

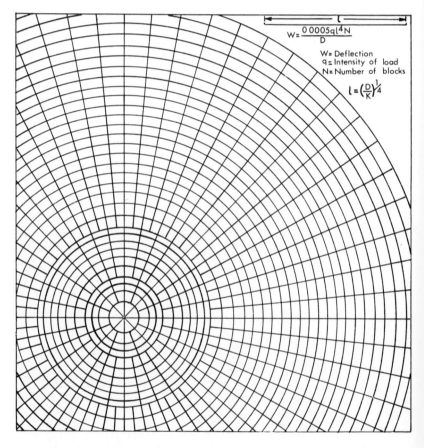

$$W = \frac{0.0005 q l^4 N}{D}$$

W = Deflection
q = Intensity of load
N = Number of blocks

$$l = \left(\frac{D}{K}\right)^{\frac{1}{4}}$$

F IG. 9.5. Influence chart for deflection: interior load. Subgrade assumed to be a dense liquid. (From [10].)

Although these charts may be used for any rigid type of pavement of uniform thickness, they are mainly intended for airport pavements. With these charts, the deflections can be obtained for any distribution of load that might be transmitted by aeroplane landing gears, if the following procedure is followed:

1. Draw the imprint of tyre or tyres on transparent paper, to a scale that depends on the properties of the slab and its supporting sub-grade. The appropriate values of these properties are determined and combined in one value, defined previously as the radius of relative

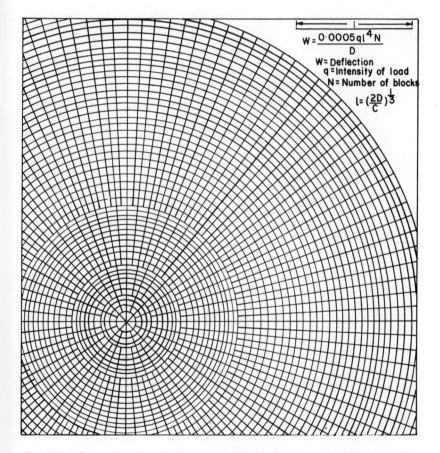

FIG. 9.6. Influence chart for deflection: interior load. Subgrade assumed to be an elastic solid. (From [10].)

stiffness, *l*. The obtained value of *l*, when compared with *l* as measured from the charts, will give the scale to be used.

2. Place the drawing on the appropriate chart in a position that depends on the location of the load with respect to the point for which the deflection is desired.

3. Count the blocks of the chart covered by the diagram showing the tyre imprints. The desired deflection is then obtained as a product of the intensity of loading, *q*, a factor expressing properties of subgrade and slab, and the number of blocks covered by the diagram.

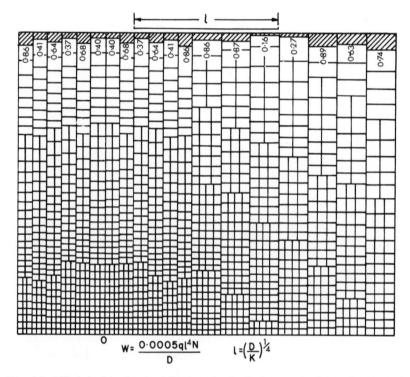

$$W = \frac{0.0005ql^4N}{D} \qquad l = \left(\frac{D}{K}\right)^{1/4}$$

FIG. 9.7. Influence chart for the deflection of point 0, due to a load near the edge and assuming a dense liquid subgrade. Deflection at the edge (shaded blocks count only as fractions). (From [10].)

The charts are for four different cases, classified according to the subgrade assumption, as follows: (a) interior load: dense-liquid subgrade, (b) interior load: solid subgrade, (c) edge load: dense-liquid subgrade, and

(d) load at 0·5 *l* from the edge: dense liquid subgrade. Figures 9.5 and 9.6 are for the case of interior load, for the two assumptions of dense-liquid and solid subgrade, respectively. Figures 9.7 and 9.8 are for the cases of edge and near-edge deflections respectively.

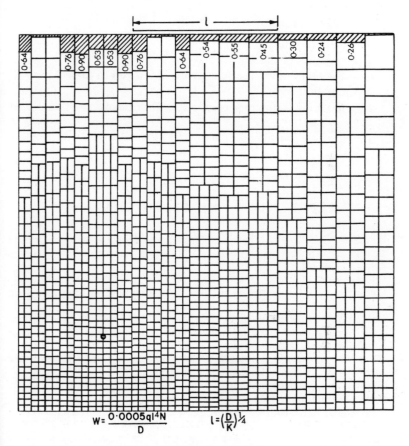

$$W = \frac{0 \cdot 0005 q l^4 N}{D} \qquad l = \left(\frac{D}{K}\right)^{\frac{1}{4}}$$

FIG. 9.8. Influence chart for the deflection at point 0, due to a load near the edge and assuming a dense liquid subgrade. Deflection at 0·5 *l* from the edge (shaded blocks count only as fractions). (From [10].)

For the assumption of a dense-liquid subgrade:

$$l = \sqrt[4]{\frac{E_c h^3}{12(1 - v^2)K}} = \sqrt[4]{\frac{D}{K}} \qquad (9.22)$$

For the assumption of a solid subgrade:

$$l = \sqrt[3]{\frac{2D}{C}} \qquad (9.23)$$

where $D = \dfrac{E_c h^3}{12(1 - v^2)}$ is the flexural rigidity of the slab;

h = slab thickness;

E_c = modulus of elasticity of the concrete;

v = Poisson's ratio of the concrete and is assumed to be 0.15 in all computations;

K = modulus of subgrade reaction, for the case of dense liquid; and

$C = \dfrac{E}{1 - v^2}$ is the rigidity of assumed solid subgrade, with E and v as elastic constants of an elastic solid.

REFERENCES AND BIBLIOGRAPHY

1. Burmister, D. M., 'The Theory of Stresses and Displacements in Layered Systems and Applications to the Design of Airport Runways', *Proc. Highway Research Board*, **23**, 1943.
2. Cheung, Y. K. and Nag, D. K., 'Plates and Beams on Elastic Foundations—Linear and Non-Linear Behaviour', The Institution of Civil Engineers, June 1968.
3. Cheung, Y. K. and Zienkiewicz, O. C., 'Plates and Tanks on Elastic Foundations—An Application of Finite Element Method', *International Journal of Solids and Structures*, **1**, 1965.
4. Ghali, A. and Neville, A. M., *Structural Analysis—A Unified Classical and Matrix Approach*, Intext Educational Publishers, 1972.
5. Hudson, W. R., 'Discontinuous Orthotropic Plates and Pavement Slabs', Ph.D. Dissertation, University of Texas, Austin, 1965.
6. McCullough, B. F. and Boedecker, K. J., 'Use of Linear-Elastic Layered Theory for the Design of CRCP Overlays', *Highway Research Record*, No. 291, 1969.
7. Peutz, M. G. F., Jones, A. and Van Kempen, H. P. M., 'Layered Systems Under Normal Surface Loads', *Highway Research Record*, No. 228, 1968.
8. Pickett, G., 'A Study of Stresses in the Corner Region of Concrete Pavement Slabs Under Large Corner Loads', Portland Cement Association, Chicago, Illinois, 1951.
9. Pickett, G. and McCormick, F. J., 'Circular and Rectangular Plates Under Lateral Load and Supported by an Elastic Solid Foundation', *First National Congress of Applied Mechanics, Chicago, ASME, 1951*.
10. Pickett, G. and Ray, G. K., 'Influence Charts for Concrete Pavements', *Trans. ASCE*, **116**, 1951, 49–72.

11. Wang, S. K., Sargious, M. A. and Cheung, Y. K., 'Advanced Analysis of Rigid Pavements', *Transportation Engineering Journal, ASCE*, February 1972.
12. Westergaard, H. M., 'Theory of Concrete Pavement Design', *Proc. Highway Research Board*, 1927.
13. Westergaard, H. M., 'New Formulae for Stresses in Concrete Pavements of Airfields', *Trans. ASCE*, 1948.

CHAPTER 10

DESIGN AND CONSTRUCTION OF PLAIN CONCRETE PAVEMENTS

10.1 INTRODUCTION

It was mentioned in the last chapter that since the modulus of elasticity of a concrete pavement is much higher than that of the subgrade or sub-base material, a major part of the load-carrying capacity is derived from the beam action of the slab.

The stresses produced in rigid pavements can be related to the following broad categories:

1. Curling stresses due to temperature differential through the thickness of the slab; and warping stresses due to difference in moisture.
2. Frictional stresses due to uniform temperature variations.
3. Infiltration stresses resulting from the filtering down of foreign matter into the joint, or from pumping action where the subgrade material is forced up into the joint by the jetting action of pumping water.
4. Stresses due to externally applied loads.

10.2 CURLING STRESSES

When a pavement slab is subjected to a difference in temperature between the top and bottom surfaces, it tends to curl. For example, the tendency of a slab to curl upward, due to an increase in the top surface temperature of the slab, will be restrained by its own weight. Thus, tensile stresses will be produced at the centre of the slab.

The value of the temperature differential depends upon the latitude of the location, the time of day, the season of the year and the colour of the slab. The maximum temperature differential is 3°F per inch thickness of slab.

The curling stresses at the edge or interior of the slab can be calculated from eqns. (10.1) and (10.2), respectively,

$$\text{edge stresses} \quad f_{\Delta t} = \frac{\alpha_c E_c \Delta t}{2} \cdot C \tag{10.1}$$

$$\text{interior stresses} \quad f_{\Delta t} = \frac{\alpha_c E_c \Delta t}{2(1 - v_c^2)} (C_1 + v_c C_2) \tag{10.2}$$

where α_c = coefficient of thermal expansion of the concrete $\cong 5 \times 10^{-6}$ in per in per degree Fahrenheit;

E_c = modulus of elasticity of the concrete (psi);

v_c = Poisson's ratio of the concrete;

L_x and L_y = free length and width of slab, respectively (ft);

C_1 = coefficient for the direction in which the curling stress is required; this can be obtained from Fig. 10.1;

C_2 = coefficient for the perpendicular direction; this can be obtained from Fig. 10.1; and

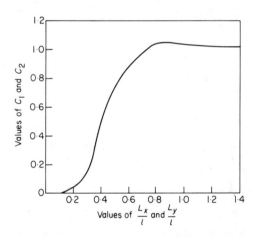

FIG. 10.1. Coefficient for curling stresses. The radius of relative stiffness, $l = \sqrt[4]{\{E_c h^3 / [12(1 - v^2) K]\}}$. (From [2].)

Δt = difference in temperature between top and bottom surfaces of the slab (°F)

= $3 \times$ slab thickness in inches.

Example

Calculate the curling stresses in a 9 in concrete pavement with 20 ft transverse joints; the lane width is 12 ft.

The modulus of elasticity of the concrete used is 4×10^6 psi and modulus of subgrade reaction, K, is 200 pci.

Solution. $l = 33 \cdot 39$ in $\Delta t = 3 \times 9 = 27°F$

$$\frac{L_x}{l} = \frac{20}{33 \cdot 39} = 0 \cdot 6 \quad \frac{L_y}{l} = \frac{12}{33 \cdot 39} = 0 \cdot 36$$

Longitudinal edge stress $= \dfrac{5 \times 10^{-6} \times 4 \times 10^6 \times 27}{2} \, 0 \cdot 89 = 240$ psi

Transverse edge stress $= \dfrac{5 \times 10^{-6} \times 4 \times 10^6 \times 27}{2} \times 0 \cdot 4 = 108$ psi

Longitudinal interior stress $= \dfrac{5 \times 10^{-6} \times 4 \times 10^6 \times 27}{2(1 - 0 \cdot 15^2)} \times (0 \cdot 89 + 0 \cdot 15 \times 0 \cdot 4)$

$= 276 \times 0 \cdot 95 = 263$ psi

Transverse interior stress $= \dfrac{5 \times 10^{-6} \times 4 \times 10^6 \times 27}{2(1 - 0 \cdot 15^2)} (0 \cdot 40 + 0 \cdot 15 \times 0 \cdot 89)$

$= 276 \times 0 \cdot 533 = 147$ psi

For longer slabs the curling stresses are higher.

10.3 FRICTION STRESSES

Uniform temperature variations cause contraction or expansion of the slab. This movement is resisted by friction between the slab and the subgrade. The minimum amount of displacement required for developing full friction is 0·06 in. Accordingly, the maximum value of coefficient of friction varies with the magnitude of slab displacement. For short slabs (less than 60 ft long), frictional stresses are not critical, and are generally neglected in the calculation. However, eqn. (10.3) can be used in calculating the stresses due to friction:

$$f_F = \frac{F\gamma L}{2 \times 144} \tag{10.3}$$

where f_F = stress due to friction in the concrete slab (psi);

γ = volume weight of the concrete used (lb/ft^3);

F = average coefficient of friction between the concrete slab and the subgrade; it is generally equal to 1·5; and

L = length of slab between transverse joints (ft).

10.4 INFILTRATION STRESSES

When foreign material, such as sand used in ice control or the like, infiltrates down into a joint during contraction, spalling may be caused at the bottom or top of the slab, when an expansion cycle starts. Infiltration can also be caused by pumping action where the subgrade material is forced up into the joint from below, by the jetting water that accumulates in a void space under the pavement. At such a location a rigid pavement will deflect, under a heavy load, but will rebound sufficiently after the removal of the load. With an increasing number of load repetitions, slab deflection will increase, the soil will go into suspension with the water, and muddy water will be ejected through the joints.

In addition to the infiltration of subgrade material into the joints, if pumping action continues a relatively large void space may be created under the slab and faulting may take place. Afterwards, a transverse crack may develop in the slab ahead of traffic with additional faulting resulting.

Also, if the subgrade contains granular dense-graded material, then in locations where voids created by repeated loading are filled with water, this material may be ejected by the deflecting slab. This type of pumping is sometimes referred to as blowing. If blowing occurs when the slab is contracted, coarse material from the base will enter the joint and may clog it sufficiently to cause a joint spall in the slab towards the traffic.

10.5 LOAD STRESSES

Wheel loads produce high bending stresses near the corners, along the edges, and to a lesser extent, at the centre of the slab. If sufficient load transfer exists across the joints, through aggregate interlock, dowel bars, or the like, the edge and corner stresses will be reduced to a large extent.

The design of rigid pavements for highways and airfields are presented separately.

10.6 CONCRETE PAVEMENTS FOR HIGHWAYS

The lane width for highways may range between 9 and 12 ft, with the 12 ft lane used in most modern highways.

Transverse cracks will develop in such pavements unless transverse joints are provided at adequate separations. The spacing of contraction cracks, L, in feet, may be predicted from the following formula:

$$L = \frac{2f_t' \times 144}{F \times \gamma} \simeq \frac{2f_t'}{F} \tag{10.4}$$

where f_t' is the concrete tensile strength in psi at a short time after casting, F is the coefficient of subgrade friction and γ is the volume weight of the concrete, assumed to be 144 lb/ft^3 in eqn. (10.4).

To avoid intermediate transverse cracks, it is recommended that transverse joints be spaced every 20 ft under average conditions. In special situations, this spacing can go down to 15 ft or up to 25 ft.

Where load transfer at the transverse joints is to depend wholly or partly upon aggregate interlock, expansion joints should be eliminated and all transverse joints constructed as contraction joints.

If dowel bars are to be used at the joints, then expansion joints can be provided at large distances apart. This distance may vary between 400 ft and 2000 ft.

10.6.1 Slab Thickness Design

Figure 10.2 shows a normal traffic distribution across pavements with 12 ft lanes and three possible design load positions (Cases I, II and III). Case I shows single- and tandem-axle loads at the transverse joint edge. Maximum flexural stresses occur at the bottom of the slab and are parallel to the joint edge. The axle loads are positioned at the point of greatest load repetition, as shown in the frequency distribution diagram of traffic on the right-hand side of Fig. 10.2.

Case II is for single and tandem axles at the outside pavement edge. Axles are normal to the pavement edge. For Case II, maximum flexural stresses are also at the bottom of the slab and are parallel to the outside edge. These stresses are somewhat higher than at the Case I position, especially for single-axle loads. However, as shown by the traffic frequency diagram in Fig. 10.2, load applications at the Case II position are rare.

Case III is the same as Case II, except that loads are 6 in inward from

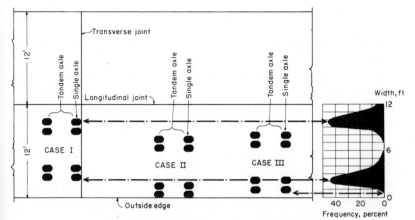

FIG. 10.2. Load positions and traffic distribution.

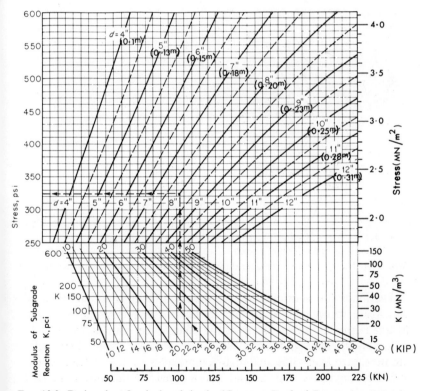

FIG. 10.3. Design chart for single-axle loads. *(Courtesy: Portland Cement Association.)*

the outside pavement edge. Edge stresses for the Case III position are less than those for either Case I or Case II.

Accordingly, design charts are developed using Westergaard's equation for load acting at a transverse joint.

The design chart shown in Fig. 10.3 is for the case of single-axle loads; that shown in Fig. 10.4 is for dual-tandem-axle loads [11].

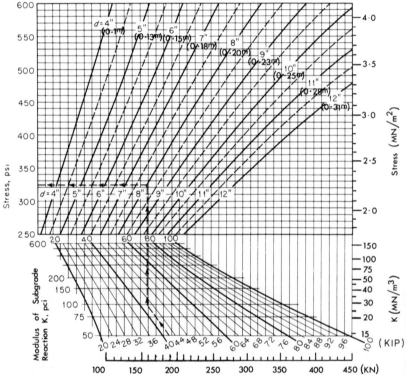

FIG. 10.4. Design chart for tandem-axle loads. *(Courtesy: Portland Cement Association.)*

The loads are multiplied by the following load safety factors (LSF), which are recommended by the Portland Cement Association. These factors represent, in fact, the dynamic effect of the vehicles.

1. For freeways and other multi-lane projects where there will be uninterrupted traffic flow and high volumes of truck traffic, LSF = 1·2.

2. For highways and arterial streets where there will be moderate volumes of truck traffic, LSF = 1·1.

TABLE 10.1 STRESS RATIOS AND ALLOWABLE
LOAD REPETITIONS
(Courtesy: Portland Cement Association)

Stress[a] ratio	Allowable repetition	Stress[b] ratio	Allowable repetition
0·51[b]	400 000	0·69	2 500
0·52	300 000	0·70	2 000
0·53	240 000	0·71	1 500
0·54	180 000	0·72	1 100
0·55	130 000	0·73	850
0·56	100 000	0·74	650
0·57	75 000	0·75	490
0·58	57 000	0·76	360
0·59	42 000	0·77	270
0·60	32 000	0·78	210
0·61	24 000	0·79	160
0·62	18 000	0·80	120
0·63	14 000	0·81	90
0·64	11 000	0·82	70
0·65	8 000	0·83	50
0·66	6 000	0·84	40
0·67	4 500	0·85	30
0·68	3 500		

[a] Load stress divided by modulus of rupture.
[b] Unlimited repetitions for stress ratios of 0·50 or less.

3. For highways, residential streets, and other streets that will carry small volumes of truck traffic, $LSF = 1·00$.

The effect of traffic volume on thickness design is considered in choosing the stress ratio corresponding to the load repetitions, as shown in Table 10.1.

10.6.2 Use of Charts for Thickness Design

The procedure for use of the design charts for pavement thickness design is exemplified in Fig. 10.5. From highway department traffic counts and loadometer data, the present average daily number of repetitions of a given single- or tandem-axle load may be determined. The expected number of such repetitions during the service life of the pavement is then readily estimated with the use of a projection factor which reflects probable increase in traffic volume for the particular route in question. With present-day design and construction practices, the service life may be taken as 40 years. The expected numbers of repetitions of particular axle loads are tabulated, as in column 6 of Fig. 10.5. Use of the design charts and Table 10.1 will then give the percentage of available fatigue resistance

CALCULATION OF CONCRETE PAVEMENT THICKNESS
(Use with Case I Single & Tandem Axle Design Charts)
Project _DESIGN ONE - A_
Type _Rural Interstate - Rolling Terrain_ No. of Lanes _4_
Subgrade k _100_ pci., Subbase _6-in. Granular Untreated_
Combined k _130_ pci., Load Safety Factor _1.2_ (L.S.F.)

PROCEDURE

1. Fill in Col. 1, 2 and 6, listing axle loads in decreasing order.
2. Assume 1st trial depth. Use 1/2-in. increments.
3. Analyze 1st trial depth by completing columns 3, 4, 5 and 7.
4. Analyze other trial depths, varying M.R.[*], slab depth and subbase type.[**]

1	2	3	4	5	6	7
Axle Loads	Axle Loads X/.2 L.S.F.	Stress	Stress Ratios	Allowable Repetitions	Expected Repetitions	Fatigue Resistance Used[***]
kips	kips	psi		No.	No.	percent

Trial depth _9.0_ in.   M.R.[*] _650_ psi   k _130_ pci

SINGLE AXLES

1	2	3	4	5	6	7
30	36.0	340	.52	300,000	3100	1
28	33.6	325	.50	Unlimited	3100	0
26	31.2	<.50	"	6200	0	
24	28.8	"	"	163,200	0	
22	26.4	"	"	639,740	0	

TANDEM AXLES

1	2	3	4	5	6	7
54	64.8	382	.59	42,000	3100	7
52	62.4	368	.57	75,000	3100	4
50	60.0	358	.55	130,000	30,360	23
48	57.6	348	.54	180,000	30,360	17
46	55.2	333	.51	400,000	48,140	12
44	52.8	318	<.50	Unlimited	150,470	0
42	50.4	"	"	171,360	0	
40	48.0	"	"	248,060	0	

Total = 64

[*]   M.R. Modulus of Rupture for 3rd pt. loading.

[**]   Cement-treated subbases result in greatly increased combined k values.

[***]   Total fatigue resistance used should not exceed about 125 percent.

FIG. 10.5. Design thickness calculation. *(Courtesy: Portland Cement Association.)*

used by each category of axle load. This is recorded in column 7. The total of column 7 is the total fatigue resistance used by the anticipated traffic in 40 years, using a 9 in pavement thickness. The procedure is then repeated using a lesser trial thickness—say 8 in, etc.—until a design thickness is determined which will utilise 100–125% of the available fatigue resistance.

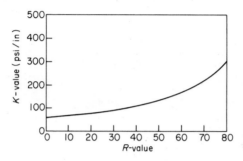

FIG. 10.6. *K* value vs. *R* value. (From [3].)

The value of *K* to be used in the design can be obtained from field experiments or from relations with other soil strength measurements such as the *R* value; this relationship is shown in Fig. 10.6.

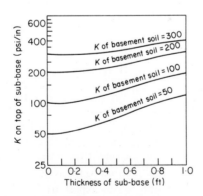

FIG. 10.7. Effect of various thicknesses of granular sub-base on *K* values. (From [3].)

Also, if a sub-base or cement treated base is used, the value of *K* at the top of these layers can be determined from Figs. 10.7 and 10.8, once the value of *K* for the subgrade is known.

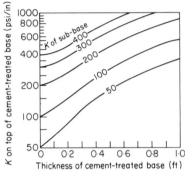

FIG. 10.8. Effect of various thicknesses of cement treated bases on K values. (From [3].)

10.6.3 Design for Case of Corner Loading

If lanes of less than 12 ft width are used, and it is expected that a large number of trucks will run over the slab corners with their wheels, then the Pickett formulae should be used:

$$\text{protected corners} \qquad f_t = \frac{3\cdot36\,P_i}{h^2}\left[1-\frac{\sqrt{a/l}}{0\cdot925+0\cdot22\dfrac{a}{l}}\right] \qquad (10.5)$$

$$\text{unprotected corners} \quad f_t = \frac{4.2\,P_i}{h^2}\left[1-\frac{\sqrt{a/l}}{0\cdot925+0\cdot22\dfrac{a}{l}}\right] \qquad (10.6)$$

where f_t = maximum tensile stress at top of slab and in the direction of the bisector of the corner angle (psi);

P_i = wheel load (including impact LSF) (lb); in the case of axles with dual wheels on each side, P_i is the load on the dual wheels including impact (lb);

h = thickness of slab (in);

a = radius of tyre imprint for a single-wheel load (in); in the case of dual wheels, this is the radius of a circle having an area equal to the tyre imprint area of one wheel plus the area of a rectangle of length equal to the centre-to-centre spacing of the dual wheels and width equal to the diameter of the tyre imprint area for one wheel.

The mixed traffic using the highway during the life period of the pavement should first be converted to an equivalent number of repetitions of a standard single-axle load. This number of load repetitions will determine the stress ratio to be used in the design, from Table 10.1.

The load on one side of the standard single axle should then be multiplied by the load safety factor (LSF) to determine the value of P_i to be used in the design.

Figures 10.9 and 10.10 can be used to determine the required thickness for the case of protected and unprotected corners, respectively, for concrete having an elastic modulus E_c equal to 5×10^6 psi.

After determining the slab thickness to satisfy the requirement of carrying the loads safely, it is recommended that the load stress plus the curling stress be checked, using eqn. (10.7):

$$f_1 + f_{\Delta t} < MR \qquad (10.7)$$

where f_1 = tensile stress due to load (psi);

$\quad f_{\Delta t}$ = tensile stress due to temperature differential in the concrete slab (psi); and

$\quad MR$ = modulus of rupture of the concrete.

This check is useful in non-reinforced concrete pavements to avoid the development of cracks in the pavement at the locations of maximum stress.

10.6.4 Typical Pavement Thicknesses in Europe

Table F.1 in Appendix F shows the thicknesses of plain concrete pavements and sub-bases which are commonly used in road construction in the different European countries.

10.7 CONCRETE PAVEMENTS FOR AIRFIELDS

The width of lane for airfield pavements can vary between 10 and 25 ft according to the width of taxiways, runways and terminal aprons, and according to the equipment available for concreting. If the longitudinal construction joints defining the lanes are spaced 25 ft apart and the pavement thickness is less than 10 in, an intermediate longitudinal joint of the dummy type should be placed.

So as to prevent transverse cracks, transverse joints spaced 20 ft apart are recommended under normal conditions for non-reinforced concrete pavements. In special situations the spacing may range between 15 and 25 ft.

Where load transfer at the joints depends wholly or partially upon aggregate interlock, expansion joints should be eliminated. If dowel bars are used in transferring the load, then expansion joints are placed every 1000 to 2000 ft apart.

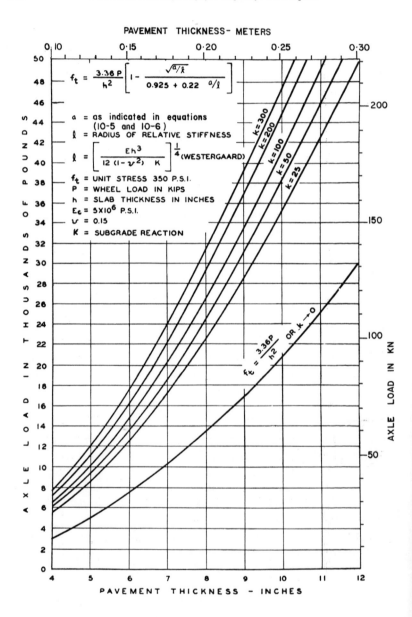

FIG. 10.9. Design chart based on eqn. (10.5) for Portland Cement Concrete Pavements: protected corners. (From [1].)

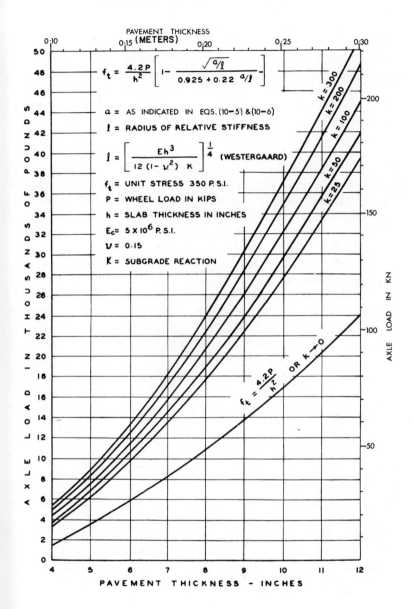

FIG. 10.10. Design chart based on eqn. (10.6) for Portland Cement Concrete Pavements: unprotected corners. (From [1].)

10.7.1 Slab Thickness Design

Analysis of the structural design of airport pavements consists in part of determining both the thickness of the pavement itself and the thicknesses of the component parts of the pavement structure.

There are a number of factors in addition to the quality of the pavement foundation which affect the ability of a pavement to provide satisfactory service. These include: (1) the magnitude and character of the aircraft loads to be carried, (2) the volume of traffic, (3) the concentration of traffic in certain areas and (4) the quality of materials used in construction.

It is general practice to design a pavement according to capacity operations of those aircraft which it is anticipated will use the airport regularly. However, for less than capacity operation, one may use a smaller thickness, equal to or greater than 0·8 times the design thickness for the critical areas. Two methods of thickness design for airfield pavements will be explained hereafter.

10.7.2 The US Federal Aviation Method

In this method [4], the gross aircraft weight for aircraft equipped with single-wheel, dual-wheel, or dual-tandem-wheel undercarriages is used. The conversion from stating pavement strength in terms of equivalent single-wheel load to terms of gross aircraft weight was made to avoid misinterpretation of the equivalent single-wheel load method of design. To convert from equivalent single-wheel load design to that based on gross aircraft weight, it was necessary to make some reasonable compromise on the effects of gear dimensions of existing civil aircraft. The 'Influence Charts for Concrete Pavements', developed by Pickett and Ray [10], were used in developing the thickness design charts based on aircraft gross weight. The following parameters were used in this work:

K = 300 pci
f_t = allowable concrete tensile stress = 400 psi (modulus of rupture should be ≥ 700 psi)
E_c = 4×10^6 psi
v = 0·15
Dual spacing = 20 and 30 in
Dual-tandem spacing = 20×45 and 30×55 in

A design curve for the dual-gear aircraft was prepared using the 20 in and 30 in spacings, respectively. This compromise curve was drawn considering the narrower spacing as favoured for lighter aircraft and the wider spacing for heavier aircraft. The same procedure was used for dual-tandem undercarriages. The curve for single-wheel-gear aircraft was developed from Pickett's influence charts; all the curves are presented in Fig. 10.11.

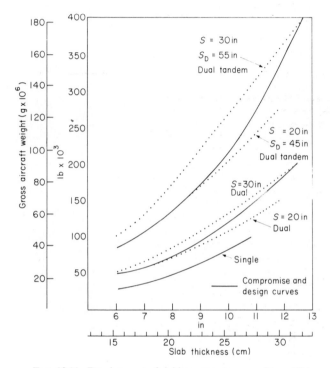

FIG. 10.11. Development of rigid pavement curves. (From [8].)

The rigid-pavement design curves presented in Fig. 10.12 can be used to determine the thickness required for the concrete slab as well as for the sub-base, once the classification of subgrade soil, the gross weight of the aircraft and the type of undercarriage assembly are known. The same design curves of Fig. 10.12 can be applied in designing the non-critical areas of airfield rigid pavements, by taking 80–90% of the pavement design thickness required for the critical areas together with the full required sub-base thickness. Traffic type and relative location of the non-critical area are the governing factors in using either 80 or 90% of the pavement thickness required for the critical areas.

It should be noted that the use of the curve of Fig. 10.12 is limited to the values of the parameters mentioned before. Also, load transfer through aggregate interlock or through dowel bars at the joints is assumed to exist.

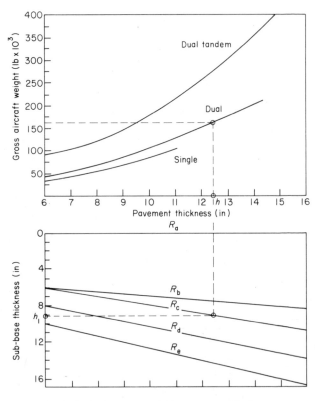

FIG. 10.12. Design curves: rigid pavement critical area.
(*Courtesy: the Federal Aviation Agency.*)

Example
The critical aircraft is of a dual-gear type, of 160 000 lb gross weight. The subgrade soil group is E-7 and poor drainage and severe frost conditions exist. Determine the thickness of the concrete pavement and sub-base for both critical and non-critical areas.

Solution. Subgrade classification is R_C.

	Critical area	Non-critical area
Thickness of concrete pavement	12·4 in, take 12·5 in	11·2 in, take 11·5 in
Thickness of sub-base	9·0 in	9·0 in

If the parameters are different from those used in plotting Fig. 10.12, the charts shown in Figs. 10.13, 10.14 and 10.15 can be used to determine

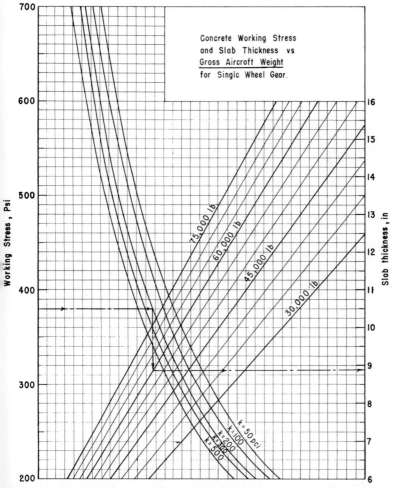

FIG. 10.13. Design curves: rigid pavements subjected to single-wheel gear loads.
(Courtesy: the Federal Aviation Agency.)

he pavement thickness for the critical area. The working stress is com-
ɔuted by dividing the concrete flexural strength by 1·75. The pavement
hickness for the non-critical areas is as mentioned before, 80–90% of
he thickness required for the critical areas.

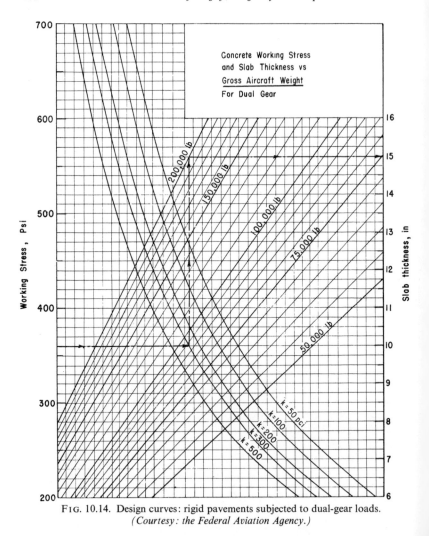

FIG. 10.14. Design curves: rigid pavements subjected to dual-gear loads.
(Courtesy: the Federal Aviation Agency.)

10.7.3 The United Kingdom Load Classification Number (LCN) Method

The load classification number system used in designing flexible pavements is also used in designing the rigid type.

The equivalent single-wheel load concept is used to convert the load on a dual- or dual-tandem-wheel assembly to an equivalent single-wheel load. For this purpose, Fig. 2.7 or Fig. 2.8, developed from the central loading condition, can be used to determine ESWL. The load classifica-

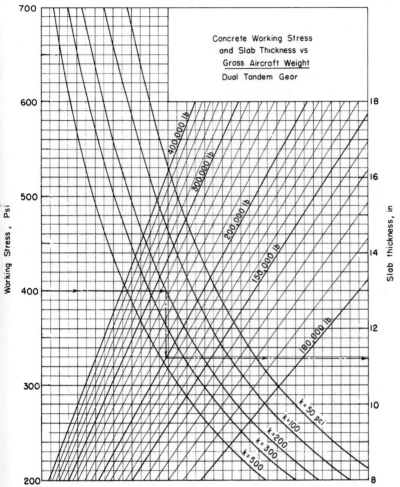

FIG. 10.15. Design curves: rigid pavements subjected to dual-tandem-gear loads.
(Courtesy: the Federal Aviation Agency.)

tion number corresponding to the value of the E S W L can be determined from Fig. 7.8, since the value of the tyre pressure under the E S W L is known.

In the case of non-reinforced, undowelled concrete pavement slabs (separate slabs), the weakest point is the corner. This method of design is based upon Teller and Sutherland's empirical modification of Westergaard's corner case expression. Teller and Sutherland's expression is:

$$f_\text{t} = \frac{3P\left[1 - \left(\dfrac{a\sqrt{2}}{l}\right)^{1\cdot2}\right]}{h^2} \tag{10.8}$$

where P = applied load (lb);
$\quad\ \ h$ = slab thickness (in);
$\quad\ \ l$ = radius of relative stiffness (in);
$\quad\ \ a$ = radius of contact area (in); and
$\quad\ \ f_\text{t}$ = flexural stress at slab corner (psi).

In eqn. (10.8) there is no direct relationship, independent of other variables, between P and a. Hence, in order to introduce the LCN system into rigid-pavement design, all calculations are made for $a = 6$ in. The calculation of thickness of pavement required for a particular LCN is made by determining what wheel load, acting on a circular contact area of radius 6 in, is equivalent to the LCN in question, and then using Teller and Sutherland's expression with these two values inserted. The value of $a = 6$ in was selected because designs based upon this assumption agreed closely with the United Kingdom results obtained from plate-bearing tests on concrete slabs of varying thicknesses, on a variety of subgrades.

To simplify design procedure, the chart shown in Fig. 10.16 has been prepared. This chart was drawn in the following manner:

1. For each of the LCNs shown on the chart, the wheel load associated with a circular contact area of radius 6 in was calculated.
2. Assuming $K = 50$ lb/in^2/in
   $$v = 0\cdot15$$
   $$E = 5 \times 10^6 \text{ lb/in}^2$$
   $$a = 6 \text{ in}$$
   the flexural stresses associated with pavements of thicknesses from 8 to 24 in were calculated for the wheel load associated with each LCN. A straight plot of flexural stress against pavement thickness for each LCN resulted in the curves shown.
3. The same calculations were made for
   $$K = 300 \text{ lb/in}^2/\text{in}$$
   $$\text{and } K = 1\,000 \text{ lb/in}^2/\text{in}$$
   and it was found that curves of almost identical shape resulted. It was therefore possible to incorporate the three design charts into one diagram by providing alternative scales of pavement thicknesses for the alternative values of K.
4. The Equivalent K Line for construction with a base consisting of 4 in of compacted lean concrete, is entirely empirical. It has been

derived from the results of plate-bearing tests carried out on experimental slabs of various thicknesses of concrete; these were laid on a layer of lean concrete over a natural clay formation.

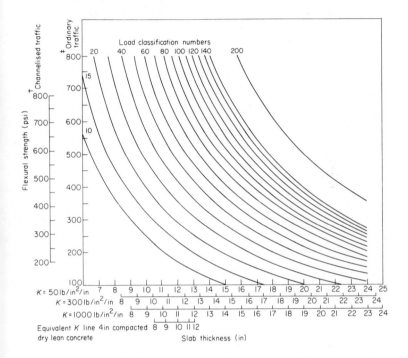

FIG. 10.16. Design chart for rigid pavements: single-slab concrete. Flexural strength on this scale is equal to 0·90 of the flexural strength obtained for the concrete from laboratory tests. († Ordinary traffic use is related to approximately 10 000 repetitions of load during the pavement life. ‡ Channelised traffic use is related to approximately 40 000 repetitions of load during the pavement life.) *(Courtesy: ICAO.)*

Notes on use of design chart. Air Ministry General Specification calls for a minimum concrete strength of 350 psi at 28 days. This minimum will rise to about 470 psi in 6 months and about 520 psi in 1 year. On the assumption that the pavement will be at least a few months old before first being brought into use, the flexural strength of 450 psi is used in all Air Ministry designs in use (this strength is theoretically produced at an age of 120 days). The ageing of the concrete will produce an increase of flexural strength and will introduce a factor of safety into the pavements which after some years will approach 1·5, the flexural strength of the

concrete having reached about 675 psi. During the pavement life, repeti-
tional loading will take place, but providing the loads only create stresses
less than the flexural strength of the concrete at the time, they will not stop
the build-up of the concrete strength with age. With a factor of safety of
1·5, a pavement will be able to withstand about 10 000 repetitions of load

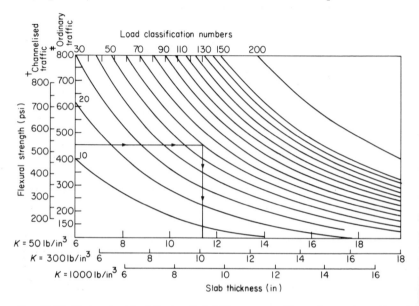

FIG. 10.17. Design chart for fully dowelled rigid pavements. Flexural strength on this scale
equals 0·80 of the flexural strength obtained for the concrete from laboratory tests.
(† Ordinary traffic use is reduced to approximately 10 000 repetitions of load during the
pavement life. ‡ Channelised traffic use is related to approximately 40 000 repetitions of
load during pavement life.) *(Courtesy: ICAO.)*

during its life, and designs for normal traffic usage therefore are based on
this number of load repetitions. If a pavement life is required to be such
that the number of load repetitions to be carried is increased from 10 000
to 40 000 the flexural strength used in the original design will have to
include a factor of safety higher than 1·5. Based on evidence provided by
the US Cement and Concrete Association, the factor of safety for 40 000
load repetitions will require to be about 1·8. On this basis the design flexural
strength to be used for 40 000 repetitions of load has been adjusted from
that required for 10 000 repetitions of load in the approximate ratio of
1·5 to 1·8. (The flexural strength of 450 psi used for 10 000 load repetitions
will require to be reduced to 385 psi for 40 000 repetitions of load.)

Although the use of non-reinforced undowelled concrete slabs is still standard practice, it is likely that dowelled joints may often be adopted in some countries. To simplify the interpretation of load plate testing of such pavements, Fig. 10.17 has been prepared. The chart has been drawn in the same way as that of Fig. 10.16, except that the calculations have been made from Westergaard's 'centre case' expression:

$$f_t = \frac{0 \cdot 275 \, P}{h^2} (1 + v) \log_{10} \frac{Eh^3}{Kb^4} \tag{10.9}$$

where f_t = flexural stress at the centre of the slab (lb/in^2);

$b = (\sqrt{1 \cdot 6 \, a^2 + h^2} - 0 \cdot 675 \, h)$ when a is less than $1 \cdot 724 \, h$; and

$b = a$ when a is greater than $1 \cdot 724 \, h$.

When using Fig. 10.17, it is important to make sure that appropriate load transfer devices will be used at all joints.

As in the case of highways, one should check that the stress due to load plus the curling stress due to temperature variations through the slab thickness does not exceed the modulus of rupture of the concrete used. This check is useful if cracks are to be avoided in plain concrete pavements at the locations of maximum stress.

In Figs. 10.16 and 10.17, it is recommended that values for the flexural strength on the vertical scale be used which are equal to 0·8–0·9 times those obtained for the concrete from laboratory tests. The factor 0·90 could be applied to Fig. 10.16, while the factor 0·80 is recommended for Fig. 10.17.

10.7.4 The Portland Cement Association Method

The method proposed in the 1973 edition of the Portland Cement Association 'Design of Concrete Airport Pavement' [12] is based on knowledge of the performance of existing pavements, full-scale loading tests and theoretical studies of pavement stresses and deflections.

In this method, it is recommended that the 90 day flexural strength of the concrete or 110% of 28 day strength be used, if 90 day test results are not available. This is justified by the fact that concrete strength increases with time, and the number of stress repetitions at any one spot by the full design load will be very small during the first few months after paving. Since variations in modulus of elasticity, E_c, and Poisson's ratio, v, of the concrete have slight effect on thickness design, the values used in this design procedure are $E = 4 \times 10^6$ psi and $v = 0 \cdot 15$.

The flexural stresses used in the design procedure are those at the interior of a slab. When the slab edges at all longitudinal and transverse joints are provided with adequate load transfer, the paved area acts as a continuous large slab and the interior loading case can be applied.

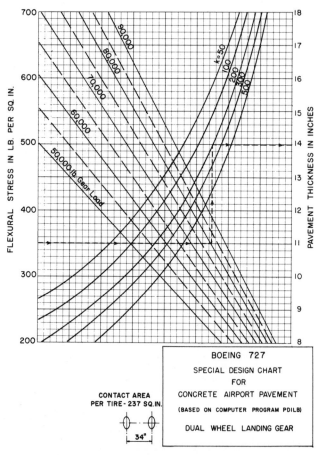

FIG. 10.18. Design chart for Boeing 727. *(Courtesy: Portland Cement Association.)*

At free edges or at edges where load transfer is doubtful, load stresses are somewhat greater than those for the interior load condition. For this reason the slab thickness at undowelled, butt joints—expansion joints at intersections of runways and taxiways—is increased to compensate for the absence of load transfer and thus keep load stresses at these slab edges within safe limits.

The outside edges of runways, taxiways or aprons do not require thickening, since aircraft wheels rarely, if ever, travel close to the outside edges. However, where future expansion of the pavement is anticipated, some means of load transfer is built into the slab edges or the edge is thickened to provide for loads at these edges.

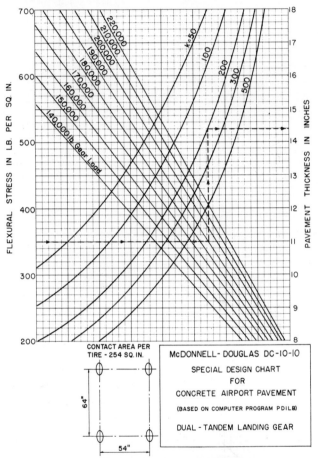

FIG. 10.19. Design chart for McDonnell-Douglas DC-10-10.
(Courtesy: Portland Cement Association.)

The design charts shown in Figs. 10.18, 10.19 and 10.20 are for Boeing 727, DC-10-10 and DC-8-62 (or 63) aircraft, respectively. Charts for other civil and military aircraft are available from the Portland Cement Association.

Careful attention is required for correct use of the charts. It is important to make sure that gear load, wheel spacings and tyre contact area shown in the applicable chart correspond to those of the aircraft for which the design of pavement is made. Load stress in these charts is based on gear

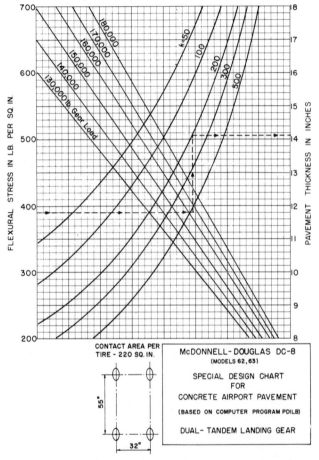

FIG. 10.20. Design chart for McDonnell-Douglas DC-8 (62, 63).
(Courtesy: Portland Cement Association.)

load rather than on gross weight of aircraft. For most aircraft, the gear load can be estimated from the gross aircraft weight with the assumption that 92–95% of the weight is on the main gears.

In Figs. 10.18, 10.19 and 10.20, a dashed-and-arrowed example line, representing the design loading reported by the aircraft manufacturers at the time of chart publication, is shown. Additional load lines, above and below the example, have also been included; those above represent future, heavier versions of the aircraft that may be developed; those below are for aircraft operated at less than the maximum design load, such as those flying into smaller airports. It is possible to interpolate between load lines or curves for K if intermediate values are used.

The safety factor, which is the ratio of modulus of rupture to working stress, used for airport pavement design depends on the expected frequency of traffic operations and their channelisation on runways, taxiways and aprons.

Estimating future traffic is undoubtedly one of the most important factors in airport pavement design and needs gathering of data from several sources on expected future operating and load conditions. When a specific forecast is made of the mixed aircraft that will operate during the design life of the pavement, fatigue methods [12] can be used for a more detailed assessment of traffic effects. If detailed information is not available, the ranges of safety factors given in Table 10.2 are recommended.

TABLE 10.2 SAFETY FACTORS FOR AIRFIELD PAVEMENTS

Installation	*Safety factor*
Critical areas: Apron, taxiways, runway ends for a distance of 1 000 ft, and hangar floors	1·7–2·0
Non-critical areas: Runways (central portion)	1·4–1·7

The allowable working stress used in the design charts is obtained by dividing the concrete modulus of rupture by the appropriate safety factor.

The lower safety factors for the central portion of runways can be permitted because most runway traffic consists of fast-moving loads that are partly airborne. In addition, the aircraft wheel loads are distributed transversely over a wide pavement area, so that the number of stress repetitions on any spot is small compared with that on taxiways.

Recent observations, however, have indicated some failures in the central portion of a few runways. If the runway pavement is rough, higher stresses can be expected in the concrete during the take-off operations. Both theoretical and experimental studies are being carried out by several organisations to investigate the effect of pavement roughness on concrete stresses during aircraft take-off operations.

On airfields with a large number of operations by planes having critical loads, safety factors near the top of the suggested range should be used. On airfields with only occasional operations by planes with critical loads, safety factors near the bottom of the range should be used.

Once the working stress for a specific aircraft is determined by dividing the modulus of rupture of the concrete by the safety factor corresponding to the volume of operations of this aircraft, the design chart for the specific aircraft can be used to determine the pavement thickness. The process should then be repeated for other aircraft of critical loads, using appropriate safety factors for the level of operations expected for these aircraft. The design thickness is selected for the most critical condition.

For heavy-duty runways serving large volumes of traffic, a keel-section design, where the centre section of the pavement is thicker than the outside pavement edges, can be used.

Example

A new runway and taxiway are to be designed to serve frequent operations of aircraft of which the B727 and DC-10 produce the critical loading conditions. In addition, the runway is expected to carry occasional operations of DC-8-63 aircraft.

The modulus of subgrade reaction, K, of the existing subgrade is 170 pci. A sub-base layer of 6 in thickness is used to prevent pumping of the subgrade and to increase K to 200 pci.

The 90 day modulus of rupture of the concrete used is expected to be 700 psi and it is required to determine the pavement thickness.

Solution. Table 10.2 is used to select appropriate values for the factor of safety as listed in columns 4 and 7 of Table 10.3. Working stresses are computed as shown in columns 5 and 8 of Table 10.3, and the required slab thicknesses are determined from Figs. 10.18, 10.19 and 10.20. These thicknesses are listed in columns 6 and 9 of Table 10.3.

Based on the values obtained, a slab thickness of 14·5 in is selected for the taxiway and runway ends, while 13·0 in is required for the central portion of the runway.

Pavement stresses induced by other less critical aircraft in the traffic forecast are next determined from the design charts for these specific aircraft. If stresses for these less critical aircraft are less than half the modulus of rupture, these aircraft will not cause fatigue in the pavements and the slab thicknesses established above are adequate.

TABLE 10.3 EXAMPLE CALCULATIONS FOR THICKNESS DESIGN
(Design K value: 200 pci. Design MR = 700 psi)

Aircraft	Gear load (lb)	Operations	Pavement facility						
			Taxiway and runway ends			Runway, central portion			
			Safety factor	Working stress (psi) (MR ÷ Column 4)	Slab thickness (in)	Safety factor	Working stress (psi) (MR ÷ Column 7)	Slab thickness (in)	
(1)	*(2)*	*(3)*	*(4)*	*(5)*	*(6)*	*(7)*	*(8)*	*(9)*	
B-727	80 000	Frequent	2·0	350	14·0	1·7	412	12·5	
DC-10	190 000	Frequent	2·0	350	14·4	1·7	412	12·7	
DC-8-63	165 000	Occasional	1·8	389	14·1	1·5	467	12·4	

Keel Section for Runways

For major airports serving large volumes of traffic, the central portion of runways may be considered a critical traffic area for which a high safety factor is appropriate.

A keel section design can produce substantial savings in these areas. A keel section means a thickened pavement in the centre portion of a runway tapered to thinner pavement at the outside runway edge, as shown in Fig. 10.21. The justification of the reduction in thickness for slabs near the outside runway edge is that very few, if any, aircraft travel close to the edge, especially on the wide runways (200 ft) often specified for major facilities.

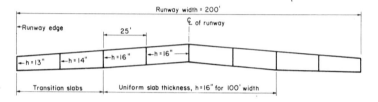

F1G. 10.21. Keel section design for runway.

The following safety factors can be used in thickness design of runways with keel section:

1. A high safety factor (usually 2·0 for runways with a large volume of traffic) for the centre portion of the runway, to obtain a thicker pavement down the middle of the runway, at least 75 ft wide. Uniform pavement thickness should be used for all slabs wholly or partly within this area (Fig. 10.21).

2. An intermediate safety factor (about 1·7) is used for the area outside the keel section to determine the lesser thickness (20–25% less than thickness of keel section).

3. The normal design procedure with low safety factor appropriate for the infrequent number of operations is used to determine minimum thickness for the outer slabs of the pavement.

Thicknesses of the transition slabs are selected to avoid any abrupt change in grade and slab thickness and to meet minimum grade requirements on the subgrade.

Extra strengthening at longitudinal joints of keel sections is not usually required, since the thickness of these sections will be greater than that required by the normal procedure.

REFERENCES AND BIBLIOGRAPHY

1. ACI Committee 325, 'Recommended Practice for Design of Concrete Pavements', ACI Manual of Concrete Practice, Part 1, 1968.
2. Bradbury, R. D., 'Design of Joints in Concrete Pavements', *Proc. Highway Research Board*, 1932.
3. Estep, A. C. and Wagner, P. I., 'A Thickness Design Method for Concrete Pavements', *Highway Research Record*, No. 239, 1968.
4. Federal Aviation Administration, 'Airport Paving', Advisory Circular, AC 150/5320-6A, USA Department of Transportation, Reprinted September 1971.
5. Fordyce, P. and Yrjanson, W. A., 'Modern Design of Concrete Pavements', *ASCE Transportation Journal*, August 1969.
6. Friberg, B. F., 'Frictional Resistance under Concrete Pavements and Restraint Stresses in Long Reinforced Slabs', *Proc. Highway Research Board*, 1954.
7. Hennes, R. G. and Ekse, M., *Fundamentals of Transportation Engineering*, Second Edition, McGraw-Hill, 1969.
8. International Civil Aviation Organisation, 'Aerodrome Physical Characteristics —Part 2', *Aerodrome Manual*, Second Edition, 1965.
9. Pickett, G., 'A Study of Stresses in the Corner Region of Concrete Pavement Slabs under Large Corner Loads', Portland Cement Association, Chicago, Illinois, 1951.
10. Pickett, G. and Ray, G. K., 'Influence Charts for Rigid Pavements', *Trans. ASCE*, 1951.
11. Portland Cement Association, 'Thickness Design for Concrete Pavements', Illinois, 1966.
12. Portland Cement Association, 'Design of Concrete Airport Pavement', Bulletin prepared by R. G. Packard, E B050.03P, Illinois, 1973.
13. Wang, S. K., Sargious, M. A. and Cheung, Y. K., 'Effect of Openings on Stresses in Rigid Pavements', *ASCE Transportation Journal*, May 1973.
14. Westergaard, H. M., 'Analysis of Stresses in Concrete Pavements due to Variations of Temperature', *Proc. Highway Research Board*, 1926.
15. Westergaard, H. M., 'Theory of Concrete Pavement Design', *Proc. Highway Research Board*, 1927.
16. Westergaard, H. M., 'Stresses in Concrete Runways of Airports', *Proc. Highway Research Board*, 1939.
17. Westergaard, H. M., 'New Formulae for Stresses in Concrete Pavements of Airfields', *Trans. ASCE*, 1948.
18. Yoder, E. J., *Principles of Pavement Design*, Wiley, 1959.

CHAPTER 11

PAVEMENT JOINTS, DOWEL BARS AND REINFORCING STEEL

11.1 INTRODUCTION

Uniform temperature changes cause shortening or lengthening of rigid slabs. If the slab is cooled, it will tend to contract, and tensile stresses will result. This may cause the slab to crack at its midpoint. Stresses due to volume changes in the concrete result only when the volume changes are restrained by subgrade friction.

Stresses due to warping can be exceedingly high and may cause transverse or longitudinal cracking of the concrete slab. The restraining force in this case is the weight of the concrete.

If a rigid-pavement slab is restrained against movement, cracking or 'blowup' may result from the induced stresses. One method of counteracting this is to use transverse and longitudinal joints, spaced in such a manner as to relieve the stresses. Also, reinforcing steel can be used to minimise the detrimental effect of cracking.

11.2 TYPES OF JOINTS

Joints for rigid pavements can be divided into the following four basic groups: (1) contraction joints, (2) expansion joints, (3) construction joints and (4) hinge or warping joints. Each type of joint mentioned above has a distinct function and is used only at certain sections of the pavement. Figure 11.1 shows typical details of the joints generally used in conventional concrete pavements.

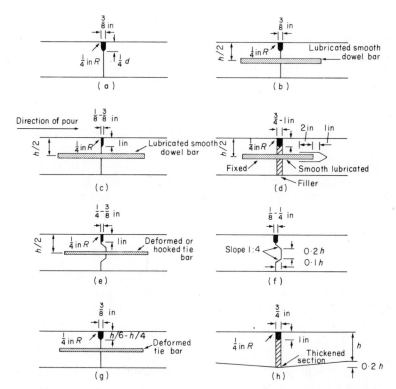

FIG. 11.1. Typical types of joints. (a) Dummy-groove contraction. (b) Dummy-groove contraction, dowelled. (c) Butt construction. (d) Expansion. (e) Keyed longitudinal tied. (f) Keyed hinge or warping. (g) Tied longitudinal warping. (h) Longitudinal expansion.

11.2.1 Contraction Joints

Figure 11.1(a) shows a conventional dummy-groove contraction joint. For this joint, a groove is cut or formed at the pavement surface to make certain that cracking will occur at this location. Load transfer at this joint is accomplished by grain interlock of the cracked lower portion of the slab—but only if the joints are kept quite tight by omitting expansion joints and using contraction joint spacing not exceeding 20–25 ft.

Contraction joints are transverse joints intended to relieve only tensile stresses resulting from contraction and warping of the concrete. Observations of performance of existing unreinforced concrete pavements have shown that allowable joint spacing for crack control increases as slab thickness increases. As a rough guide, the joint spacing (in feet) for unreinforced pavements should not greatly exceed twice the slab thickness (in inches). In addition, adequately designed slabs are not likely to develop an intermediate crack if the length-to-width ratio does not exceed 1·25. If the pavement is reinforced, longer joint spacing is allowed. However, it is desirable to limit joint spacing of pavements having distributed steel to about 30 or 40 ft for pavements less than 12 in thick and 50 ft for thicker pavements. For these longer joint spacings, dowels are required in all transverse joints because the joints open wider and the load transfer by aggregate interlock is doubtful. Observations of performance of un-reinforced concrete pavements have shown that for a distance of about 100 ft back from each free end and 60 ft back from each expansion joint the joints will gradually open to a point where the aggregate interlock may not be effective. For this reason, dowels must be placed across con-traction joints of unreinforced pavements near the free ends and near any expansion joint.

Dowel bars are generally of a standard design that varies from place to place; they are generally spaced at mid-depth of the slab. The dowel bar transfers the load across the joint, and it is necessary to lubricate one-half of the dowel bar to permit freedom of sliding, because the slabs move relative to one another in the longitudinal direction.

The joint groove can be formed by sawing, or it may be formed by placing a metal or fibre strip in the uncured concrete and then removing the strip as soon as initial set of the concrete takes place. If sawed joints are used, sawing should be done as soon as possible after pouring, but not so soon as to leave marks on the pavement or to displace pieces of aggregate.

One method of forming dummy-groove joints is to place a small metal bar having the size of the required joint and then remove the bar as soon as feasible.

11.2.2 Expansion Joints

Expansion joints must be constructed with a clean break throughout the depth of the slab to permit expansion to take place (Fig. 11.1(d)). The clear distance across the joint is usually $\frac{3}{4}$ in, although in some cases it may be desirable to permit a $1–1\frac{1}{2}$ in opening. Since the joint has no aggregate interlock, it is necessary to provide some type of load transfer. This is best accomplished by means of dowel bars which must be smooth and lubricated on at least one side. An expansion cap must also be provided to allow space for the dowel bar to move during the expansion process.

Expansion joints for highways are highly susceptible to pumping action, so some countries have discontinued their use. The omission of expansion joints tends to hold the interior areas of the pavement in restraint and limits crack and joint opening, thus adding to the aggregate interlock effectiveness.

In airport pavements, expansion joints may not be needed either transversely or longitudinally if the contraction joints are spaced as outlined before. If used in airports, expansion joints are generally spaced 1000–2000 ft apart and are either dowelled or provided with thickened edges. Also, expansion joints must be provided between concrete pavements and all buildings or other fixed structures. They are sometimes required at intersections of runways, taxiways and aprons.

The structural adequacy of an expansion joint is determined to a large extent by its load-transfer device. If adequate load transfer is provided, deflection of the forward slabs is minimised, which in turn tends to reduce pumping. Also, it is necessary to maintain the joints periodically, and in some cases to replace the filler material in the joint.

Common types of fillers are fibrous and bituminous materials, cork and preformed neoprene strips. Care should be taken to prevent infiltration of shoulder and subgrade material through the filler of the expansion joints, otherwise spalling of the concrete edges will occur during the expansion of the pavement slabs.

It is essential to seal the joints periodically to prevent infiltration of surface water. Resealing of joints is best accomplished during a cool period, when the joint has opened, thus permitting of placement of a bituminous seal.

High expansion forces can result when certain concrete aggregates are used. Field studies have indicated that the frequency of occurrence of blowups is a function of the source and type of coarse aggregate. Thus, if due caution is used in selecting the coarse aggregate, distress due to blowups can be minimised. On most construction, this relieves the necessity of using expansion joints.

Furthermore, the magnitude of the longitudinal compressive stresses generated in concrete pavements when expansion joints are eliminated depends upon the temperature of the concrete during construction and upon the air temperature during the concrete hardening process.

Studies made in Germany have indicated that when expansion joints are omitted, the longitudinal compressive stresses resulting from high temperature will be in the order of 1400 psi (100 kgf/cm^2), if the pavement is constructed at low temperatures. No blowups are expected from these stresses if short slabs less than 17 ft (5·1 m) long are selected, provided the slabs are thicker than 7 in (18 cm) and the contraction joints are not clogged with sand. This conclusion applies to the prevailing climatic conditions in Germany.

In the USA and Canada, this system of eliminating the expansion joints has been in use for a number of years, especially in airfield pavements where the slab thickness is generally of the order of 12 in (30 cm). The spacing between transverse contraction joints is usually 20 ft (6 m).

However, in countries with severe weather conditions, especially in those with very hot Summers, extensive studies should be made to determine whether or not it is feasible to eliminate expansion joints in their concrete pavements.

11.2.3 Construction Joints

Construction joints are usually of the butt type and contain dowel bars for transferring the load across the joint. Construction joints are used at the transition from old to new construction, such as at the end of a day's pour. In some cases a keyed construction joint such as that indicated in Fig. 11.1(f) is used. The butt type is perhaps the most common for highway work; keyed longitudinal construction joints are quite often used on airfields. Keyed longitudinal construction joints are not usually tied with tie bars except on the extreme outer slabs. If this type of joint is used, it is common practice to pour alternate lanes, forming the key by means of special metal plates or wood strips fastened to the forms. If the key joints are not made properly, cracks may develop in the weakened slab parallel to the joint.

11.2.4 Hinge or Warping Joints

Hinge and/or warping joints are employed mainly in longitudinal joint construction. The type of joint used depends primarily upon the method of pouring the concrete slabs. If lane-at-a-time construction is to be used, keyed joints are generally built. In some cases, the keyed joints are tied together with tie bars to make sure that the key transfers load properly. This is accomplished by disjointing the tie bar at the pavement form by means of bolt and nut arrangements, or making holes through the shuttering.

If two-lane construction is used on highways, the most convenient type of longitudinal warping joint is of the dummy-groove type, where tie bars are placed at intervals of about 3 ft to make certain that grain interlock is maintained. It is important to note that load transfer across a tied joint is brought about solely by grain interlock. Thus, the tie bars must be firmly anchored and bonded to prevent movement.

11.3 DOWEL BARS AND TIE BARS

Dowel bars are plain bars embedded in a concrete pavement to provide vertical shear resistance across a joint. They should offer no restraint to the opening or closing of the joints at any time. They should also be made corrosion-resistant; this is done by painting the bonded part with a paint that does not break the bond but prevents corrosion. The unbonded part is generally greased so that it can move freely, and at the same time the grease prevents corrosion.

There are various methods of calculating stresses in dowel bars and determining the dowel group action. Since these methods rely upon certain values that can vary to a large extent, only tables recommending minimum dowel requirements will be provided here (Tables 11.1, 11.2).

TABLE 11.1 RECOMMENDED MINIMUM DOWEL REQUIREMENTS FOR CONTRACTION OR EXPANSION JOINTS IN HIGHWAY CONSTRUCTION

Pavement thickness (in)	Dowel diameter (in)	Dowel spacing (in)	Dowel length (in)
6	$\frac{3}{4}$	12	18
7	1	12	18
8	1	12	18
9	$1\frac{1}{4}$	12	18
10	$1\frac{1}{4}$	12	18

TABLE 11.2 MINIMUM DOWEL REQUIREMENTS FOR LONGITUDINAL BUTT JOINTS, TRANSVERSE CONTRACTION, EXPANSION AND CONSTRUCTION JOINTS IN AIRFIELD CONSTRUCTION

Pavement thickness (in)	Dowel diameter (in)	Dowel spacing (in)	Dowel length (in)
6	$\frac{3}{4}$	12	18
7	$\frac{3}{4}$	12	18
8	1	12	18
9	1	12	18
10	1	12	18
11	$1\frac{1}{4}$	15	20
12	$1\frac{1}{4}$	15	20
13	$1\frac{1}{4}$	15	20
14	$1\frac{1}{4}$	15	20
15	$1\frac{1}{4}$	15	20
16	$1\frac{1}{2}$	15	24

TABLE 11.3 RECOMMENDED MAXIMUM SPACING FOR ½ AND ⅝ IN TIE BARS FOR HIGHWAYS

Type and grade of steel	Working stress (psi)	Pavement thickness (in)	Tie bars ½ in diameter				Tie bars ⅝ in diameter			
			Over-all length (in)	Spacing (in) Lane width (ft)			Over-all length (in)	Spacing (in) Lane width (ft)		
				10	11	12		10	11	12
Structural-grade billet or axle steel	22 000	6	20	46	42	38	24	48	48	48
		7		39	36	33		48	48	48
		8		34	31	28		48	48	45
		9		30	28	25		48	43	40
		10		27	25	23		43	39	36
Intermediate-grade billet or axle steel	27 000	6	24	48	48	47	27	48	48	48
		7		48	44	40		48	48	48
		8		42	38	35		48	48	48
		9		37	34	31		48	48	48
		10		34	31	28		48	48	44
Rail steel or hard-grade billet or axle steel	33 000	6	27	48	48	48	33	48	48	48
		7		48	48	48		48	48	48
		8		48	47	43		48	48	48
		9		46	42	38		48	48	48
		10		41	37	34		48	48	48

Tie bars are deformed bars embedded in a concrete pavement at a joint, generally a longitudinal one, and designed to hold abutting slabs together by direct bond with the surrounding concrete. Table 11.3 shows the recommended maximum spacing of tie bars in highway construction.

In airport construction, the recommended minimum diameter of tie bars is $\frac{5}{8}$ in. The recommended range of tie bar length is 30–36 in at a spacing of 30 in centre to centre.

11.4 TEMPERATURE STEEL REINFORCEMENT

Welded wire-fabric or bar-mat reinforcement may be used in rigid-pavement slabs to control temperature cracks. The reinforcement is meant to ensure intimate contact of slab parts, so as to produce a degree of load transfer across the crack through aggregate interlock. Accordingly, a wheel load near the crack will be supported by the slabs on both sides with correspondingly smaller concrete flexural stress and deflection at the crack. In addition, the narrower cracks that occur when reinforcement is used minimise water seepage to the subgrade soil and the infiltration of soil into the crack.

In most cases, the reinforcement is assumed to add nothing to the structural capacity of the pavement, and so will not change the design thickness of the concrete pavement. Using steel reinforcement, however, allows of placing of the transverse joints at larger distances, 40–100 ft apart. The AASHO Road Test has indicated that the performance of a 15 ft long non-reinforced concrete pavement is the same as that of a 40 ft long concrete pavement with temperature reinforcement.

The area of steel is calculated to overcome the maximum subgrade friction at mid-length of the slab. When the temperature drops, a pavement slab tends to shorten. This contraction is resisted by the subgrade. The frictional resistance, which increases from the ends to a maximum value at mid-length of the slab, must be overcome by the tensile resistance of the steel crossing any transverse crack. The area of steel can be calculated from eqn. (11.1).

$$A_s = \frac{FwL}{2f_s} \tag{11.1}$$

where A_s = is the cross-sectional area of steel required per foot of slab width (in^2);

w = is the weight of slab (psf);

$L =$ is the length of slab (ft);

$f_s =$ is the allowable steel stress (psi), generally taken as 0·67 times the yield stress of the steel; and

$F =$ is the coefficient of resistance between the slab and subgrade, generally taken as 1·5.

The cross-sectional area of steel in the transverse direction, per foot of slab length, is generally between 20 and 30% of the area, A_s, of longitudinal steel. This area should also be checked to make sure that it will develop a tensile force greater than or equal to the strength of the tie bars per foot length of pavement at longitudinal joints of multi-lane pavements.

The minimum size of reinforcing steel in heavy-duty pavements is $\frac{1}{4}$ in diameter for longitudinal bars and 0·225 in diameter for transverse bars if welded wire-fabric mats are used. If deformed bar mats are used, the minimum sizes are $\frac{3}{8}$ in and $\frac{1}{4}$ in diameter for the longitudinal and transverse bars, respectively.

For welded wire fabrics the spacing of the longitudinal wires should not exceed 6 in and the transverse wires 12 in. If bar mats are used, the spacing of the longitudinal bars should not exceed 15 in, while that of the transverse bars should be no greater than 30 in.

The reinforcement should be placed at a depth not less than 2 in nor more than $\frac{1}{3} h$ from the top of the slab. Figure 11.2 shows a longitudinal and a transverse cross-section of a lightly reinforced concrete pavement.

11.5 JOINT SEALANTS

Joint sealants are used in all joints to keep out damaging material. They must be capable of withstanding repeated extension and compression as the pavement expands and contracts with temperature and moisture change. In order to maintain an effective seal, the joint width must be made large enough so that subsequent joint movement will not put undue strain on the sealants. This means that some joints must be sawed wider at the top for a depth equal to joint width and not less than $\frac{1}{2}$ in to form a reservoir for the sealant material. For poured joint sealants, the shape of this reservoir has a critical effect on the sealant's capacity to withstand extension and compression and a joint shape as nearly square as possible is desired. The joint sealant should also be $\frac{1}{8}$ in below surface.

In addition to the conventional sealing compounds, like the bituminous and rubberised sealing materials used in joints, preformed neoprene compression joint seals are taking their place in pavement construction.

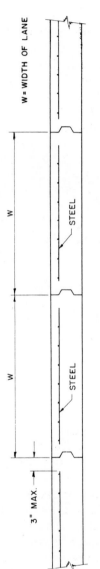

TRANSVERSE CROSS-SECTION OF PAVING LANES

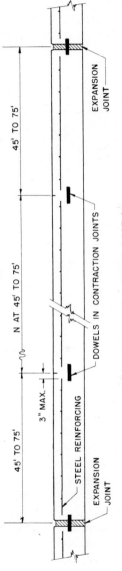

LONGITUDINAL CROSS-SECTION

FIG. 11.2. Details of reinforced concrete pavement.

In airport pavements, any type of joint sealant used must be fuel resistant. Also, because of the jet blast, it is important to avoid having any loose material in these joints.

The effectiveness of in-service performance of a preformed neoprene compression joint seal is dependent to some degree upon proper installation. Generally, the following principles of positioning in contraction joints apply (see Fig. 11.3).

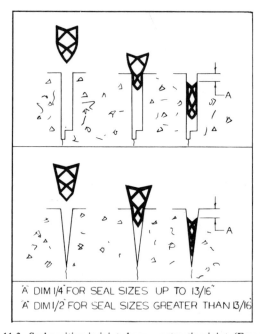

A DIM 1/4" FOR SEAL SIZES UP TO 13/16"
A DIM 1/2" FOR SEAL SIZES GREATER THAN 13/16

Fig. 11.3. Seal position in joint shape: contraction joint. (From [10].)

1. The seal configuration should be positioned in the joint shape with its vertical axis reasonably parallel to the joint interfaces.

2. The top corners of the seal should be in reasonable contact with the joint interfaces.

3. The top of the seal, at the time of installation, should be no higher than $\frac{1}{4}$ in below the riding surface of the pavement for contraction joint seals $\frac{13}{16}$ in (uncompressed width) and smaller. For larger seals, such as $1\frac{1}{4}$ in contraction seals (uncompressed width) and down to but not including $\frac{13}{16}$ in sizes, the top surface should be positioned at approximately $\frac{1}{2}$ in below the riding surface of the pavement at the time of installation. While

this might appear to be somewhat low in the joint, long-term experience has shown that typical attrition to the top edges of joints, as well as the inevitable minor edge ravelling from joint sawing, justifies these setting heights as practical. To obtain their full life, these rubber seals should never under any condition of slab movement be touched by traffic, studded tyres, snowplough, etc., since attrition will take its toll. The top portions can actually sustain wear not unlike the heels of shoes.

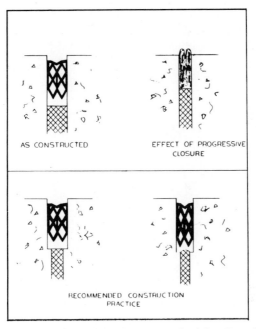

AS CONSTRUCTED

EFFECT OF PROGRESSIVE CLOSURE

RECOMMENDED CONSTRUCTION PRACTICE

FIG. 11.4. Seal position in joint shape: expansion joint. (From [10].)

Pavement expansion joints are normally wider, usually anywhere from $\frac{3}{4}$ to $1\frac{1}{2}$ in wide, and since they exist to relieve compressive stress, they require different construction practices to reflect movement phenomena peculiar to their specific application. When used in a line of contraction joints, they can and usually do progressively close, in addition to reflecting normal thermal volume change. The construction practice recommendations shown in Fig. 11.4 are easily obtained in the field and preclude the possibility of seal extrusion. Generally, the same three positioning rules used in contraction jointing will apply to expansion joints as well. Due to the complete absence of control of joint geometry, plus a tendency towards durability loss from the old hand-edging process, a recent strong trend is in evidence towards the sawing of expansion joints.

Since most longitudinal joints are tied together with a variety of devices, keys, etc., it is logical to assume that a zero movement condition prevails. Based on widespread photographic evidence from thousands of miles of concrete pavements now in service throughout North America and abroad, it is suggested that there may well be a number of categories of movement phenomena that can and actually do occur at longitudinal joints. This movement permits of the entry of free water, solids and chemicals harmful not only to concrete, but also to the highly corrodible metals used in tie bars and hook bolts, and to the hydraulically vulnerable sub-base as well.

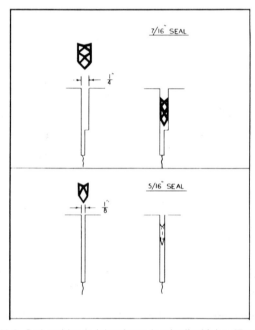

FIG. 11.5. Seal position in joint shape: longitudinal joint. (From [10].)

The rules for positioning transverse joints apply equally to longitudinal joints. Since the stroke of movement is not nearly so dynamic as with transverse joints, smaller seal widths and, in non-freeze–thaw areas, narrower joint widths, have appeared to offer successful sealing solutions. Figure 11.5 illustrates $\frac{1}{8}$ in saw cuts and small $\frac{5}{16}$ in seals currently in widespread use on longitudinal joints. The $\frac{1}{4}$ in saw cut with $\frac{7}{16}$ in seal is also used on longitudinal joints in freeze–thaw areas, in general.

It would be extremely difficult if not impossible to insert a compression seal into most joint openings, without lubricating the joint interfaces. In addition to this, a priming agent is required to establish continuity of the seal with the joint interfaces. Because it is desirable to bond compression seals in place, a lubricant that is also an adhesive, with an approximate 3–5 min pot life, is used. Insufficient as well as excessive lubricity can be troublesome in the installation practice. A sufficient amount of the lubricant–adhesive should be used to negate the forces of frictional refusal as differentiated from mechanical refusal.

The practice of bevelling the edges of saw cuts and setting the compression seals near the bottom of the bevel began in Switzerland and is now spreading throughout Europe. Figure 11.6 illustrates the principle. Long-term comparisons of performance in the field of both conventional saw cuts and bevelled edge saw cuts have proved the latter markedly superior.

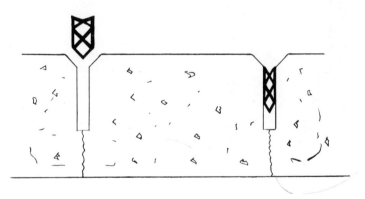

FiG. 11.6. Detail of bevelled saw cuts showing position of compression seal: Swiss–German system, slab length 7·8 m. (From [10].)

Automatic and semi-automatic methods of machine installation are presently in various stages of use, both in North America and in Europe. They may be categorised in principle as follows: (1) mechanical compress–eject, (2) vacuum compress–eject, (3) two-phase pull down, (4) combination vacuum compress–two-phase pull down, (5) progressive set of rolling wheels, and (6) compress–eject hand roller.

The development of reliable automatic or semi-automatic machines has been relatively costly and slow, primarily because no two paving projects appear to have exactly the same joint geometry, thermal movement, seal configuration, construction personnel, etc. The causes of seal refusal are many, and any abrupt change in the joint geometry, whether it is due to small spalls, excessive cavitation of interfaces or differential ravelling, etc., will still cause the most sophisticated installing machine to balk.

REFERENCES AND BIBLIOGRAPHY

1. ACI Committee 325, 'Structural Design Considerations for Pavement Joints', *Journal of the American Concrete Institute*, July 1956.
2. ACI Committee 325, 'Recommended Practice for Design of Concrete Pavements', *ACI Manual of Concrete Practice*, Part 1, 1968.
3. Friberg, B. F., 'Frictional Resistance under Concrete Pavements and Restraint Stresses in Long Reinforced Slabs', *Proc. Highway Research Board*, 1954.
4. *Highway Research Record*, No. 320, 'Joint Sealants, Paint and Pipe', 1970.
5. Pachuta, J. M. and Smith, D. J., 'Development of Effective Poured-in-Place Systems for Concrete Joints', *Highway Research Record*, No. 287, 1969.
6. Portland Cement Association, 'The Design of Concrete Pavements for City Streets', Illinois, 1963.
7. Schuster, F. O., 'The Behaviour of Concrete Pavements on Stabilized Soils', *Proc. 2nd European Symposium on Concrete Roads, Bern, 1973.*
8. Stewart, P. D. and Shaffer, R. K., 'Investigation of Concrete Protective Sealants and Curing Compounds', *Highway Research Record*, No. 268, 1969.
9. Teller, L. W. and Cashell, H. D., 'Performance of Doweled Joints under Repetitive Loading', *Highway Research Board Bulletin 217*, 1958.
10. Watson, S. C., 'Installation of Preformed Neoprene Compression Joint Seals', *Highway Research Record*, No. 287, 1969.
11. Yoder, E. J., *Principles of Pavement Design*, Wiley, 1959.

CHAPTER 12

CONTINUOUSLY REINFORCED CONCRETE PAVEMENTS

12.1 INTRODUCTION

Joints in rigid pavements, which have been used with varying degrees of success, have continued to be points of potential structural weakness. They also present problems of maintenance and permit of the infiltration of water and soil. Also, they often adversely affect the riding quality of the pavement.

Careful measurement of slab movements on experimental pavements laid in Indiana in 1938 showed that 500–1000 ft at the ends of long reinforced pavement slabs moved on the subgrade, but that the central portions of the slabs did not. Rather, minute closely spaced cracks developed in the concrete. In most instances these centre sections were giving excellent performance. It therefore appeared that a heavily reinforced continuous pavement without joints should perform like these central sections. Soon after the Second World War, experimental sections to test this theory were constructed in several states in America. Figure 12.1 shows the increase in use of this type of pavement in the USA since 1958.

According to these findings, continuously reinforced concrete (CRC) pavement, regardless of length, can be defined as a pavement that has no transverse expansion or contraction joints, except at its ends and at bridges or other structures. The continuity of the longitudinal steel reinforcement is terminated only at the ends. Only transverse construction joints are permitted, at which appropriate splicing of the longitudinal reinforcement should be made. Pavements of this type, up to 40 miles long, have been built with no transverse joints. In these pavements, minute closely spaced

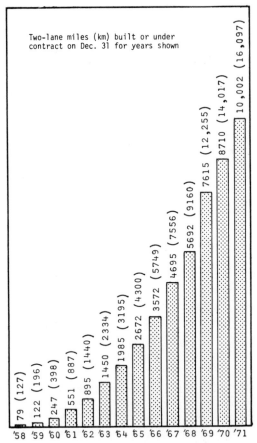

Fɪɢ. 12.1. Increase in use of CRCP. (From [7].)

cracks are allowed, since the width of these cracks is generally less than 0·012 in.

The central part of this type of pavement behaves as if completely restrained; the slab movement takes place only near the slab ends, as shown in Fig. 12.2(b).

12.2 SLAB WIDTH

The same widths used in non-reinforced concrete pavements can be applied in continuously reinforced concrete pavements. This is usually

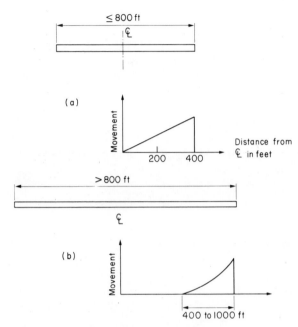

FIG. 12.2. Effect of slab length on longitudinal movement. (a) Short pavement without joints. (b) Long pavement without joints.

a lane width of 10–12 ft in highways. In airports, larger widths of up to 25 ft are generally used.

12.3 SLAB THICKNESS

Since cracks are an acceptable fact in this type of pavement, curling stresses due to temperature changes between the top and bottom surfaces of the concrete slab are not considered in the design.

The stresses due to load only determine the required thickness for the pavement.

The only corners and transverse joints for continuously reinforced concrete pavements are the ones at both ends of the slab. Slab ends are generally either thicker than the intermediate part or rest on sleeper slabs that increase their load-carrying capacity. Accordingly, the cases of corner load or load at a transverse joint are not used in the design of continuously reinforced concrete pavements.

12.4 REINFORCING STEEL DESIGN

The percentage of longitudinal steel (area of longitudinal steel times 100, divided by cross-sectional area of the concrete slab) should be enough to keep the transverse cracks very tight throughout the life of the pavement. Many continuously reinforced concrete pavements have been built with this percentage ranging as low as 0·5. Climatological factors influence the amount of reinforcement required, and the engineer should be aware of this fact. Vetter [22] found that to prevent yielding of the steel at cracks due to restrained volume changes, the minimum amount of steel should be determined by the equation:

$$p_s = \frac{f_t'}{f_y} 100 \tag{12.1}$$

The result of eqn. (12.1) is normally the minimum steel percentage required for shrinkage.

The minimum amount of steel to control restrained volume changes due to temperature was found to be:

$$p_s = \frac{f_t'}{f_y - nf_t'} 100 \tag{12.2}$$

Equation (12.2) gives a higher percentage of steel than eqn. (12.1).

In addition to omitting some climatological factors, the Vetter formulae for determining the minimum percentage of longitudinal steel, do not take into account the friction factor, F, between the slab and the subgrade. Provided F is approximately 1·5, no correction to eqn. (12.2) need be made. However, if there is reason to believe that F varies appreciably from 1·5, eqn. (12.2) should be replaced by eqn. (12.3):

$$p_s = \frac{f_t'}{f_y - nf_t'} (1·3 - 0·2F) (100) \tag{12.3}$$

In eqns. (12.1)–(12.3), the following notation has been used:
F = friction factor between the slab and its supporting medium;
E_c = Young's modulus of elasticity for concrete;
E_s = Young's modulus of elasticity for steel (may be taken as 29 000 000 psi);
f_t' = tensile strength of concrete (psi $\simeq$ 0·40 M R);
f_y = yield strength of steel (psi);

M R = modulus of rupture of concrete as measured by third-point beam
    loading at 28 days (psi);

$n = E_s/E_c$;

p_s = percentage of steel = (A_s/A_c) 100; and

A_c = cross-sectional area of concrete slab.

It must be noted that in the above equations the yield value of the steel
and the tensile failure value of the concrete are used to establish an
equilibrium situation, from which one determines the required minimum
area of steel. Using a larger percentage of steel (which one may obtain by
multiplying p_s by a safety factor) will keep cracks more tightly closed. A
safety factor of 1·35 applied to p_s is considered adequate, i.e. p_s should be
increased by 35%.

Having established the required percentage of longitudinal steel, the
longitudinal steel area may be computed by the formula:

$$A_s = \frac{Bh}{100} p_s \qquad (12.4)$$

where A_s = cross-sectional area of longitudinal steel (in²);

h = slab thickness (in); and

B = width of slab (in).

In CRC pavements with controlled longitudinal joints 15 ft or less
apart, transverse steel may serve the following purposes:

1. To maintain the specified spacing of the longitudinal steel. This is
accomplished by securing the longitudinal bars to the transverse bars, at
selected intersections, with wire ties or clips or by welding.

2. To serve as tie bars across the longitudinal joints in lieu of conventional tie bars.

3. To hold chance longitudinal cracks tightly closed. (The probability
of such cracks occurring in a properly designed and constructed pavement, with a longitudinal joint, tied or untied, separating the traffic lanes,
is remote. Chance longitudinal cracking has not been found to be a
problem in plain concrete pavements with no more than three lanes tied
together.)

4. To aid in supporting the longitudinal steel above the sub-base, in
the case of preset reinforcement.

The decision to use transverse steel will depend on which of these
purposes are required in the particular case. For example, if reinforcement
is to be embedded in the concrete by a mechanical method to ensure
proper location and spacing of the longitudinal steel without tying to the
transverse steel, then transverse bars would not be necessary for items 1

and 4 above. Thus, if conventional tie bars were used across longitudinal joints, the only purpose to be served by transverse steel would be item 3 above.

When transverse steel is required, its area may be determined in the same manner as for a jointed reinforced concrete pavement, and should not be less than the cross-sectional area of the tie bars. If the transverse steel does not extend across the longitudinal joints, tie bars between lanes should be provided, as in other concrete pavements. Furthermore, longitudinal weakened plane joints must be constructed.

The amount of transverse steel required to hold chance longitudinal cracks tightly closed can be calculated from the formula:

$$A = \frac{F\,wb}{2f_s} \qquad (12.5)$$

where F = coefficient of subgrade resistance to slab movement, usually taken as 1·5;

A = area of steel per foot length of slab (in^2);

b = width of slab between the longitudinal joints;

w = weight of slab (lb/ft^2); and

f_s = allowable working steel stress (psi), usually taken as 0·75% of yield strength.

Experience gained from the construction and observation of several thousand miles of CRC highway pavement is the basis of the following important steel reinforcement details:

1. Steel reinforcement may consist of deformed bars, deformed bars fabricated into mats, or deformed welded wire fabric.

2. It is recommended that longitudinal steel, regardless of type, have a minimum yield strength of 60 000 psi. Experience shows that the use of lower grades results in spacings between members which interfere with the placing and consolidation of concrete. Furthermore, in the case of loose bars installed on the grade, the use of a small number of bars is economical because less handling and tying are required.

3. Spacing of longitudinal steel should not exceed 8 in, and the minimum spacing should not be less than 4 in when the concrete is placed in a single course or 3 in when the concrete is placed in two courses with the steel installed between the two.

4. Size of longitudinal steel will be equal to the total required steel area divided by the number of longitudinal bars or wires, as determined from the adopted spacing.

5. Spacing of transverse steel, when required, will depend on whether the reinforcing steel consists of deformed bars or deformed wire. Deformed bars should be limited to a maximum spacing of 48 in. Currently the spacing of the transverse wires in deformed welded wire fabric is normally limited by manufacturing practice to a maximum of 16 in. In either case the size of the transverse member will depend on the required transverse steel area. It is suggested that the minimum size of transverse bar should be a number 3 bar (nominal diameter of $\frac{3}{8}$ in) and the minimum size of transverse wire should be a D-4 wire (nominal diameter of 0·225 in).

6. Reinforcement should be located in such a manner that all steel will have a minimum concrete cover of 2 in, and the longitudinal steel will not fall below the mid-depth of the slab.

7. Lap splices of individual bars or prefabricated bar mats or deformed wire mats may be located in the same transverse section, or preferably the laps of the individual pairs of bars or mats may be staggered across the width of the pavement. If not more than one-third of the longitudinal steel members are lapped within any 3 ft length of pavement, the length of the lap should be 25 times the bar or wire diameter, with a minimum of 16 in. If the laps are in the same transverse section across the entire slab, the length should be 30 times the bar or wire diameter, with a minimum of 18 in. In the case of deformed welded wire fabric, if the splice should include transverse wires, the mats should be lapped so that the end transverse wires of the two mats overlap a minimum of 4 in centre to centre. The 4 in space is desirable for placing concrete.

8. When bar mats or welded wire mats, having widths less than the full width of the slab, are placed alongside each other, the transverse wires or bars of the mats should be lapped a minimum of 20 times the bar or wire diameter.

12.5 DESIGN AND CONSTRUCTION CRITERIA

Information developed on the performance of continuously reinforced pavements suggests that careful consideration be given to the following:

1. The subgrade should receive the same preparation as that required for pavements of conventional design. Irregularly compacted and otherwise variable foundation support was determined to be the primary cause of failure of a 2000 ft section of continuously reinforced pavement.

2. A 3–6 in granular sub-base should prevent excessive edge pumping. Greater thicknesses are required to help protect against high volume

change and frost-susceptible soils. The sub-base should be well compacted to assure against densification and settlement under traffic loads.

3. The longitudinal steel should be located no lower than, and preferably above, the mid-depth of the pavement. Cores drilled at cracks in experimental pavements show that crack width is considerably greater above than below the level of the longitudinal steel. Surface widths of cracks were observed to be abnormally wide in the few very local areas where the steel was inadvertently placed in the bottom third of the pavement.

4. Care should be taken to make sure that the proper lap of welded wire fabric or deformed bars is maintained.

A few detrimental cracks have occurred in pavements where the transverse wires of adjacent sheets of welded wire fabric were not overlapped and where bar laps were in a line at right angles to the axis of the highway. In these cases deficient bond between the steel and concrete, and consequent slippage of the steel, caused excessive crack opening and breakage of the pavement. Among the contributing factors to the poor bond development were insufficient length of lap and separation of the two layers of concrete where two-lift construction was used. In several instances the bar reinforcement was not lapped at all.

5. The transverse construction joint should be designed to permit of continuation of the basic longitudinal steel. The continuing steel should be supplemented by additional steel sufficient in amount to provide adequate resistance to repeated shear and bending stresses caused by traffic loads.

6. The longitudinal steel should extend without lapping a minimum of 3 ft into the first completed portion of the pavement at a construction joint, and at least 5 ft in the following portion. The first lap, if it occurs within 20 ft beyond the construction joint in the direction of paving, should be double the lap normally specified, wherever possible. In the case of preplaced steel, where double laps are impractical, it would be desirable to install supplementary steel at all laps within 20 ft beyond the construction joint. Where welded wire fabric or prefabricated bar mats are placed between two lifts of concrete, the portions of the sheets or mats extending beyond a construction joint in the direction of paving should be supported on suitable chairs.

Under certain conditions, particularly after an undue delay in resuming paving, abnormal crack concentrations may occur in the portion of the pavement immediately following a construction joint. Experience has shown that such concentrations will have little, if any, effect on the structural integrity of the pavement, provided the cracks are held tightly closed.

7. Preferably, the concrete should be placed in one lift. If two-lift construction is specified, the use of two pavers should be specified also, to shorten the time interval between the first and second lift of concrete. As mentioned earlier, separation of the two layers of concrete was a contributing factor to several lap failures in a pavement reinforced with bars. The separation was the direct result of the first layer of concrete obtaining a partial set before placing the second layer.

8. Full-width pavement vibration should be mandatory. This operation promotes proper bond between the concrete and the relatively large amount of reinforcing steel, and in the case of two-lift construction also minimises the occurrence of a plane of separation.

9. Concrete patches, used to restore continuity at localised failed areas, should be at least 6 ft wide, 3 ft each side of the failed area. The performance of patches 6 ft or greater in width has been generally excellent, except where the supporting foundation was poor.

12.6 FACTORS AFFECTING CRACK WIDTH AND SPACING IN CRC PAVEMENTS

Steel percentage has been singled out as perhaps the most important variable in the design of continuously reinforced concrete pavements.

In a study by the Texas Highway Department [16], other factors were found to influence the behaviour and performance of an experimental continuous pavement. It was found that a steel percentage of 0·5 and 0·6 performed satisfactorily, but cement type, time of placement, concrete properties and other environmental factors had a marked influence on the behaviour of the test project.

The behaviour of the pavement is explained through the use of the following variables:

1. Reinforcing steel strains
2. Concrete strains
3. Crack width
4. Slab temperature gradient
5. Crack pattern development
6. Variations in crack pattern
7. Steel strain due to wheel loads

Related to these measurable variables are the following parameters:
1. Weather conditions
2. Slab thickness
3. Subgrade and sub-base
4. Curing of concrete
5. Type of steel
6. Type of cement
7. Concrete properties
8. Age
9. Temperature (slab and air)
10. Crack spacing
11. Axle loads
12. Steel percentage
13. Concrete strains

The average crack spacing is obtained by dividing the length of slab under consideration by the number of cracks within the slab.

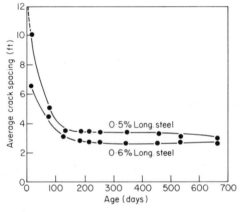

FIG. 12.3. Typical relationship between age and average crack spacing. (From [16].)

Effect of age. Figure 12.3 shows the effect of age on the average crack spacing. As has been shown by other investigations, crack spacing decreases with age at a very fast rate during the early age of the slab (3–4 months), reaching a constant value at about 4 months.

Indications are that these pavements will attain their final crack spacing within a year.

Steel percentage. Figure 12.3 can also be used to compare the effect of steel percentage on the average crack spacing of a slab. There is an inverse relation between the average crack spacing and the percentage of longitudinal steel, but the relationship changes with time. After more than 2 years, the difference in average crack spacing between a pavement with 0·5% (number 5 round bars at 7½ in centres located at mid-depth of slab) and a pavement with 0·6% (number 5 round bars at 6½ in centres located at mid-depth of slab) of longitudinal reinforcement is negligible.

At very early ages, the pavement with low percentage will present a larger average crack spacing. This crack spacing after 6 months is about 25% larger than that of the pavement with the high steel percentage.

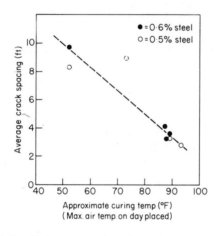

FIG. 12.4. Typical relationship between crack spacing and curing temperature. (From [16].)

Curing temperature. Figure 12.4 shows an inverse relation between the average crack spacing and the curing temperature. It is apparent that a 0·1% difference in steel percentage is not very significant in influencing the crack pattern.

Effect of cement type. Longitudinal forces in the pavement are shared by both concrete and steel reinforcement according to their relative stiffness. But once the concrete cracks (Fig. 12.5), the steel should resist all the force.

Figure 12.6 shows the effect of cement type on steel stresses during the early curing period. The dashed portion of each of the relations is the change in the steel stress conditions at the time of pavement cracking. The change in stress magnitude in both cases is noteworthy. On cracking, the stress in the steel goes from approximately 500 to 4500 psi with type I cement, a difference of 4000 psi. With the type III cement, the stress in the steel goes from 2000 to 20 000 psi, a difference of 18 000 psi. The cracking in concrete with a type III cement is of an explosive nature.

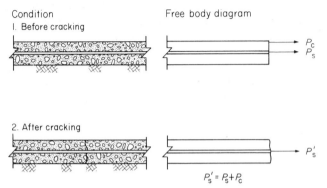

FIG. 12.5. Force conditions before and after concrete cracking. (From [16].)

As background for an explanation of the extreme difference in stress levels, the following observations should be made from Fig. 12.6. First the failure in type I occurs at approximately 14 h, whereas the failure in type III does not occur until 33 h. The basic difference between type I and

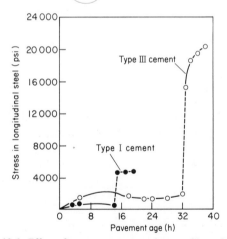

FIG. 12.6. Effect of cement type on steel stress. (From [16].)

type III cement is that type III attains its strength much more rapidly than type I, especially during the first 72 h. At the time of fracture, the type I cement has attained a strength of only 40 psi and the type III has attained a strength of 220 psi. This means that approximately 2500 lb of force were dropped into each steel bar when the type I cement was used, whereas up to 11 000 lb of force were dropped into each bar with type III. Type III cement is not recommended, since it results in wider cracks due to higher steel strains.

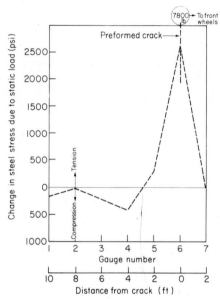

FIG. 12.7. Effect of static load on steel stress distribution: 0·5%. (1. Stresses shown are for 7800 lb wheel load directly over gauge No. 6. 2. Stresses shown are differential stresses only. Temp. stresses deleted.) (From [16].)

Effect of temperature. The following formulae have been found to give an approximation for the steel stresses, if the slab temperature T and the crack spacing X are known.

For steel percentage of 0·5,

$$S_s = 32\,410 + 9850X - (391 + 99X)\,T \tag{12.6}$$

For steel percentage of 0·6,

$$S_s = 22\,000 + 8140X - (194 + 84·7X)\,T \tag{12.7}$$

Once the crack has occurred, stress concentrations in the steel may be as high as 3 or 4 times the magnitude of steel stresses in the interior.

Each individual slab segment is a fluctuating body. These segments expand and contract with temperature changes, although the centre of the slab segment is fixed in relation to the ground. Any opening movement that occurs at a crack must be balanced by a closing movement in the area between cracks. A continuous pavement may be considered similar to a jointed pavement except that the movement is severely restricted by the steel.

Effects of loads. Figure 12.7 shows the longitudinal distribution of stress due to a wheel load being placed directly over the crack.

At a distance of approximately 3 ft from the crack, the steel stress goes from tension to compression. Steel stresses at interior points are minor when compared with the stress at the crack, being less than $\frac{1}{6}$ of the magnitude.

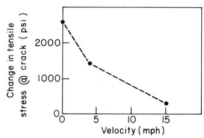

FIG. 12.8. Effect of velocity of load on steel stresses at crack: 0·5%. (Curve shown is for the effect of rear wheels.) (From [16].)

For dynamic loading, Fig. 12.8 shows that the stress obtained at the crack is inversely proportional to the velocity of the wheel load. In other words, the stress at the crack decreases as the velocity increases.

12.7 CRACK WIDTH

Figure 12.9 shows a typical relation between crack width and the slab's mid-depth temperature.

The following expressions have been found to give approximate values for crack width if the temperature of the slab is known.

For 0·5% steel,

$$\Delta X = (63 \cdot 1 - 0 \cdot 64\ T) \times 10^{-3} \tag{12.8}$$

For 0·6% steel,

$$\Delta X = (50 \cdot 3 - 0 \cdot 52\ T) \times 10^{-3} \tag{12.9}$$

where ΔX = crack width (in); and
T = temperature of slab at mid-depth when pavement age is less than 20 days.

Crack width obtained by the use of these equations is strictly the magnitude at the surface.

A fundamental concept in the design of continuously reinforced concrete pavements is the determination of the optimum percentage of longitudinal steel. Sufficient steel must be provided to hold the cracks tightly closed and to prevent the steel from being overstressed. Steel stresses are expected to be maximum at the transverse cracks.

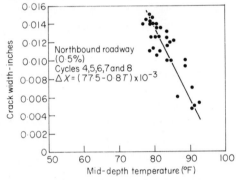

FIG. 12.9. Relationship between crack width and temperature. (From [16].)

It is preferable to have cracking patterns with a small average crack spacing, since shorter slab segments will result in smaller fluctuations in the crack widths.

The initial 10 days of pavement life is the period when the concrete properties are undergoing the greatest rate of change. The strength, the modulus of elasticity and the thermal coefficient are increasing at various rates as the concrete ages.

12.8 DESIGN OF CRC PAVEMENTS FOR HIGHWAYS

12.8.1 Recommendations

By the end of 1971, about 10 000 miles of equivalent two-lane highway of the continuously reinforced concrete pavement type had been either built or placed under contract in the USA. In general, these pavements have performed in a satisfactory manner.

The USA Bureau of Public Roads recommends the following minimum criteria for construction of continuously reinforced concrete roads.

1. Heavy-duty traffic: 8 in slab thickness, 0·6% continuous longitudinal steel (minimum yield point = 60 000 psi), either deformed bars or welded wire fabric, placed $2\frac{1}{2}$ in from surface but never below mid-depth of the slab and 3–6 in granular sub-base to prevent edge pumping.

2. Relatively little truck traffic: 7 in slab thickness, 0·7% steel and 3–6 in granular sub-base.

All projects in service conforming to these design minima have shown generally satisfactory performance when properly constructed.

The above-mentioned recommendations are based on the truck loads used in the USA. In countries where heavier truck loads are utilised, such as in Europe, thicker pavements should naturally be used. Because this type of pavement is heavily reinforced in the longitudinal direction, transverse cracks develop at close intervals throughout the entire length. Crack intervals of from 2 to 10 ft have been reported, and investigators have noted that crack intervals are related to the percentage of longitudinal steel. The longitudinal reinforcement must be in sufficient amount and distributed throughout the width of the pavement so that these transverse cracks are kept very narrow and shear interlock will prevail between the slab sections on either side of a crack.

12.8.2 General Concepts of Thickness Design

It has been generally accepted that the concrete slab must resist stresses and deflections produced by applied wheel loads. The function of the longitudinal steel is to keep cracks in the concrete tightly closed so that there is effective load transfer across these cracks. Accordingly, designers can apply the same design methods used for jointed pavement by assigning a very high degree of load transfer to represent the situation for continuously reinforced concrete pavement.

Thickness design and longitudinal steel design must be considered simultaneously in a continuously reinforced concrete pavement. If too small an amount of steel is used, transverse cracks will open an excessive amount and aggregate interlock will be lost, resulting in appreciable slab deflections and ultimate slab deterioration and failure. Measurements made on experimental pavements with approximately 0·6% longitudinal steel showed average surface crack widths ranging from 0·003 to 0·031 in. It should be noted that cracks which may be that wide at the upper surface generally measure much less at mid-depth and near the bottom of the slab.

12.8.3 Thickness Design

As tensile and flexural stresses accumulate in the concrete pavement, cracks form and the stresses are redistributed. When the cracks are held tightly closed by longitudinal steel, there is some shear transfer across the crack, but the moment transfer is somewhat reduced, since these cracks act as hinges. Thus, deflection may become critical, as well as load stress, in the design of a continuously reinforced pavement. Extensive load-deflection tests have shown the deflection of 8 in CRCP to be no greater than that of 10 in jointed pavement.

As failures in concrete pavements in the AASHO road test were influenced greatly by deflections and resulting pumping, the empirical expressions developed from this test reflected these items, and were extended to apply to CRCP design [3]. Subsequently, modifications were introduced to permit of more versatility in materials and some latitude in load transfer at cracks. The modified procedure is selected because it has adequately served as the design procedure for continuously reinforced concrete pavements in service.

The results of the AASHO road test were used to develop a design equation for the thickness of concrete pavements based on certain fixed and certain variable parameters. Design parameters varied in the tests were axle load relations, frequency of load application and slab thickness. Parameters held as constant as possible were modulus of elasticity of concrete, E_c; modulus of rupture of concrete, MR; modulus of subgrade reaction, K; and slab continuity. The environmental conditions may also be accepted as a constant parameter since only one geographical site was included in the test.

Analyses were based on the mathematical model:

$$\log W_t = \log \rho + \frac{H_t}{\beta} \qquad (12.10)$$

or

$$H_t = \beta \left(\log W_t - \log \rho \right) \qquad (12.11)$$

in which H_t = a function (the logarithm) of the ratio of loss in serviceability at time t to the total potential loss taken to the point where $p = 1.5$;

p = serviceability (an index number established in AASHO road test) at end of time period;

ρ = a function of design variables and load variables that denotes the expected number of axle load applications from the time of construction to the time the pavement reaches a serviceability index p of 1.5;

W_t = axle load applications at time t; and

β = a function of design and load variables that influences the shape of the p–W performance curve.

The term H_t was further defined as:

$$H_t = \log \frac{(c_0 - p_t)}{(c_0 - 1.5)} \qquad (12.12)$$

where p_t = present serviceability index value at time t; and

c_0 = a constant related to initial condition of pavements in the AASHO road test, and taken as 4.5 for reinforced concrete pavements.

Evaluation of the terms of eqn. (12.10) by the techniques used at the AASHO road test gave the following values for ρ and β:

$$\log \rho = 5.85 + 7.35 \log (h+1) - 4.62 \log (L_1 + L_2) + 3.28 \log L_2 \qquad (12.13)$$

$$\beta = 1.00 + \frac{3.63 (L_1 + L_2)^{5.20}}{(h+1)^{8.46} L_2^{3.52}} \qquad (12.14)$$

where h = thickness of the concrete slab (in);

L_1 = load on one single axle or one tandem axle set (kips); and

L_2 = axle code number (one for single axle and two for tandem axle).

Equivalency factors were obtained from the AASHO road tests to relate all axle loads to a common denominator. The common denominator chosen was the 18 000 lb single-axle load, so L_1 in the design equation became 18 and L_2 became 1.

Spangler's equation, because of its simplicity of form, was used for convenience to include the fixed parameters of the AASHO road test in a design equation. An almost linear relation was noted when values of load stress in the AASHO road test slab were plotted against stress computed by Spangler's equation. This relation was expressed as:

$$\log W_p = f + b \log FS \qquad (12.15)$$

where $FS = MR/\sigma$;

MR = modulus of rupture of concrete;

σ = maximum tensile stress in the concrete (psi), computed from the Spangler equation

$$\sigma = \frac{JL_1}{h^2}\left(1 - \frac{a_1}{l}\right);$$

J = a coefficient dependent on load transfer characteristics of slab continuity, and taken as 3·2 at the AASHO road test;

a_1 = distance (in) from the corner of slab to centre of load L_1, and equals $a\sqrt{2}$ where a is radius of circle equal in area to the loaded area under half the axle load;

$$l = \sqrt[4]{\frac{E_c h^3}{12(1 - v_c^2)K}};$$

b = slope of the log W_p – log (M R/σ) curve;

f = a constant; and

W_p = number of applications of a given load to reach a terminal serviceability.

A relationship was then obtained which expressed b in terms of p for 18 000 lb single-axle loads for a range of p between 1·5 and 2·5.

The expressions given above were then used to obtain a relationship between the physical properties of the concrete used in the AASHO road test and the physical properties which might be encountered in design. These relationships were then substituted into eqn. (12.15) and the following equation was obtained:

$$\log W_p' = \log W_p + b_p (\log \mathrm{FS}' - \log \mathrm{FS}) \qquad (12.16)$$

where W_p = number of load applications required to reach a given serviceability p for a given rigid slab from the AASHO road test equation;

FS = M R/σ ratio for the AASHO road test slabs;

M R = modulus of rupture of AASHO road test slabs;

σ = tensile stress calculated for AASHO road test slabs;

W_p' = the number of load applications to reach a given serviceability p for a similar rigid slab with different physical properties as described by FS′;

FS′ = M R′/σ' ratio for the modified properties;

M R′ = modulus of rupture for slabs to be designed (third-point loading);

σ' = tensile stress calculated for slabs to be designed; and

b_p = 4·22 − 0·32p.

Two important assumptions were then made. These assumptions were presented by the AASHO design committee as:

1. The variation in load applications W_t required to reach a certain FS level for different loads was properly evaluated by the AASHO road test equation and was adequately expressed by the equivalence factors used in the design guide.

2. Any change in FS due to variations in the physical constants (E_c, K, h, MR) will have the same effect as varying thickness h; this relationship was given by eqn. (12.15).

The final AASHO road test equation was developed by incorporating eqn. (12.10), developed from theoretical considerations, in eqn. (12.16), which represented results obtained from the AASHO road test. The resulting equation was:

$$\log W'_p = 7\cdot35 \log (h+1) - 0\cdot06 + \frac{H_t}{\beta'} + b_p \log \frac{MR'\sigma}{MR\sigma'} \qquad (12.17)$$

where β' is the value of β when $L_1 = 18$ and $L_2 = 1$.

In the AASHO road test slabs the values for the physical properties of the concrete and subgrade soil were determined as:

$E_c = 4\cdot2 \times 10^6$ psi (static test at 28 days);
$K = 60$ psi/in (30 in diameter plate); and
$MR = 690$ psi (28 days, third-point loading).

By including these values for E_c, K and MR in the design equation, substituting values for σ and σ' determined from the Spangler equation and computing the term b_p as given previously, the following equation resulted:

For $p = 2\cdot5$,
$$\log W'_t = 7\cdot35 \log (h+1) - 0\cdot06 + \frac{H_t}{\beta'} +$$
$$+ 3\cdot42 \log \left[\frac{MR' (h^{0\cdot75} - 1\cdot132)}{690 \left(h^{0\cdot75} - \dfrac{18\cdot416}{Z^{0\cdot25}} \right)} \right] \qquad (12.18)$$

where W'_t = number of cumulative 18 kip axle load applications to time t; and
$Z = E_c/K$.

It was realised that there were factors such as climatic and geological differences between the AASHO road test site and other geographic regions which must be taken into consideration in any design. Equation (12.18) also reflects the specific values of concrete properties of different qualities of concrete. By assuming Poisson's ratio as 0·20, the modified equation becomes:

$$\log \Sigma L = -9\cdot483 - 3\cdot837 \log \left[\frac{J}{MRh^2}\left(1 - \frac{2\cdot61a}{Z^{0\cdot25}h^{0\cdot75}}\right)\right] + \frac{H_t}{\beta'} \qquad (12.19)$$

where ΣL = the number of accumulated equivalent 18 kip single-axle loads;

J = a coefficient dependent on load transfer characteristics or slab continuity;

MR = third-point loading modulus of rupture of concrete at 28 days (psi);

h = nominal thickness of concrete pavement (in);

$Z = E_c/K$;

E_c = modulus of elasticity for concrete (psi);

K = modulus of subgrade reaction (psi/in);

a = radius of equivalent loaded area = 7·15 in for AASHO road test under 9 kip wheel representing half the axle load;

$$H_t = \left[\log \frac{4\cdot5 - p_t}{3}\right]; \text{ and}$$

$$\beta' = 1 + \left[\frac{(1\cdot624)\ 10^7}{(h+1)^{8\cdot46}}\right].$$

The AASHO subcommittee on rigid-pavement design in effect used a safety factor of 1·33 of the concrete strength to obtain a working stress. This essentially is a 'pavement life' term, which actually reduced the logarithm of the predicted load applications by a factor of 0·935. Performance studies on pavements would be the best basis on which to base such a factor, and designers in one state have found that the factor in that locality should be 0·896 to more nearly match results in that state. If an average of these two factors (0·9155) is used, and p_t is taken as 2·5, eqn. (12.19) becomes

$$\log \Sigma L = -8\cdot682 -$$
$$-3\cdot513 \log \left[\frac{J}{MRh^2}\left(1 - \frac{2\cdot61a}{Z^{0\cdot25}h^{0\cdot75}}\right)\right] - \frac{0\cdot1612}{\beta'} \qquad (12.20)$$

12.8.4 Use of ACI Committee 325 Nomograph in Thickness Design

In eqn. (12.10) the J term, which appeared in the basic Spangler equation, needs to be evaluated. This is the continuity or load transfer term. In the

AASHO road test this factor was fixed at the value of 3·2, which value appears in the basic Spangler equation. The above equation, which may be applied to rigid reinforced concrete pavements using a factor $J = 3·2$, may also be applied to continuously reinforced concrete pavements if some other value of J is used. One investigator has suggested a value for J of 2·2. Equation (12.20), with a value for J of 2·2, is recommended for pavement thickness determination. A nomograph of eqn. (12.20), along with an example solution using a value of $J = 2·2$, is presented in Fig. 12.10.

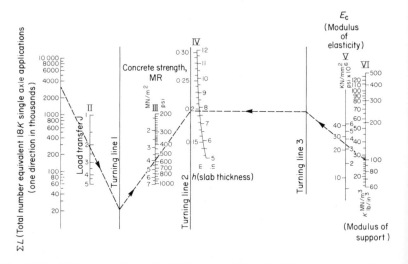

FIG. 12.10. Thickness design method. (Use compatible values for concrete properties on Lines III–V.) (*Example:* Given: $\Sigma L = 3\ 650\ 000$ applications; $J = 2·2$; $k = 100$ psi/in; $MR = 580$ psi; $E_c = 3\ 400\ 000$ psi. Answer, $h = 8·0$ in.) (*Courtesy: ACI Committee.*)

It should be noted that one of the most difficult values to be fixed in this and any other design procedure is the number of load applications to be applied to the pavement during its life. This value should be established by statistical prediction procedures from studies of past traffic counts and loadmeter station data. It should also be noted that the nomograph of Fig. 12.10 is so arranged that the designer may vary the value of J from 2·2 to any other value he chooses, within the range of 1–5. This can be done simply by passing a line through the appropriate point on the second column from the left.

The nomograph in Fig. 12.10 may be used in the following manner:

1. Mark ΣL, the total equivalent 18 kip single-axle loads expected during the life of the proposed pavement, on Scale I.

2. Mark J value on Scale II.

3. Mark subgrade support value on Scale VI.

4. Select a concrete strength and mark it on Scale III.

5. Mark E_c on Scale V. The value of E_c must be compatible with the concrete strength.

6. Connect the points on Scales I and II, projecting this line to a point on Turning line 1.

7. Pass a line through this turning point and through the value chosen on Scale III, extending the line to Turning line 2.

8. Move to the right of the nomograph and pass a line through the values chosen on Scale VI and V, extending the line to Turning line 3.

9. Connect these points on Turning lines 2 and 3, and read the value where the line passes through Scale IV. This is the required thickness.

This relatively simple-to-operate nomograph provides flexibility for the designer to choose his own value of J, the continuity value. High values of J should be associated with poor load transfer characteristics. Until more field research data are available on this subject, a value for J of 2·2 may be considered reasonable.

For values of ΣL less than 1×10^6, the nomograph gives small pavement thickness. A thickness less than 6 in is not recommended. The design procedure for thickness of pavement is semi-empirical, and is an extrapolation from equations obtained for jointed rigid pavements. The Spangler equation was originally derived for the corner loading condition on jointed rigid pavements. Its use as a mathematical form in this design procedure was not in relation to a corner loading situation, but because it has a simple mathematical form and could be easily adjusted to fit values obtained in the AASHO road test. Table 12.1 can be used in determining 18 kip single-axle applications (ΣL).

12.8.5 Other Methods for Thickness Design

The thickness of a CRC pavement can also be evaluated using the conventional method of design. For this purpose it is important to determine the equivalent number of 18 kip load applications during the design life of the pavement. Therefore, traffic counts, loadmeter data and statistical predictions are needed to determine the number of repetitions for the different categories of axle loads during that period. Table 12.1 can be applied to determine the required number of equivalent 18 kip single-axle load applications and Table 10.1 can be used to determine the corresponding stress ratio. Using the equivalent single-wheel load concept explained in Chapter 2, after assuming a reasonable thickness for the slab, will give

TABLE 12.1 EIGHTEEN KIP AXLE EQUIVALENCY FACTORS
(Courtesy: ACI Committee)

Single axles		Tandem-axle groups	
Weight (lb)	Factor	Weight (lb)	Factor
1 000	0·000 03	10 000	0·012 6
2 000	0·000 20	11 000	0·018 2
3 000	0·000 76	12 000	0·025 6
4 000	0·002 14	13 000	0·035 3
5 000	0·004 97	14 000	0·047 5
6 000	0·010 1	15 000	0·062 9
7 000	0·018 8	16 000	0·081 8
8 000	0·032 4	17 000	0·105
9 000	0·052 6	18 000	0·133
10 000	0·081 7	19 000	0·166
11 000	0·122	20 000	0·206
12 000	0·176	21 000	0·253
13 000	0·248	22 000	0·308
14 000	0·341	23 000	0·371
15 000	0·458	24 000	0·444
16 000	0·604	25 000	0·527
17 000	0·783	26 000	0·622
18 000	1·00	27 000	0·729
19 000	1·26	28 000	0·850
20 000	1·57	29 000	0·986
21 000	1·93	30 000	1·137
22 000	2·34	31 000	1·305
23 000	2·82	32 000	1·49
24 000	3·36	33 000	1·70
25 000	3·98	34 000	1·92
26 000	4·67	35 000	2·16
27 000	5·43	36 000	2·43
28 000	6·29	37 000	2·72
29 000	7·24	38 000	3 03
30 000	8·28	39 000	3·37
31 000	9·43	40 000	3·74
32 000	10·70	41 000	4·13
33 000	12·09	42 000	4·55
34 000	13·62	43 000	5·00
35 000	15·29	44 000	5·48
36 000	17·12	45 000	5·99
37 000	19·12	46 000	6·53
38 000	21·31	47 000	7·11
39 000	23·69	48 000	7·73
40 000	26·29	49 000	8·38
41 000	29·12	50 000	9·07
42 000	32·20	51 000	9·81
43 000	35·53	52 000	10·59
44 000	39·15	53 000	11·41
45 000	43·07	54 000	12·29
46 000	47·30	55 000	13·22
47 000	51·87	56 000	14·20
48 000	56·79	57 000	15·24
49 000	62·09	58 000	16·33
50 000	67·78	59 000	17·50
		60 000	18·72
		61 000	20·02
		62 000	21·39
		63 000	22·83
		64 000	24·35
		65 000	25·96
		66 000	27·65
		67 000	29·43
		68 000	31·30
		69 000	33·27
		70 000	35·34

the value of the load and tyre imprint area that can be used in the design. By applying Westergaard's equation for the central loading case, the required thickness can be determined.

12.9 DESIGN OF CRC PAVEMENTS FOR AIRFIELDS

The use of continuously reinforced concrete in the construction of airfield pavements is relatively new. Many state highway departments in the USA have utilised continuous reinforcement in the construction of highway pavements with good results. In general, continuously reinforced concrete pavement offers the following advantages to airports:

1. A pavement requiring a minimum of surface maintenance, since there are no transverse joints needing cleaning and sealing.

2. A pavement fully unitised, with no structural discontinuity due to transverse joints.

3. A pavement that provides a smooth riding surface free of transverse joints. Pavement roughness is a critical factor as aircraft become larger, use higher tyre pressures and operate at higher speeds.

4. A pavement with longer service life than with a conventional pavement.

5. A pavement that eliminates pumping of soil and water, because it is without joints.

6. A pavement with a low average annual cost, considering first cost, accumulative maintenance cost and service life.

7. A pavement that will allow for better airport operations with less delays and hazards to aircraft and the flying public.

8. A pavement with high salvage value when rehabilitation becomes necessary.

12.9.1 Thickness Design

No nomographs, to the author's knowledge, have yet been developed to determine the thickness of airfield pavements. However, since continuously reinforced concrete pavements have no transverse joints, and aircraft do not move on airfields with their wheels close to the free edge, the central loading condition can be used in the thickness design.

The charts developed by the Portland Cement Association [14] for the design of airfield pavements can be used for this purpose. These charts are based on Westergaard's solution for the central loading case and thus may be considered adequate for the design of continuously reinforced concrete pavements. In his solution, Westergaard considered a Winkler's type of foundation whereby the subgrade is regarded as consisting of a group of springs.

The two charts shown in Figs. 12.11 and 12.12 can be used for the case of aircraft with single-wheel assemblies. Figures 12.13 and 12.14 are for

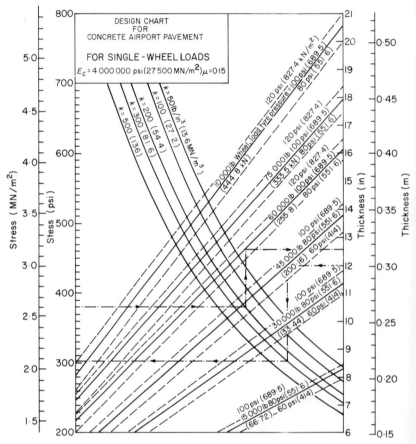

FIG. 12.11. Rigid airfield pavement subject to single-wheel loads.
(Courtesy: Portland Cement Association.)

the case of aircraft with dual-wheel assemblies, while Fig. 12.15 is for the case of dual-tandem-wheel assemblies. All the charts are plotted for a concrete having modulus of elasticity E equal to 4 000 000 psi.

To use these charts, an appropriate value for the allowable tensile stress of the concrete should first be determined. This value depends upon the number of load repetitions expected during the life period of the pavement. Multiplying the modulus of rupture of the concrete by the stress ratio, which can be obtained from Table 10.1, once the number of load repetitions

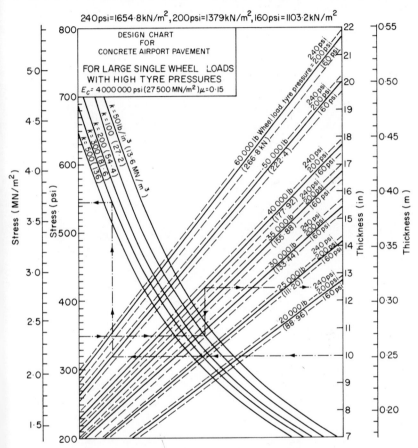

FIG. 12.12. Rigid airfield pavement subject to large single-wheel loads.
(*Courtesy: Portland Cement Association.*)

is known, will give the allowable stress. If the design is based on an un-restricted number of load repetitions, then the allowable tensile stress will be equal to half the modulus of rupture of the concrete.

These charts can be applied to determine the required thickness for a given load, tyre pressure and subgrade strength, if the allowable tensile stress of the concrete is known. They can also be used to check the stresses in a pavement of known thickness under a specified load and tyre pressure. The arrowed lines in the figures indicate the method of using the figures for the two cases.

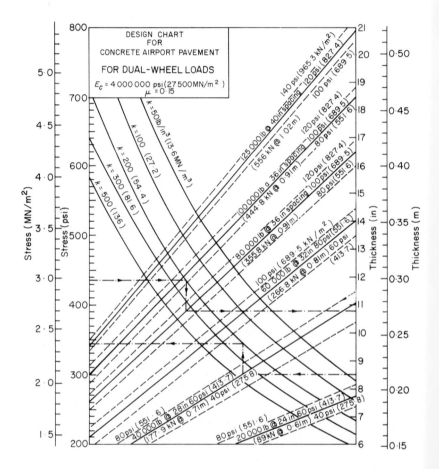

FIG. 12.13. Rigid airfield pavement subject to dual-wheel loads.
(Courtesy: Portland Cement Association.)

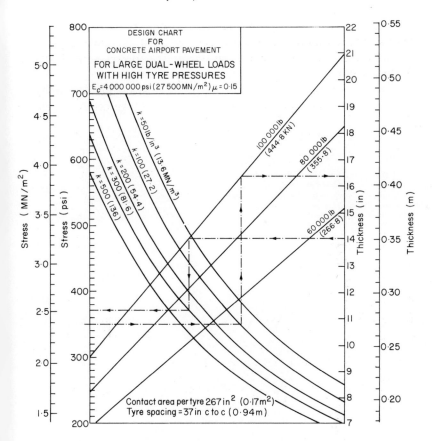

FIG. 12.14. Rigid airfield pavement subject to large dual-wheel loads.
(Courtesy: Portland Cement Association.)

12.10 PAVEMENTS WITH CONTINUOUS REINFORCEMENT AND ELASTIC JOINTS

12.10.1 Method of Design and Construction

In CRC pavements, the continuing function of the large amounts of continuous steel in the longitudinal direction is insignificant compared with its initial early-age objective. This initial objective is to promote fine cracks across the pavement section. In a mature pavement, crack widths at different temperatures seem to be directly related to crack spacing and temperature but not to the amount of steel. This indicates that fine cracks might be promoted by less steel through the use of transverse weakened planes. The result would be a substantial saving in steel quantities and possibly equally satisfactory performance. Elastic-joint pavements are intended to meet this design objective.

The term 'elastic joint' may be defined as a transverse joint across which steel reinforcement ties the two pavement elements together. As the elastic

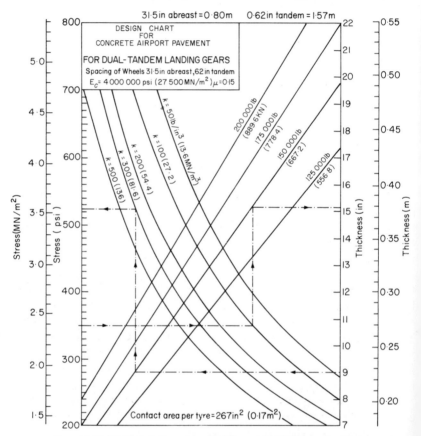

FIG. 12.15. Rigid airfield pavement subject to dual-tandem-wheel loads.
(Courtesy: Portland Cement Association.)

joint opens with decreasing temperatures, the steel is strained in tension. Overstressing the continuous steel at low temperatures can be avoided by treating the parts nearest the joint (30–40 in on each side) by asphalt coating to prevent bonding to the concrete in that part.

In the experimental sections using this type of pavement, constructed in Sweden [13], the joints consisted of ⅛ in bitumen coated hard masonite crack starters, 2 in deep placed in cut grooves. A 0·2 in deep, 45° notch was sawed in the surface at each transverse joint line 3 weeks after casting. Figure 12.16 shows the details of these joints.

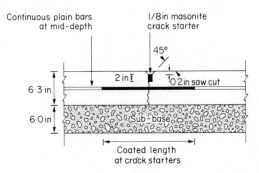

FIG. 12.16. Details of elastic joints and continuous reinforcement.

The attempt was made in Sweden to obtain the benefits of continuous-pavement designs without using the large amounts of continuous re-inforcement in the longitudinal direction. This has been tried by using elastic joints with continuous steel reinforcement crossing the joints at mid-depth to form transverse hinges across the pavement.

An initial short highway test section, constructed in Sweden, indicated that 0·2% continuous steel and elastic joints spaced 15–20 ft apart could be used. The steel bars should be coated with asphalt to prevent bond, for a short distance of about 30–40 in on each side under the elastic joints. These results were reached after trying continuous reinforcement of varying magnitudes, consisting of ½–⅝ in plain bars of 75 000 psi yield strength, spaced 12–16 in centre to centre. The 6·3 in test slabs were resting on a 6 in granular sub-base.

The highway test section constructed in 1964 was found in excellent condition when inspected after 3 years of use. No random cracks or other deteriorations could be seen.

REFERENCES AND BIBLIOGRAPHY

1. ACI Committee 325, Sub-committee VII, 'Continuous Reinforcement in Highway Pavements', *Journal of the American Concrete Institute*, December 1958.

2. ACI Committee 325, Sub-committee VII, 'Second Progress Report—Continuously Reinforced Concrete Pavements', *Journal of the American Concrete Institute*, November 1962.

3. ACI Committee 325, Sub-committee VII, 'A Design Procedure for Continuously Reinforced Concrete Pavements', *Journal of the American Concrete Institute*, June 1972.

4. Antonelli, F. A., 'Continuous Reinforcement in Airport Pavements', ASCE National Meeting on Transportation Engineering, Washington, D.C., July 1969.

5. Dechamps, Y. *et al.*, 'Large Scale Construction of Continuously Reinforced Concrete Pavements', *Proc. 2nd European Symposium on Concrete Roads, Bern, 1973*, 97–115.

6. Friberg, B. F., 'Frictional Resistance under Concrete Pavements and Restraint Stresses in Long Reinforced Slabs', *Proc. Highway Research Board*, 33, 1954, 167–182.

7. Highway Research Board, 'Continuously Reinforced Concrete Pavement', National Cooperative Highway Research Program 16, Washington, D.C., 1973.

8. Hughes, P. C., 'Evaluation of Continuously Reinforced Concrete Pavement', Final Report, Invest. No. 184, Const. Div., Office of Materials, Minn. Dept. of Highways, 1970.

9. Majidzadeh, K. and Talbert, L. O., 'Performance Study of Continuously Reinforced Concrete Pavements', Final Report, Ohio Dept. of Highways, September 1971.

10. McCullough, B. F. and Ledbetter, W. B., 'LTS Design of Continuously Reinforced-Concrete Pavements', *Proc. ASCE*, **86**, HW4, December 1960.

11. McCullough, B. F. and Treybig, H. J., 'A State-wide Deflection Study of Continuously Reinforced Concrete Pavement in Texas', *Highway Research Record*, No. 239, 1968.

12. Morgan, F. D., 'Preliminary Report on Continuously Reinforced Concrete Pavement in Oregon', *Highway Research Record*, No. 112, 1966.

13. Persson, B. O. E. and Friberg, B. F., 'Concrete Pavements with Continuous Reinforcement and Elastic Joints', *Highway Research Record*, No. 291, 1969.

14. Portland Cement Association, 'Design of Concrete Airport Pavement', Illinois, 1955.

15. Schwartz, D. R., Continuously Reinforced Concrete Pavement Performance in Illinois', *Proc. 24th Annual Ohio Highway Engineering Conference, Ohio State University, Columbus, April 1970*.

16. Shelby, M. D., and McCullough, B. F., 'Determining and Evaluating Stresses of an In-Service Continuously Reinforced Concrete Pavement', *Highway Research Record*, No. 5, 1963.

17. Spangler, M. G., 'Stresses in the Corner Region of Concrete Pavements', Bulletin No. 157, Iowa Engineering Experiment Station, Iowa State University, Ames, 1942.

18. Sternberg, F., 'Performance of Continuously Reinforced Concrete Pavement, 1–84, Southington', Final Report, Conn. State Highway Dept., June 1969.

19. Starkey, C. E. and Parish, A. S., 'Maryland's Two Experimental Continuously Reinforced Concrete Pavement Projects—I-83-Baltimore/Harrisburg Expressway', 1970 Final Report. Md. State Highway Adm., August 1971.

20. Teng, T. C. and Coley, J. O., 'Continuously Reinforced Concrete Pavement Observation Program', Report No. 1, Mississippi State Highway Department, Jackson, Miss., 1968.
21. Treybig, H. J. *et al.*, 'Effect of Transverse Steel in Continuously Reinforced Concrete Pavement', *Highway Research Board Special Report 116*, 1971.
22. Vetter, C. P., 'Stresses in Reinforced Concrete due to Volume Changes', *Trans. ASCE*, **98**, 1933, 1039–1080.
23. Wire Reinforcement Institute, 'Continuously Reinforced Concrete Pavements', Washington, D.C., 1964.

CHAPTER 13

EXPANSION JOINTS AND END ANCHORAGE FOR CONTINUOUSLY REINFORCED CONCRETE PAVEMENTS

13.1 INTRODUCTION

It has been noted that the free end of a continuously reinforced concrete pavement, regardless of its length, may undergo 1 in or more of longitudinal movement during a normal year. This movement is approximately confined to the end 300 ft of the pavement, the central portion of the pavement being fully restrained. Observations on the performance of continuously reinforced concrete pavements have indicated that while the spacing of the cracks is greater near the free end, the structural capacity of this slab is not impaired, because the width of the cracks is no greater than that in the restrained central portion of the pavement.

It is strongly recommended that adequate terminal provision be made to prevent bridge abutments from being damaged. Similar consideration should be given to locations where continuously reinforced concrete pavement is joined with other types of pavement. It is suggested that the continuously reinforced concrete pavement be terminated at least 20 ft from the bridge when bridge approach slabs are to be constructed.

Two methods of handling the expansion at pavement ends have been used with success. These are the terminal joint method and the end anchorage method.

13.2 TERMINAL JOINTS

Figure 13.1 shows one type of these joints, where a wide-flange I-Beam is cast in a reinforced concrete sleeper slab providing continuity at the joint.

The continuously reinforced concrete pavement is cast on the sleeper slab and is separated by a bond breaker such as Mylar. The inside faces of the WF beam are greased, and a void adjacent to the web is created by inserting expansion material 1–2 in thick. The end of the continuously

SUGGESTED BEAM SIZE

Pavement depth (in)	Beam size	Embedment in sleeper slab (in)
7	W10 × 54	3
8	W12 × 58	4
9	W12 × 58	3
10	W14 × 61	4

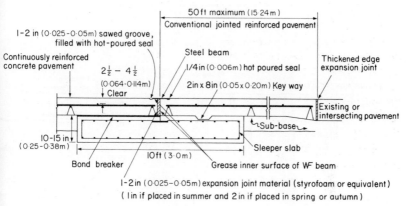

FIG. 13.1. Detail of wide-flange beam terminal joint.

reinforced concrete pavement is allowed to expand towards the wide flange beam, compressing the expansion material. Vertical plates are welded at the ends of the wide flange beam at the edge of pavement, in order to prevent foreign material from entering the void, and thus causing damage to the wide flange beam as a result of pavement expansion.

13.3 END ANCHORS

Some engineers are not entirely satisfied with the concept of free end and prefer to have the entire length in complete restraint. The concept requires that some form of anchorage be provided either at the end or distributed gradually along the active length near the end. Concrete piles (Fig. 13.2) or transverse trench lugs (Fig. 13.3) can be used for this purpose. These anchorages are usually designed to resist the full tensile force of the continuous reinforcement at yield point. However, provision should be made for some expansion space between the last anchor and the bridge approach slab or other pavement type (Fig. 13.4). Research has indicated that where end anchorages have been used, longitudinal movements plus growth of up to 1 in have been noted.

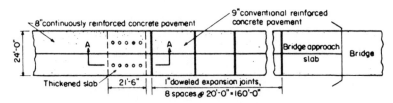

PLAN OF 10-PILE END ANCHORAGE - MISSISSIPPI

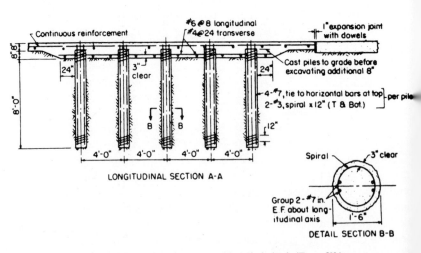

LONGITUDINAL SECTION A-A

DETAIL SECTION B-B

FIG. 13.2. End anchorage used in Mississippi. (From [2].)

Due to inherent uncertainties in assessing the soil resistance, anchor stiffness, forces involved, effect of frost, etc., trial installations have varied. Figures 13.5 and 13.6 show some modifications for future consideration.

End anchors allow of the use of continuously reinforced concrete pavements of short lengths, as in the case of industrial driveways or the

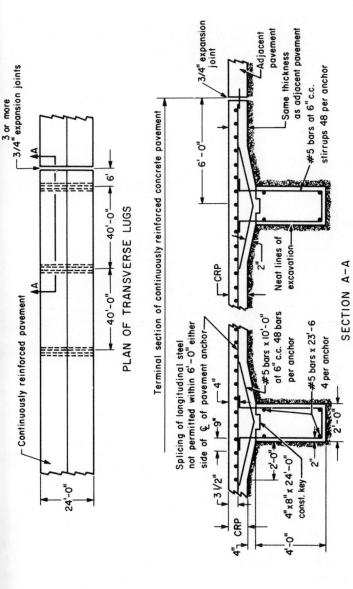

FIG. 13.3. Lug-type terminal anchor. (From [3].)

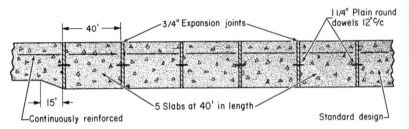

FIG. 13.4. Relief section at junction of continuously reinforced pavement and standard pavement. (From [3].)

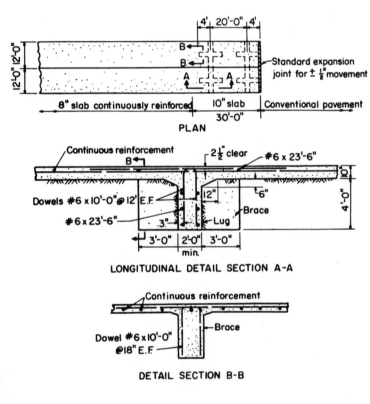

FIG. 13.5. Knee-braced multiple-lug anchorage: suggested design. (From [2].)

like. Anchors at right angles are suitable for this purpose, shown in Fig. 13.6. The end movement of a CRC pavement with terminal anchorage may

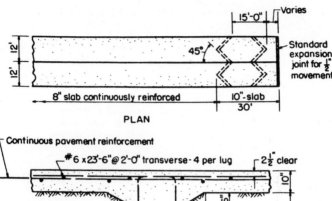

PLAN

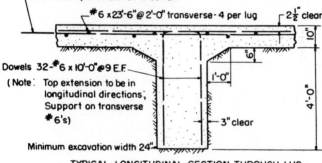

TYPICAL LONGITUDINAL SECTION THROUGH LUG

FIG. 13.6. Chevron pattern multiple-lug anchorage: suggested design. (From [2].)

be predicted using the following empirical equation suggested by McCullough [8]:

$$\Delta x = \frac{0.012\,53 \left(\dfrac{L}{|G|+1}\right)^{0.107} \cdot \Delta T}{F^{2.027}(N+1)^{0.312}} \tag{13.1}$$

where Δx = total movement for a given temperature change experienced at an expansion joint (in);

ΔT = change in air temperature for a given period (°F);

$|G|$ = absolute value of grade (%);

L = length of slab contributing to end movement (ft), and is generally assumed 500 ft maximum;

F = sub-base coefficient of friction, and can be estimated from Table 13.1; and

N = number of rigid lugs.

The values of the coefficient of sub-base friction, given in Table 13.1, may be relatively high and should be used with care.

TABLE 13.1 SUB-BASE COEFFICIENTS FOR USE IN
EMPIRICAL EQUATION

Sub-base	Sub-base coefficient of friction	Sub-base	Sub-base coefficient of friction
Surface treated	2·65	Crushed river gravel	1·93
Lime stabilised	2·13	Crushed limestone	1·93
Asphalt stabilised	1·96	Cement stabilised	1·90
Round river gravel	1·95	Crushed sand stone	1·35

13.4 LUG-TYPE TERMINAL ANCHOR

13.4.1 Design of Anchor Beams

A continuously reinforced concrete pavement slab may expand or contract due to changes in temperature, moisture, chemical reaction or mechanical stress. The movement of the slab relative to the subgrade creates frictional stresses that tend to restrain this movement. However, these frictional stresses do not prevent the development of the slab strain completely.

Movements as high as 4 in have been observed at the end of a few continuously reinforced concrete slabs. The infiltration of foreign material into the shrinkage cracks causes pavement growth with time. The large movements must either be restrained by adequate anchorage or accommodated by expansion joints.

Expansion joints capable of accommodating such large movements are rather elaborate and require considerable maintenance.

Figure 13.3 shows a lug-type terminal anchor commonly used to reduce the movement. In this type of anchor the beams extend the full width of the pavement. It is essential that the thickness and reinforcement in the anchor, and the pavement slab in the anchorage area, be adequate to prevent shear failure. The subgrade should be thoroughly compacted in the usual manner and the excavation for lugs then made. The concrete for the lugs should be placed without any side forms. Under normal conditions, restraint of the end movement is not increased significantly by increasing the number of lugs over three.

R. A. Mitchell [10] suggested a procedure for design of end anchors of continuously reinforced concrete pavements. The assumptions used for the derivation of some of the formulae have been correlated with field observations only to a very limited degree. Figure 13.7 shows the assumed slab-anchor system.

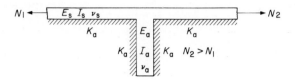

FIG. 13.7. Assumed elastic system.

The symbols v, E, I and K represent Poisson's ratio, elastic modulus of the concrete, moment of inertia of the concrete member and modulus of reaction of the subgrade soil, respectively. The subscripts s and a identify the properties of the slab and anchor, respectively.

It is assumed that the weight and stiffness of the system are sufficient to prevent significant vertical movement of the joint. Because $N_2 > N_1$, the slab is displaced to the right. As the top of the anchor is deflected horizontally, rotation is partially restrained by the slab and a bending moment is induced at the joint.

The slab may be considered as an infinitely long beam of unit width, with axial load N and transverse moment M_0, as shown in Fig. 13.8.

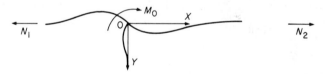

FIG. 13.8. Infinite beam on elastic foundation with transverse applied moment.

An anchor system should be designed to accommodate a selected unit strain. This strain may be induced by changes in temperature, moisture or chemical state of the pavement slab.

The end movement of the slab should be assumed equal to the range of the expansion joint to be provided. The greater the allowed end movement, the smaller the required anchor forces.

The following properties of the subgrade soil must be determined or assumed:

1. coefficient of vertical subgrade reaction under the slab, K_s;
2. coefficient of horizontal subgrade reaction throughout the anchor depth, K_a;

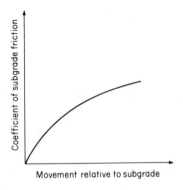

FIG. 13.9. General nature of subgrade friction–movement relationship.

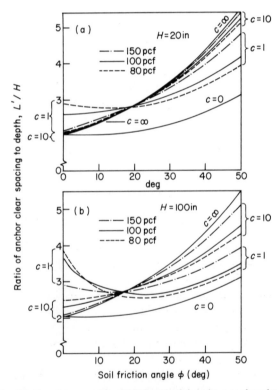

FIG. 13.10. Theoretical optimum spacing–depth ratio for various anchor depths and soil properties. (From [10].)

3. angle of internal friction, ϕ;
4. effective cohesion, c; and
5. effective unit weight of soil, $\bar{\gamma}$.

A curve of subgrade friction coefficient vs. movement must be assumed, as shown in Fig. 13.9.

Concrete properties such as Poisson's ratio and elastic modulus must be determined or assumed. The selection of anchor dimensions and shape require consideration of construction methods. Rectangular trenches can be dug with a variety of trenching machines. The deeper an anchor is, the more resistance it can develop. One should check that the anchor dimensions are sufficient after the maximum bending moment and shear have been determined. It is recommended that the first anchor unit be located at a distance from the free end equal to the anchor depth.

The required clear spacing between the remaining anchor units, L, should be determined with the following formulae or by the use of Fig. 13.10:

for
$$H > \frac{2c}{\bar{\gamma}} \tan(45° + \phi/2),$$

$$L = \frac{\dfrac{\bar{\gamma}H}{2c}(A^2 - B^2) + 2(A+B) - \dfrac{2c}{H\bar{\gamma}}}{\dfrac{\bar{\gamma}\tan\phi}{c} + \dfrac{1}{H}} \tag{13.2}$$

for
$$H \leqslant \frac{2c}{\bar{\gamma}} \tan(45° + \phi/2),$$

$$L = \frac{\dfrac{\bar{\gamma}H}{2c}A^2 + 2A}{\dfrac{\bar{\gamma}\tan\phi}{c} + \dfrac{1}{H}} \tag{13.3}$$

where $A = \tan(45° + \phi/2)$;
$B = \tan(45° - \phi/2)$;
$H = $ anchor depth;
$\phi = $ angle of internal friction;
$c = $ effective cohesion; and
$\bar{\gamma} = $ effective unit weight of the soil
$= (\gamma - \gamma_\omega)$, for saturated soil
$= \gamma$, for unsaturated soil.

At failure, it is assumed that the block of soil between two anchors is sheared off from the subgrade, as shown in Fig. 13.11.

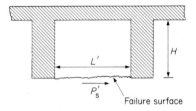

FIG. 13.11. Simplified failure mechanism.

Along the bottom failure surface, a shear force P'_s is developed. It is recommended that 65% of this value be used as the maximum resisting force developed by a single anchor:

$$P'_s = L'c + L'\bar{y}H \tan \phi \qquad (13.4)$$

$$\text{limiting anchor resistance} = 0.65 P'_s = P \qquad (13.5)$$

To determine the shear and moment strength required in the anchor, the conditions shown in Fig. 13.12 are assumed.

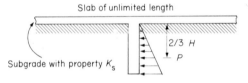

FIG. 13.12. Assumed condition for moment and shear analysis.

P, as calculated before, is then considered equal to the resultant of the subgrade horizontal pressure, which is assumed to increase linearly with depth:

maximum moment acting in the anchor,

$$M_0 = \frac{2}{3} PH \qquad (13.6)$$

maximum shear,

$$V = P = 0.65 P'_s \qquad (13.7)$$

It is recommended that the slab be thick enough and sufficiently reinforced to withstand double the moment and shear given by eqns. (13.8) and (13.9):

moment, $$M = M_0 e^{-\beta_s x} \cos \beta_s x \qquad (13.8)$$

where $\beta_s = \sqrt[4]{\dfrac{K_s(1-v)^2}{4E_s I_s}}$; and

x = distance from anchor axis along the slab.

The moment drops to zero at a distance $x = \pi/2\beta_s$ from the axis of the anchor to either side. The moment is maximum at $x = 0$, and equal to M_0.

maximum $M = M_0$:

shear, $$V = M_0 \beta_s e^{-\beta_s x}(\cos \beta_s x + \sin \beta_s x) \qquad (13.9)$$

Shear is maximum at the anchor and drops to zero at a distance $x = \frac{3}{4}(\pi/\beta_s)$. Maximum $V = M_0 \beta_s$. Figure 13.13 shows a plot of β_s vs. slab thickness t_s for $E_s = 4 \times 10^6$ psi and $v = 0.15$.

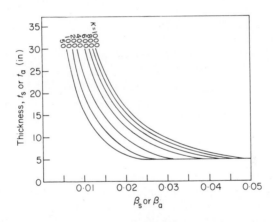

FIG. 13.13. Relationship between β and element thickness t. (From [10].)

If the distance between anchors is less than $2(\pi/2\beta_s)$, the moment on the slab should be assumed to be zero midway between the anchors. That is:

for $L < \pi/\beta_s$,

$$M = 0 \text{ at } x = L/2 \qquad (13.10)$$

It is recommended that reinforcing steel, required for moment and shear induced by the anchor, be provided, in addition to the normal continuous-pavement reinforcing steel. Slab thickness should be sufficient to resist loads induced by the anchor or traffic loads, whichever is greater. Reinforcing steel should be checked for bond, minimum spacing and cover.

13.4.2 Longitudinal Movement Analysis

After design of the anchors and end slab, the following analysis should be performed:

1. Determine δ_1 at point 1 (Fig. 13.14):

$$\delta_1 = \delta_0 + \varepsilon H \tag{13.11}$$

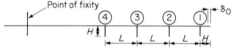

FIG. 13.14. Diagram of anchor system.

where δ_1 = movement at point 1;
δ_0 = allowed end movement;
ε = total strain to be restrained; and
H = anchor depth.

2. Determine factor C from eqn. (13.12) or from Fig. 13.15:

$$C = \left[\left(\frac{E_s t_s^3}{3(1-v^2)} \right)^3 K_s \right]^{\frac{1}{4}} \tag{13.12}$$

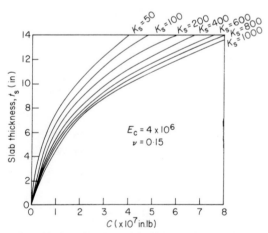

FIG. 13.15. C factor vs. slab thickness. (From [10].)

The ratio K_a/C can then be calculated and used to determine the value of R from Fig. 13.16. The factor β_a can be obtained from Fig. 13.13.

Once R has been determined, the value of P can be computed from eqn. (13.13):

$$P = RHK_a\delta_1 \tag{13.13}$$

This value of P should be compared with $0.65 P_s'$. The smaller of the two should be used as the anchor resistance force P_1.

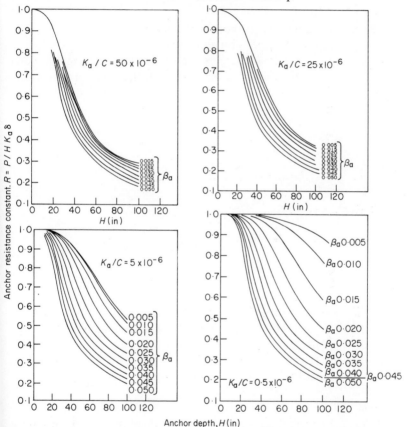

FIG. 13.16. Resistance R constant vs. anchor depth for various K_a/C ratios. (From [10].)

3. Determine movement δ_2 at point 2, Fig. 13.14:

$$\delta_2 = \delta_1 - \varepsilon L + \frac{P_1 L}{A_{s1} E_s} \tag{13.14}$$

where A_{sl} = cross-sectional area of the slab;
  E = modulus of elasticity of concrete, E_s is for the slab and E_a is for the anchor beam; and
  L = distance between anchors.

4. Repeat step 2 to determine P_2.
5. Determine δ_3:

$$\delta_3 = \delta_2 - \varepsilon L + (P_1 + P_2)\frac{L}{A_{sl}E_s} \qquad (13.15)$$

6. Repeat for additional anchors as necessary.
7. Plot a graph of δ vs. position along slab, as in Fig. 13.17.

For a satisfactory design the curve must become horizontal at some point at or above the horizontal axis, $\delta = 0$. If the curve becomes horizontal far above the axis, the system is over-designed. The total axial force

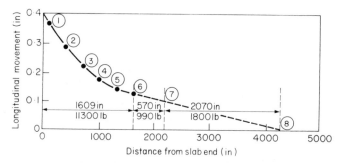

FIG. 13.17. Longitudinal movement of slab.

acting on the slab at some point beyond the last anchor must be equal to or greater than $AE\varepsilon$. The cross-sectional area of the concrete, A_{sl}, at this point may be different from that used in step 5.

Between the last anchor and the point of fixity, friction forces act on the slab and should be considered in the analysis.

Example

The following calculations are for a system where the expansion joint is 1 in wide.

  Data.

  Allowed end movement, $\delta_0 = 0.4$ in
  Anchor depth, $H = 84$ in
  Anchor thickness, $t_a = 24$ in (rectangular anchor)

Soil properties.

Horizontal coefficient, K_a = 200 psi/in

Vertical coefficient, K_s = 100 psi/in

Internal friction angle, ϕ = 32°

Cohesion, c = 10 psi

Effective unit weight, $\bar{\gamma}$ = 120 pcf = 0·0695 pci

Concrete properties.

Elastic modulus, $E = 4 \times 10^6$ psi

Poisson's ratio, v = 0·15

Thermal coefficient, $\alpha = 7 \times 10^{-6}/°\text{F}$

Unit weight, γ_c = 150 pcf = 0·0868 pci

A strain due to a pavement temperature increase of 50°F is assumed. Moisture and chemical strain are assumed zero. Thus,

$$\varepsilon = 50 \times 7 \times 10^{-6} = 3·5 \times 10^{-4}$$

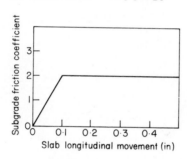

FIG. 13.18. Assumed subgrade friction curve. (From [10].)

The subgrade friction curve shown in Fig. 13.18 is assumed. Anchor spacing check:

$$\frac{2c}{\bar{\gamma}} \tan (45° + \phi/2) = \frac{2 \times 10}{0·0695} \tan \left(45° + \frac{32°}{2} \right) = 512 > 84$$

Therefore, we use the following equation:

$$L = \frac{\dfrac{\bar{\gamma}H}{2c} A^2 + 2A}{\dfrac{\bar{\gamma} \tan \phi}{c} + \dfrac{1}{H}}$$

$$A = \tan \left(45° + \frac{32°}{2} \right) = 1·804$$

and

$$L' = \frac{\dfrac{0.0695 \times 84(1.804)^2}{2 \times 10} + 2(1.804)}{\dfrac{0.0695}{10}(0.625) + \dfrac{1}{84}} = 281 \text{ in}$$

Adding the anchor thickness of 24 in gives a spacing $L = 305$ in, centre to centre.

$$P'_s = L(c + \bar{\gamma}H \tan \phi)$$
$$= 305\,[10 + (0.0695)(84)(0.625)] = 4180 \text{ lb}$$

maximum shear $= 0.65 \times P'_s = 2720$ lb

maximum anchor moment:

$$M_0 = \frac{2}{3}PH = \frac{2}{3}(2720)(84) = 152\,000 \text{ in-lb}$$

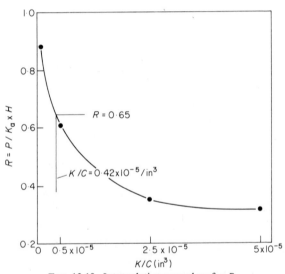

FIG. 13.19. Interpolation procedure for R.

On the slab:

maximum moment $= M_0 = 152\,000$ in-lb

maximum shear $\quad = 152\,000\,(0.0125)$
$\quad\quad\quad\quad\quad\quad\quad = 1900$ lb

The value $\beta_s = 0.0125$ is read from Fig. 13.13 assuming a 14 in thick slab. For this thickness, double reinforcement will be required to resist the moment near the anchor.

From Fig. 13.15, with $t_s = 14$ in, $C = 4.75 \times 10^7$ in-lb:

$$\frac{K_a}{C} = \frac{200 \text{ lb/in}^2}{4.75 \times 10^7 \text{ in-lb}} = 0.42 \times 10^{-5} \text{ in}^{-3}$$

From Fig. 13.13, $\beta_a = 0.010$/in. For each of the four values of K_a/C plotted in Fig. 13.16, one point is plotted in Fig. 13.19. The values are read from Fig. 13.16 for $H = 84$ in and are interpolated for $\beta_a = 0.010$.
From Fig. 13.19, the value of R is 0.65:

$$R = \frac{P}{H K_a \delta} = 0.65$$

Thus,

$$P = (0.650)(84)(200)\delta = 10\,900\,\delta \text{ lb/in}$$

Longitudinal movement.
$\delta_0 = 0.4$ in (allowed movement at the end)
At first anchor (distance H from the end):
$\delta_1 = \delta_0 - \varepsilon H = 0.400 - 0.000\,35\,(84) = 0.371$ in
$P_1 = 10\,900\,(0.371) = 4050$ lb $> 0.65\,P_s'\,(2720 \text{ lb})$
Therefore,
use $P_1 = 0.65\,P_s' = 2720$ lb
At the second anchor, located at 305 in from the first anchor:

$$\delta_2 = \delta_1 - \varepsilon L + \frac{P_1 L}{A_{sl} E_s}$$

$$= 0.371 - 0.000\,35\,(305) + 2720\,\frac{305}{10 \times 4 \times 10^6}$$

$$= 0.371 - 0.107 + 0.020 = 0.285 \text{ in}$$
$$P = 10\,900\,(0.285) = 3100 > 0.65\,P_s'$$
use $P_2 = 0.65\,P_s' = 2720$ lb

$$\delta_3 = \delta_2 - \varepsilon L + (P_1 + P_2)\frac{L}{A_{sl} E_s}$$

$$= 0.285 - 0.107 + (5440)\,\frac{305}{10 \times 4 \times 10^6} = 0.219 \text{ in}$$
$$P = 10\,900 \times 0.219 = 2390 < 0.65\,P_s'$$
use $P_3 = 2390$ lb

$$\delta_4 = \delta_3 - \varepsilon L + (P_1 + P_2 + P_3)\frac{L}{A_{sl}E_s}$$

$$= 0.321 - 0.107 + (7830)(76 \times 10^{-7}) = 0.172 \text{ in}$$

$$P = 10\ 900\ (0.172) = 1950 < 0.65\ P_s'$$

$$P_4 = 1950 \text{ lb}$$

$$\delta_5 = \delta_4 - \varepsilon L + (P_1 + P_2 + P_3 + P_4)\frac{L}{A_{sl}E_s}$$

$$= 0.172 - 0.107 + (9780)(76 \times 10^{-7}) = 0.139 \text{ in}$$

$$P_5 = 10\ 900 \times 0.139 = 1520 \text{ lb} < \frac{P_s'}{2}$$

$$\delta_6 = 0.139 - 0.107 + (11\ 300)(76 \times 10^{-7}) = 0.118 \text{ in}$$

The longitudinal movements of points 1 through 6 are shown in Fig. 13.17.

The required total restraining force is:

$$\Sigma P = A_{sl}E_s\varepsilon = 10(4 \times 10^6)(3.5 \times 10^{-4})$$

$$= 14\ 000 \text{ lb}$$

Since a thinner pavement is assumed outside the region of the anchor, the corresponding area has been used to calculate the required ΣP.

Anchors 1 through 5 develop a resistance of 11 300 lb. The line from point 6 to point 8 is a straight extension of the line from 5 to 6. The theoretical movement would actually fall above line 6 to 8. Assuming that the movement is as indicated and that the subgrade friction coefficient is as shown in Fig. 13.18, the subgrade resistance between points 6 to 8 can be determined:

$$P_{6-7} = 2(0.0868)(10)(570) = 990 \text{ lb}$$

$$P_{7-8} = 1 \times (0.0868)(10)(2070) = 1800 \text{ lb}$$

Thus, the total restraining force at point 8 is:

$$11\ 300 + 990 + 1800 = 14\ 090 \text{ lb}$$

Because subgrade friction would actually deflect curve 6 to 8 upward and thus induce even greater subgrade friction, it is safe to assume that five anchor units are sufficient. A more precise incremental analysis between point 6 and some point of fixity beyond 8 might even indicate that four anchor units would be sufficient.

If the number of anchor units exceeds three, then it would be more effective to repeat the calculations using anchor units of larger depth.

REFERENCES AND BIBLIOGRAPHY

1. ACI Committee 325, Sub-committee VII, 'Continuous Reinforcement in Highway Pavements', *Journal of the American Concrete Institute*, December 1958.
2. ACI Committee 325, Sub-committee VII, 'Second Progress Report—Continuously Reinforced Concrete Pavements', *ACI Manual of Concrete Practice, Part 1*, 1968.
3. ACI Committee 325, Sub-committee VII, 'A Design Procedure for Continuously Reinforced Concrete Pavements', *Journal of the American Concrete Institute*, June 1972.
4. Antonelli, F. A., 'Continuous Reinforcement in Airport Pavements', ASCE National Meeting on Transportation Engineering, Washington, D.C., July 1969.
5. Dechamps, Y. *et al.*, 'Large Scale Construction of Continuously Reinforced Concrete Pavements', *Proc. 2nd European Symposium on Concrete Roads, Bern, 1973*, 97–115.
6. Highway Research Board, 'Continuously Reinforced Concrete Pavements', National Cooperative Highway Research Program 16, Washington, D.C., 1973.
7. McCullough, B. F., 'A Field Survey and Exploratory Excavation of Terminal, Anchorage Failures on Jointed Concrete Pavements', *Texas Highway Dept., Res. Rep. 39-1*, March 1965.
8. McCullough, B. F., 'Evaluation of Terminal Anchorage Installation on Rigid Pavements', *Highway Research Record*, No. 362, 1971.
9. McCullough, B. F., and Sewell, T. F., 'Parameters Influencing Terminal Movement on Continuously Reinforced Concrete Pavement', *Texas Highway Dept., Res. Rep. 39-2*, August 1964.
10. Mitchell, R. A., 'End Anchors for Continuously-Reinforced Concrete Pavements', *Highway Research Record*, No. 5, 1963.
11. Shelby, M. D. and Ledbetter, W. B., 'Experience in Texas with Terminal Anchorage of Concrete Pavement', *Highway Research Board Bulletin 332*, 1962.

CHAPTER 14

DESIGN OF PRESTRESSED CONCRETE PAVEMENTS

14.1 INTRODUCTION

Experimental work and actual construction of prestressed concrete pavements have progressed in Europe since the Second World War.

A considerable improvement in concrete pavement design can be achieved by prestressing. Conventional concrete pavements are designed on the basis of low concrete flexural tensile strength without effectively utilising the high strength of concrete in compression. The precompression in the concrete due to prestressing is cumulative with the inherent flexural strength of the material, to produce an increase in stress range in the flexural zone.

In addition to this, early studies on experimental prestressed test slabs cast at Orly Airport near Paris, and tested by static load tests, indicated the following.

Prestressed pavement slabs have a considerably greater load-carrying capacity than is indicated by computations based on increasing the flexural strength by the amount of the prestress, and by assuming failure when the first crack due to positive moment occurs under the load.

These tests and later ones, together with theoretical studies, led to the following concept of failure.

As successive increments of load are applied, the slab deforms elastically to the point at which the stress due to the maximum moment beneath the loaded area exceeds the sum of the applied prestress and flexural strength of the concrete. At this point, a crack under the load forms a plastic hinge in the bottom of the slab. With the formation of the plastic hinge under the loaded area, the moments in the slab are redistributed so that with additional increments of load there is no further increase in positive moment, but a substantial increase in radial moment, some distance away from the loaded area. Tensile cracking occurs in the top surface of the slab when the maximum negative radial moment is equal in magnitude to the positive moment which produces the initial cracking at the bottom surface of the slab. The tensile cracking in the top surface of the slab forms visible circumferential cracks. When this condition is reached, failure is considered to have occurred (Fig. 14.1).

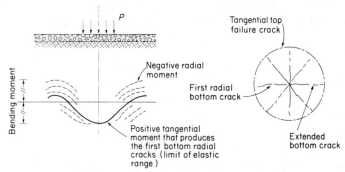

FIG. 14.1. Prestressed concrete pavements failure concept.

When the loading is increased beyond this point, vertical shear failure occurs and the load punches through the slab.

If the load under which the bottom cracks develop is removed, the force of the prestress closes the crack and the pavement regains its rigidity. Prestressed pavements can be designed to carry traffic loadings in this phase of behaviour well beyond the purely elastic range. This is the characteristic that gives prestressed pavements a potential structural advantage over conventional pavements.

In designing a prestressed concrete pavement, both the slab thickness and the prestressing cables required to produce the prestressing force should be determined. In addition, it is strongly recommended that the vertical stresses or deflection in the layers underneath the pavement be checked. Because prestressed pavements are generally thin, it is important to make sure that the loads carried by the pavement can also be carried safely by the underlying layers.

The design procedures are based on a combination of theory and experience, in varying degrees. Today's prestressed concrete pavements are the results of the accumulated experience of pavement engineers gained by: (1) study and appraisal of existing pavements, (2) observation of experimental road tests carrying normal traffic, (3) accelerated controlled traffic tests on existing or specially constructed pavement sections, (4) laboratory experimentation, and (5) theoretical and rational analysis.

The lengths of prestressed concrete pavement slabs between expansion joints should be taken as long as practically possible to reduce the number of expansion joints. The slab length is essentially a function of end pre-stress applied, subgrade friction restraint, required minimum prestress for the mid-length location, weather conditions and capacity of concreting plant. However, exaggeration in slab lengths, unless based on very accurate studies of the weather conditions and subgrade friction, may lead to serious cracks in the central part of the slab. The most critical time for prestressed pavements is that period between casting and applying the prestressing forces. For this reason, the concrete has to be protected against evaporation by spraying the surface with a curing compound within $\frac{1}{2}$–2 h after casting. Besides, portable protection tents should be used to keep the concrete free from cracks until the full prestressing force, or a part of it, is applied to the concrete. Suitable slab lengths range between 300 and 700 ft.

First, a slab thickness is assumed, generally 4–6 in for highways and 5–8 in for airports, provided that sufficient cover above the cables is provided for in both cases. When the critical stresses in the pavement are calculated according to this assumed thickness, the thickness can be changed if found unsuitable, and the procedure of determining the stresses is repeated.

14.2 STRESSES IN PRESTRESSED CONCRETE PAVEMENTS

Critical stresses in these pavements are a combination of:
1. curling stresses due to difference in temperature between the top and bottom of the slab;
2. frictional stresses between the subgrade and the pavement;
3. external load stresses.

In the design of prestressed pavements, the sum of the tensile stresses due to the above-mentioned three main factors must not exceed the allowable flexural tensile strength of concrete plus the prestressing stresses.

The fundamental formula for design of prestressed pavements is:

$$f_t + f_p > f_{\Delta t} + f_F + F_L \qquad (14.1)$$

where f_t = allowable concrete flexural stress

$$= \frac{\text{modulus of rupture of concrete}}{\text{factor of safety}};$$

f_p = compressive stress in concrete due to prestressing;
$f_{\Delta t}$ = curling stress due to difference in temperature between top and bottom surfaces of the concrete;
f_F = stresses due to subgrade friction; and
f_L = stresses due to traffic loads.

14.3 CURLING STRESSES

Curling stresses can be calculated from the formula

$$f_{\Delta t} = \pm \frac{\alpha E_c \Delta t}{2(1-v)} \qquad (14.2)$$

where α = coefficient of thermal expansion $\simeq 6 \times 10^{-6}$ in/in/°F;

$\dfrac{\Delta t}{h}$ = temperature gradient = 3·0°F/in, when concrete is warmer at top surface;

h = slab thickness (in);
E_c = static modulus of elasticity of the concrete used (psi); and
v = Poisson's ratio of the concrete $\simeq 0.16$.

14.4 STRESSES DUE TO SUBGRADE FRICTION

When a pavement slab is exposed to uniform temperature variations, it will tend to expand or contract. This movement is resisted by the friction between the slab and the subgrade.

While slab expansion due to increase in temperature causes frictional compressive stresses in the slab, as shown in Fig. 14.2(a), slab contraction due to decrease in temperature causes frictional tensile stresses that reduce the effect of prestressing, as shown in Fig. 14.2(b).

The frictional stresses in slabs of lengths less than 700 ft increase linearly towards the centre. The maximum stress is at mid-length and can be calculated from eqn. (14.3):

$$\max . f_F = \frac{F\gamma L}{2 \times 144} \text{ psi} \qquad (14.3)$$

where F = coefficient of subgrade friction;

L = total length of slab between expansion joints (ft); and

γ = unit weight of concrete $\simeq 145 \text{ lb/ft}^3$.

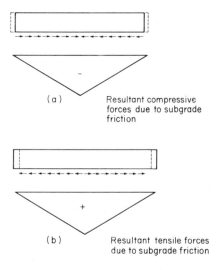

(a)   Resultant compressive forces due to subgrade friction

(b)   Resultant tensile forces due to subgrade friction

FIG. 14.2. Forces due to subgrade friction. (a) Expansion of pavement slab due to increase in temperature. (b) Contraction of pavement slab due to decrease in temperature.

For long slabs, the frictional forces increase linearly in a similar way to that shown in Fig. 14.2, until a certain length from the ends. Beyond that length, the frictional forces increase at a smaller rate towards the centre, taking a parabolic shape as shown in Fig. 14.3.

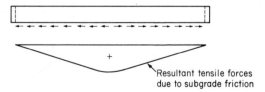

Resultant tensile forces due to subgrade friction

FIG. 14.3. Forces due to subgrade friction in long pavement slabs.

The maximum frictional stress at slab mid-length can be calculated from eqn. (14.4):

$$\max. f_F = f_p - \bar{\alpha} E_c w_a e^{-\bar{\alpha}\left(\frac{L}{2} - x_s\right) \times 12}$$

$$(14.4)$$

where f_p = compressive stress in concrete due to prestressing (psi);

w_a = initial displacement beyond which the friction factor between the pavement and subgrade becomes constant $\simeq 0.03$ in;

$$\bar{\alpha} = \sqrt{\frac{t_a}{E_c h w_a}}; \quad t_a = F\gamma h \text{ (psi); and}$$

$$x_a = \frac{h}{t_a}(f_p - \bar{\alpha}E_c w_a) \text{ (ft).}$$

14.5 STRESSES DUE TO TRAFFIC LOADS

In prestressed concrete pavements, sleeper slabs are generally cast underneath the transverse slab ends, thus increasing to a large extent the bearing capacity of the corners and transverse edges. Also, if transverse prestressing is used, which is generally the case, then enough load transfer will be generated at the intermediate longitudinal edges. For these reasons, the case of interior loading can be used in calculating the stresses due to traffic.

Westergaard's formula for an interior load acting on elliptical area, which is given in eqn. (14.5), can be used:

$$\begin{matrix} f_x \\ f_y \end{matrix} = \frac{P}{h^2}\left[0.275(1+v)\log_{10}\frac{E_c h^3}{K\left(\frac{a+b}{2}\right)^4} \mp 0.239(1-v)\frac{a-b}{a+b}\right] \quad (14.5)$$

where f_x = tensile stress at the bottom of slab in the x direction, the x-axis being taken in the direction of traffic;

f_y = tensile stress at the bottom of slab in the transverse direction;

P = maximum single load used for the design of the pavement;

a, b = semi-axes of the ellipse representing the foot print of a tyre, a being in the direction of the x axis; and

K = modulus of subgrade reaction.

14.6 PRESTRESSED CONCRETE PAVEMENTS FOR HIGHWAYS

14.6.1 Design Procedure

As explained before, a higher flexural tensile stress can be allowed in prestressed concrete pavements, especially since the traffic loads for which a factor of safety is needed form only one component of the maximum stress used in the design.

The allowable flexural stress can vary between 0·67 the modulus of rupture of the concrete, for the case of heavy traffic with a high percentage of trucks, to 0·80 the modulus of rupture, for the case of light traffic with a small percentage of trucks. This can be indicated as follows:

$$f_t = \frac{MR}{FS} \tag{14.6}$$

where f_t is the allowable flexural stress in the concrete;

MR is the modulus of rupture of the concrete;

FS is a factor of safety; and

FS = 1·5 for prestressed concrete pavements carrying heavy traffic with a high percentage of trucks;

= 1.25 for prestressed concrete pavements carrying light traffic with a small percentage of trucks.

The procedure for designing a prestressed concrete pavement for a highway is as follows.

1. Assume a reasonable thickness for the pavement slab. This thickness generally varies between 4 and 6 in.

2. Choose a suitable length for the pavement slab according to the environmental conditions in the area. The length commonly used varies between 300 and 700 ft.

3. Calculate the curling stresses using eqn. (14.2) and the frictional stresses using eqn. (14.3), for slabs up to 700 ft long, or eqn. (14.4) for longer slabs.

4. Determine the value of the maximum single-axle load to be used in the analysis. Use the equivalent single-wheel load concept to find out the value of the equivalent single-wheel load that should be used in the design.

5. Calculate the maximum stress, f_x, due to load in the longitudinal direction using eqn. (14.5).

6. Add the three stresses mentioned in (3) and (5). Use eqn. (14.1) to determine the prestressing stress, f_p.

7. Check that the prestressing stress does not exceed 800 psi, preferably 650 psi only. Also, check that the prestressing stress remaining at slab mid-length after deducting the frictional stresses is greater than or equal to 150 psi.

8. If the conditions in (7) are not satisfied, choose another thickness and repeat the process.

9. To calculate the transverse prestressing, determine the maximum stress, f_y, due to load in the transverse direction. Add this stress to the curling stress calculated before and use eqn. (14.1) to determine the transverse prestressing stress after neglecting f_F, which is the frictional stress in the transverse direction.

14.7 PRESTRESSED CONCRETE PAVEMENTS FOR AIRFIELDS

The procedure explained before for highways can also be used for airfields, except that a smaller factor of safety can be used with the concrete modulus of rupture. Also, a larger thickness is generally assumed for airfields where heavier aircraft loads are expected. The thickness commonly used varies between 5 and 8 in.

14.7.1 Design Charts
To simplify the design procedure, charts have been developed for designing prestressed concrete pavements for airfields. These charts can be easily used by pavement engineers.

The following practical points have been considered in plotting these charts in a way that covers a wide range of practical cases. Suitable values are chosen for the different variables. The single-wheel load used varies between 15 kips and 100 kips. The tyre pressure ranges from 60 to 100 psi for loads up to 45 kips and from 80 to 120 psi for higher loads. The thickness of pavement slab varies between 4 in and 9 in. The range for the modulus of subgrade reaction is between 100 pci and 500 pci. The sum of the allowable concrete flexural stress plus concrete stress due to prestressing varies between 700 psi and 1700 psi. Slab length can be any value between 300 ft and 700 ft. Two values for the coefficient of subgrade friction are arbitrarily chosen as $F = 0.6$ and $F = 0.8$.

A computer program is set to give the values of curling stresses, stresses due to friction and stresses due to load. In determining the curling stresses as well as the stresses due to load, one value for the modulus of elasticity of the concrete, $E_c = 4 \times 10^6$ psi, is considered. This value corresponds to an f_c' approximately equal to 4400 psi. If a concrete of higher quality is used, then slightly higher, but still tolerable, stresses than those obtained from the charts will occur.

The case of interior loading is considered for calculating the stresses due to traffic. This may be justified for the following reasons.

1. Sleeper slabs of suitable width are generally cast underneath the prestressed concrete slab ends to provide continuity for the pavement. These sleeper slabs allow of higher bearing capacity for the corners and cross edges of the pavement.

2. In airports, it rarely happens that the aircraft wheels go very close to the longitudinal free edge of the pavement. In addition, there is enough load transfer at the intermediate longitudinal edges due to transverse prestressing to consider these edges as non-critical.

The allowable concrete flexural stress may be taken as high as 80–100% the modulus of rupture of the concrete. The situation in which the three critical stresses used in the design add together can only occur on few occasions in practice. In addition, the prestressing of a concrete pavement changes the criterion for failure from a bottom tension crack to a top circular crack. The failure load is at least double the load that produces the first bottom crack. For this reason, a factor of safety between 1 and 1·25 may be enough for f_t.

The allowable stress in the prestressing steel, f_{se}, after losses, should not exceed $0·6 f_s'$, where f_s' is the specified minimum ultimate tensile strength of the prestressing steel.

Although the charts are plotted for slab length ranging between 300 ft and 700 ft, they can be easily used for any other length outside this range simply by extending the inclined lines at the bottom and the top of the charts until they meet the horizontal line corresponding to the new required length.

If the coefficient of subgrade friction, F, has any value between 0·6 and 0·8, then the corresponding stress, $f_t + f_p$, can be obtained by interpolation. If the value of F is outside this range, say equal to 0·50, then the stress $f_t + f_p$ can also be obtained by considering a fictitious slab length equal to the given length multiplied by 0·50/0·60 together with a value of $F = 0·60$.

Also, if the actual value of the tyre pressure is outside the range given in the charts, then a new load corresponding to a tyre pressure within this range can be obtained from the approximate relation:

$$\frac{P_1}{P_2} = \left(\frac{A_1}{A_2}\right)^{0·27} \tag{14.7}$$

where P_2 = a new load which, if applied on an area A_2, will produce approximately the same stress as that produced by the load P_1 when applied on the area A_1;

A_1 = the contact area on which the load P_1 should be applied to give the actual value of the tyre pressure; and

A_2 = the contact area on which, if a new load P_2 is applied, will produce a tyre pressure within the range given in the charts.

Once the stress $f_t + f_p$ corresponding to a certain thickness is determined from the charts and the value of f_t is defined according to the quality of the concrete and the factor of safety used, the required concrete longitudinal prestressing stress, f_p, can be easily determined.

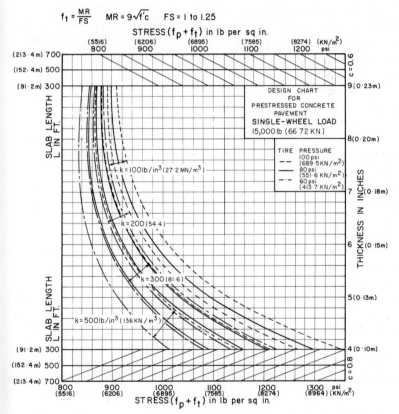

$f_t = \dfrac{MR}{FS}$ $MR = 9\sqrt{f'c}$ $FS = 1 \text{ to } 1.25$

FIG. 14.4. Design chart for prestressed pavement under 15 kip single-wheel load. (From [18].)

In prestressed pavements it may be necessary to have at least 150 psi effective prestress retained at mid-length after deduction of the frictional stresses to provide a margin of safety against the formation of cracks in this part of the slab.

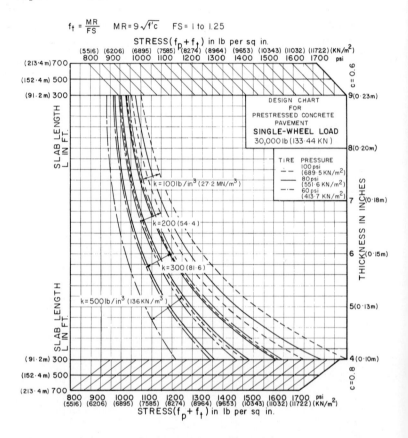

FIG. 14.5. Design chart for prestressed pavement under 30 kip single-wheel load. (From [18].)

A transverse prestressing of 0·6–0·7 the longitudinal prestressing for slabs 700 ft long, and of 0·70–0·80 the longitudinal prestressing for slabs of 300 ft long, may be used.

14.7.2 Use of Design Charts for the Case of a Single-Wheel Load

A separate design chart (Figs. 14.4–14.9) is plotted for each single-wheel load ($P = 15, 30, 45, 60, 75$ and 100 kips). For other loads, interpolation can be used. In each chart, the slab thickness in inches is plotted against the total concrete stress, f_t+f_p, for different values of the modulus of subgrade reaction K ranging from 100 pci to 500 pci. In addition, for each value of K, three values for the tyre pressure are considered. The length of pavement slab between expansion joints, as well as the values of the coefficient of subgrade friction, are taken into account by the inclined lines at the top and bottom of each chart.

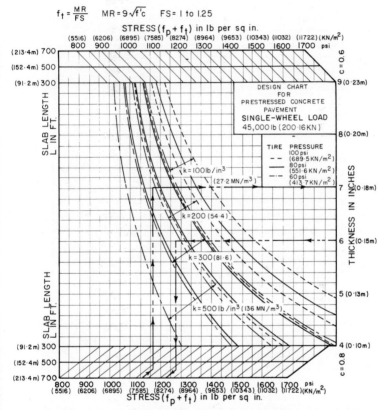

FIG. 14.6. Design chart for prestressed pavement under 45 kip single-wheel load. (From [18].)

Once the load, tyre pressure and modulus of subgrade reaction are given, a reasonable preliminary value for the thickness can be assumed. Enter the appropriate chart at the right with the assumed thickness and proceed horizontally to the left to meet the appropriate curve of K corresponding to the given tyre pressure. Proceed vertically, upward if

$F = 0.6$ and downward if $F = 0.8$, to meet the horizontal line representing the slab length; then move parallel to the inclined lines to read the value of the total stress, $f_t + f_p$, in psi.

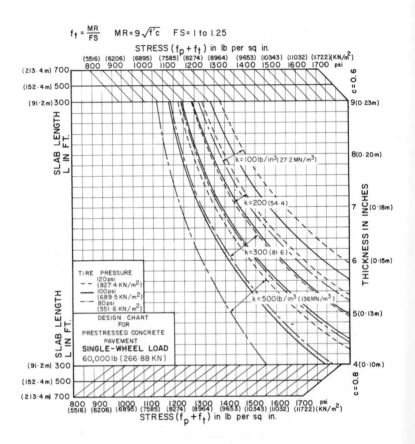

FIG. 14.7. Design chart for prestressed pavement under 60 kip single-wheel load. (From [18].)

Example

Required: The thickness of a prestressed concrete pavement slab and the prestressing stress in the concrete.

Given: $P = 45$ kips; tyre pressure $= 80$ psi; $f'_c = 4500$ psi; $K = 400$ pci; $F = 0.8$; $L = 600$ ft.

Solution.

1. Modulus of rupture of concrete = $9\sqrt{f'_c}$ = 600 psi

$$f_t = \frac{MR}{FS}.$$

Assuming the factor of safety in this case = 1·2, then

$$f_t = \frac{600}{1\cdot2} = 500 \text{ psi}$$

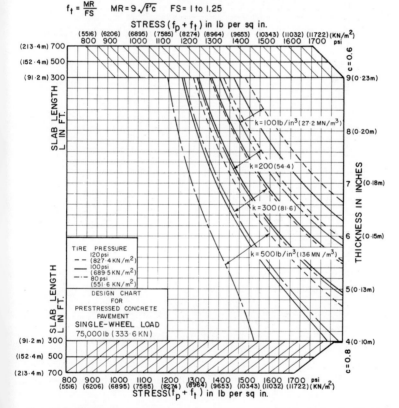

$f_t = \dfrac{MR}{FS}$ MR = $9\sqrt{f'c}$ FS = 1 to 1.25

FIG. 14.8. Design chart for prestressed pavement under 75 kip single-wheel load. (From [18].)

2. Assume a thickness of 6 in for the slab. Enter the design chart in Fig. 14.6 on the right margin at a thickness of 6 in and proceed left hori-

zontally to the desired curved line for $K = 400$ and tyre pressure = 80 psi (interpolate between the two full lines for $K = 300$ and $K = 500$).

3. Proceed vertically downward, since $F = 0.80$, to meet the horizontal line representing $L = 600$, then parallel to the inclined lines (as shown by the arrowed line in Fig. 14.6) to the lower edge of the chart. Read the required value of $f_t + f_p$, which is 1225 psi in this example.

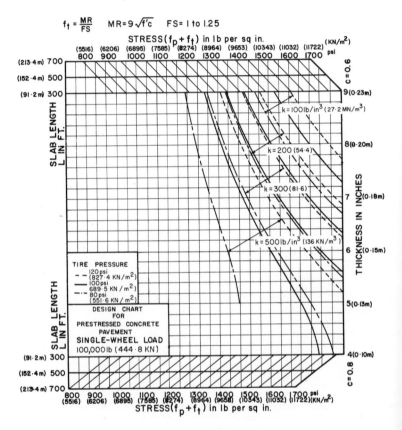

FIG. 14.9. Design chart for prestressed pavement under 100 kip single-wheel load. (From [18].)

4. The required prestressing stress in the concrete is:

$$f_p = 1225 - 500 = 725 \text{ psi}$$

Should a smaller prestressing stress, say 625 psi, be recommended, fo any reason, then either a smaller slab length or a bigger slab thickness should be used. If the slab length is kept the same, then the required new slab thickness can be determined as follows:

1. Enter the design chart in Fig. 14.6 on the bottom margin at the new value $f_t + f_p = 500 + 625 = 1125$ psi, and proceed parallel to the inclined lines to meet the line $L = 600$.

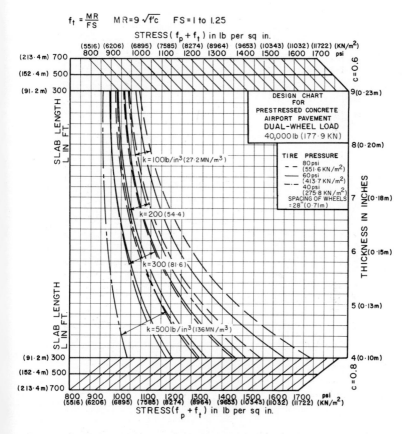

FIG. 14.10. Design chart for prestressed pavement under 40 kip dual-wheel load. (From [19].)

2. Proceed upward vertically to meet the desired curve line for $K = 400$ and tyre pressure 80 and then horizontally to the right. Read the required thickness on the right edge of the chart, which is 7 in in this example.

14.7.3 Use of Design Charts for Airfield Pavements Under Dual- and Dual-Tandem-Wheel Loading

Five charts are plotted for the case of dual-wheel undercarriage (Figs. 14.10–14.14) and two charts are plotted for the case of dual-tandem-wheel

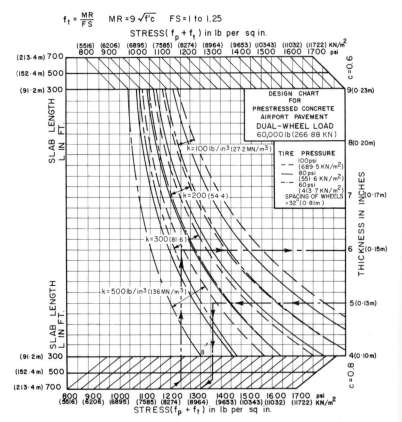

FIG. 14.11. Design chart for prestressed pavement under 60 kip dual-wheel load.
(From [19].)

undercarriages (Figs. 14.15 and 14.16). In plotting these charts, the equivalent single-wheel load concept is used to convert aircraft multiple-wheel-gear loads to equivalent single-wheel loads. They can be easily used to determine the thickness and the corresponding required prestressing stress. They also allow the choice between different possible alternatives in trying to achieve optimal solutions.

The load indicated on each chart for dual-wheel undercarriages represents the actual maximum load on one dual-wheel landing gear.

The charts for dual-tandem undercarriages are plotted for an actual maximum load on one dual-tandem-wheel landing gear ranging between 125 and 200 kips. Only one value for the contact area per tyre is considered; this is taken equal to 267 in². One of the two charts is plotted for a coefficient of subgrade friction $F = 0.60$ and the other for $F = 0.80$.

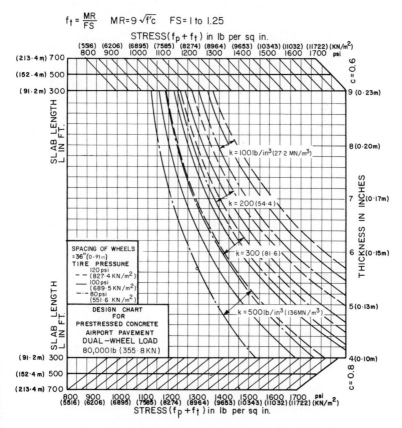

FIG. 14.12. Design chart for prestressed pavement under 80 kip dual-wheel load. (From [19].)

Once the load on the landing gear, the tyre pressure and the modulus of subgrade reaction are given, a reasonable preliminary value for the thickness can be assumed. The corresponding value for $f_t + f_p$ can be determined from the charts. If the value of f_t is estimated, according to the quality of the concrete used and according to the chosen value for the factor of safety, then the required value for the prestressing stress, f_p, can

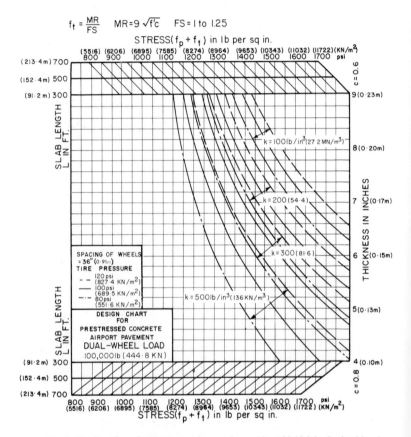

FIG. 14.13. Design chart for prestressed pavement under 100 kip dual-wheel load. (From [19].)

be determined. Should another value for f_p be preferred, then the thickness required to correspond to the new value of $f_t + f_p$ can easily be obtained.

Examples

1.. The design aircraft has dual-wheel landing gears.

Given: P (on one landing gear assembly) = 60 kips
       f'_c = 4500 psi K = 400 pci
       F = 0·8 L = 600 ft
       tyre pressure = 80 psi
       spacing of wheels centreline to centreline = 32 in

Required: The thickness of prestressed concrete pavement slab and the prestressing stress in the concrete for an airfield with the given data.

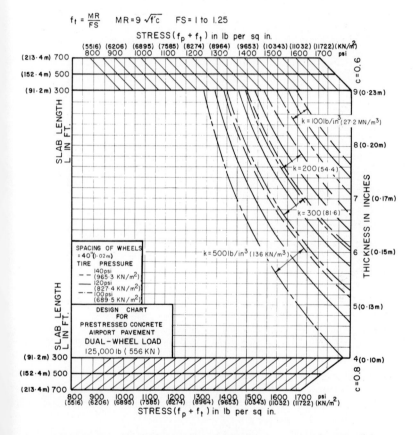

FIG. 14.14. Design chart for prestressed pavement under 125 kip dual-wheel load. (From [19].)

Solution. Modulus of rupture of concrete = $9\sqrt{f'_c}$ = 600 psi.
Assuming a factor of safety equal to 1·2,

$$f_t = \frac{MR}{FS} = \frac{600}{1\cdot2} = 500 \text{ psi}$$

Assume a slab thickness of 5 in and enter the design chart in Fig. 14.11 on the right margin at that thickness. Proceed left horizontally to the curve for $K = 400$ pci and tyre pressure $= 80$ psi (interpolate between the two full curves for $K = 300$ pci and $K = 500$ pci).

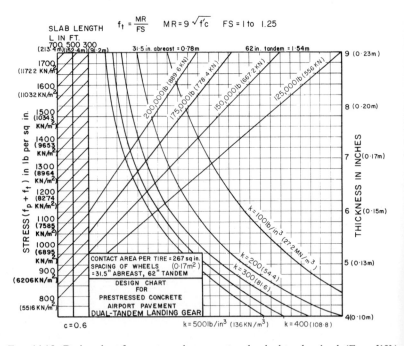

FIG. 14.15. Design chart for prestressed pavement under dual-tandem load. (From [19].)

Since $F = 0\cdot8$, proceed vertically downward to meet the horizontal line representing $L = 600$ ft, then parallel to the inclined lines (as shown by the arrowed line in Fig. 14.11) to the lower edge of the chart. This gives a value of 1320 psi for $f_t + f_p$.

The prestressing stress required in the concrete for this case is:

$$f_p = 1320 - 500 = 820 \text{ psi} > 800 \text{ psi}$$

Should a prestressing stress of, say, 700 pci be recommended, then the required new slab thickness can be determined as follows:

Enter the design chart in Fig. 14.11 on the bottom margin at the new value $f_t + f_p = 500 + 700 = 1200$ psi, and proceed parallel to the inclined lines to meet the horizontal line corresponding to $L = 600$ ft.

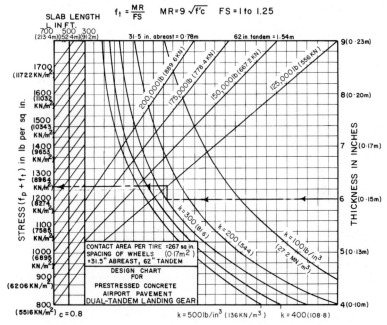

FIG. 14.16. Design chart for prestressed pavement under dual-tandem load. (From [19].)

Proceed upward vertically to meet the curve for $K = 400$ pci and tyre pressure 80 psi and then horizontally to the right. The required new thickness, which is 6 in in this example, can be read on the right edge of the chart.

2. The design aircraft has dual-tandem landing gears.

Given: P (on one landing gear assembly) $= 125$ kips
     $f'_c = 5500$ psi $K = 300$ pci
     $F = 0.8$ $L = 700$ ft
     contact area per tyre $= 267$ in^2
     spacing of wheels $= 31.5$ in abreast, 62 in tandem

Required: The thickness of prestressed concrete pavement slab and the prestressing stress in the concrete for an airfield with the given data.

Solution. Modulus of rupture of concrete $= 9\sqrt{f_c'} = 670$ psi. Assuming a factor of safety equal to 1·1, then

$$f_t = \frac{MR}{FS} = \frac{670}{1·1} \simeq 600 \text{ psi}$$

Assume a slab thickness of 6 in and enter the design chart in Fig. 14.16 on the right margin at that thickness. Proceed left horizontally to the curve for $K = 300$ pci. Proceed vertically to meet the inclined line for $P = 125\,000$ lb and horizontally to the left to meet the vertical line $L = 700$ ft (as shown by the arrowed line in Fig. 14.16). The value of $f_t + f_p$ can be read on the left margin of the chart, and is equal to 1250 psi. The required compressive stress in the concrete due to prestressing is

$$f_p = 1250 - 600 = 650 \text{ psi}$$

REFERENCES AND BIBLIOGRAPHY

1. ACI Committee 325, Sub-committee VI, 'Proposed Design for Experimental Prestressed Pavement Slab', *Journal of the American Concrete Institute*, April 1968.
2. ACI Committee 325, Sub-committee VI, 'Prestressed Pavement—A World View of its Status', *ACI Manual of Concrete Practice*, Part 1, 1968, 325-37 to 325-46.
3. *Concrete and Constructional Engineering*, 'German Recommendations for Prestressed Concrete', **49**, June 1954, 189–192.
4. Dollet, H. and Robin, M., 'The Experimental Prestressed Concrete Road at Bourge-Servas', *Travaux (Paris)*, January 1954 (in French).
5. Franz, Gotthard, 'Fundamentals of Prestressed Slabs', *Beton und Stahlbetonban (Berlin)*, No. 4, No. 5, No. 6, 1953 (in German).
6. Friberg, B. F., 'Pavement Research, Design and Prestressed Concrete', *Proc. Highway Research Board*, **34**, 1955, 65–84.
7. Harris, Allan James, 'Prestressed Concrete Runways: History, Practice and Theory', *Airports and Air Transportation*, November–December 1956, 109–111.
8. *Highway Research Record*, No. 44, 'Study of Stresses in Prestressed Concrete Pavements at Maison-Blanche Airport', Washington, D.C., 1963.
9. *Highway Research Record*, No. 60, 'Load Tests on Experimental Prestressed Concrete Pavements', Washington, D.C., 1963.
10. *Highway Research Board Special Report 78*, 'Prestressed Concrete Pavements', Washington, D.C., 1963.
11. Kellersmann, G. H., 'A New Sandwich Construction Method for Prestressed Concrete Airfield Pavements', *Proc. 2nd European Symposium on Concrete Roads, Bern, 1973*, 191–203.
12. Leonhardt, F., *Prestressed Concrete Design and Construction*, Second Edition, Wilhelm Ernst, Berlin, 1964, 554–568.

13. Mellinger, F. M., 'Summary of Prestressed Concrete Pavement Practices', *Journal of the Air Transport Division of the ASCE*, August 1961.
14. Melville, Phillip L., 'Review of French and British Procedures in the Design of Prestressed Pavements', *Highway Research Board Bulletin 179*, May 1958, 1–12.
15. Mittelman, G., 'Construction of Prestressed Pavement at an Airport in Portugal', *Journal of the American Concrete Institute*, July 1967.
16. Osawa, Y., 'Strength of Prestressed Concrete Pavements', Journal of the Structural Division, *Proc. ASCE*, October 1962.
17. Sargious, M. and Bessadah, A., 'Strengthening Prestressed Concrete Pavement', *Journal of the American Concrete Institute*, November 1969.
18. Sargious, M. and Wang, S. K., 'Economical Design of Prestressed Concrete Pavements', *Journal of the Prestressed Concrete Institute*, July–August 1971.
19. Sargious, M. and Wang, S. K., 'Design of Prestressed Concrete Airfield Pavements under Dual and Dual Tandem Wheel Loading', *Journal of the Prestressed Concrete Institute*, November–December 1971.
20. Stott, James Pearson, 'Prestressed Concrete Roads' (with discussion), *Proc. Institution of Civil Engineers (London)*, Part 2, **4**, No. 3, October 1955, 491–538.

CHAPTER 15

PRESTRESSING TECHNIQUES, PRESTRESSING STEEL, ANCHORS AND JOINTS

15.1 INTRODUCTION

To introduce the initial prestressing stress, two distinctly different systems of design and construction can be employed.

1. The continuous type of slab wherein no prestressing steel is used. The prestressing can be produced by:
   (a) Using a hydraulic jack between abutments and the slabs to be stressed, or between the slabs themselves. The use of jacks results in gaps that are subsequently filled. The concrete of the pavements is said to be post-stressed.
   (b) Using expansive cements in the concrete pavements which should be cast between abutments designed to prevent concrete from expanding and thus producing compressive stresses in the slabs. Unfortunately, the high costs of these expansive cements in addition to the lack of experience regarding the change in the forces produced from expansion with time, dimensions, mix proportion, etc., have made their use till now only practical in the case of small repairs.

2. The individual type of slab in which stressing is accomplished through the use of high-tensile-strength cables. The prestressing steel used can be:
   (a) Pretensioned steel in which steel strands are pulled to prescribed tension between anchors placed prior to concrete casting. When the concrete has attained required strength, the strands are cut at the joints.
   (b) Post-tensioned steel in which horizontal strands or bars are

coated or enclosed in tubes, unstressed before the concrete is cast. After concrete strength development, the steel is tensioned by jacking against the concrete end faces.

15.2 SYSTEM WITH CONTINUOUS TYPE OF PRESTRESSED SLABS

This system, in which hydraulic jacks are used, has the advantage of saving the steel tendons, and thus the costs of the required pavement anchors or abutments may be absorbed for suitable pavement lengths.

In addition, this type of pavement is comparatively more easy to repair than the pavement prestressed by cables, although in many cases transverse prestressing is produced by transverse cables. The disadvantages of this system are that the slabs cannot expand freely and so the stresses vary with moisture and temperature changes; as a result, it is difficult to predict the amount of initial stressing necessary to produce the desired residual required to resist the load stresses. Also, an increase in temperature, humidity or both will increase the prestress and may produce buckling failures. Such failures were reported in Belgium when 3 in and 4 in thick pavements were post-stressed and rigidly restrained at low temperatures. As expansion was impossible, increasing temperatures caused buckling.

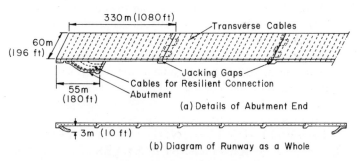

FIG. 15.1. Diagram of prestressed concrete runway at Maison-Blanche Airport, Algiers.
(Courtesy: Campenon Bernard.)

This type of prestressing technique has been used with great success at Maison-Blanche Airport in Algiers in building the two runways (Fig. 15.1). The first runway was built in 1953/54 and is 8000 ft long, 197 ft wide and 7·2 in thick, with a minimum prestress equal to 225 psi. The second runway was built in 1960/61 and is 7700 ft long, 146 ft wide and 6·4 in thick, with a minimum prestress equal to 255 psi in the longitudinal direction and 142 psi in the transverse direction, and a maximum allowed stress of 1700 psi.

The longitudinal prestressing of the pavements was obtained by post-stressing with flat jacks and elastic abutments. The jacks are 20 ft long and 6·7 in high and operate under high water pressure.

The jacks were used to counterbalance the shrinkage of the concrete immediately after placement and to induce the specified prestress later. The jacking joints were filled with concrete before putting the pavement in operation.

For the first pavement the jacking joints were constructed at 987 ft intervals, with two intermediate temporary jacking joints; these two joints were used only to counterbalance the shrinkage of the concrete. On the second runway, the intervals were reduced to 645 ft and temporary jackings were not required.

The jacks for the second runway were taken out after inducing the specified stress. However, it is possible to use them again in the same joints, if necessary, to induce higher prestress. This can be done by removing a steel plate placed to cover the gap, and inserting the jacks.

To avoid excessive curling of the slab at the joints, dowel bars and steel reinforcement were used at these joints.

The elastic end abutments consist of curved concrete slabs buried beneath the pavement with one end joining the end of the pavement at ground level and the other end about 10 ft in the ground. This buried end is designed to be restrained from any motion.

The wire tendons which give the abutment 'elasticity' pass freely through metal conduits at mid-depth of the concrete abutments and are anchored at the extremities of the pavement.

15.3 SYSTEM WITH INDIVIDUAL TYPE OF PRESTRESSED SLABS

This system, in which the pavements are stressed by prestressing cables bonded in the surrounding concrete along their length, has the advantage that the cables will provide lateral support for the concrete. Buckling due to prestressing alone will be entirely obviated, no matter how long or slender the pavement may be.

This is an important advantage for prestressed concrete in general, when the prestressing is produced by steel cables inside the concrete and not by external means.

15.3.1 Prestressing with Longitudinally Pretensioned Steel

Prestressing in which pretensioned steel is used in the longitudinal direction has an advantage: namely the direct bond between the concrete and the thin prestressing wires allows of a reasonable cover above the tendons, even for small slab thicknesses. There are, however, some difficulties associated with pretensioning. It is suitable only for longitudinal prestressing, for which purpose abutments for the anchorage of the prestressing cables must be provided at intervals of 2–7 slab lengths. The abutments for the prestressed concrete runway of an airport constructed in Portugal (Fig. 15.2) were about 2180 ft apart. The tensioning of the

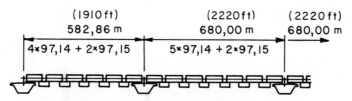

(1910 ft)	(2220 ft)	(2220 ft)
582,86 m	680,00 m	680,00 m
4×97,14 + 2×97,15	5×97,14 + 2×97,15	

FIG. 15.2. Pretensioned pavement slabs. (From [5].)

longitudinal cables between each two abutments required special prestressing jacks in the form of a hydraulic machine which stretched the cables about 10 ft in one operation. In order to avoid the risk of transmission of the prestressing forces to the concrete by direct bond at an early age of the concrete, the steel joints for both intermediate and expansion joints were provided with lashes that were clamped to the tensioned wires. After concreting and hardening of a prestressing strip, the wires were cut at the joints and the prestressing force of the cables was transmitted in this way, through the transverse steel joints, to the front surfaces of each section.

15.3.2 Prestressing with Post-Tensioned Steel

Prestressing by using post-tensioned steel (Fig. 15.3) is considered by several west European countries to be the most advantageous from the point of view of both economy and simplicity in construction. However, it is important to avoid longitudinal cracks, which may occur over the cable ducts if the ducts are large and have insufficient cover. Reinforced concrete sleeper slabs of suitable width, generally about 10 ft, are cast underneath the prestressed concrete slab ends to provide end support and continuity for the pavement. Precast concrete anchor beams, in which the anchor heads of the prestressing cables are embedded, are placed on the sleeper slabs. These anchor beams must be properly reinforced to resist the splitting tensile forces due to the transmission of the prestressing forces

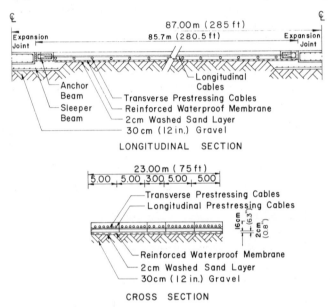

FIG. 15.3. Prestressed concrete pavements of the post-tensioned type.

at the ends. The prestressed cables, which are generally enclosed in tubes, are stretched between the anchor heads. The prestressing force is applied at the ends after casting the concrete.

15.4 THE PRESTRESSING

Once the required precompressive stresses, f_p, have been calculated in both directions, suitable cables and prestressing system can be chosen to produce the required prestressing force. Recommended spacing between the cables in the longitudinal and transverse directions is 2–4 ft.

The prestressing per cable can be obtained by multiplying the prestressing stress, f_p, by pavement thickness, h, and by the centreline-to-centreline spacing of the cables, in inches. The allowable stress in the prestressing steel, f_{se}, after losses, should not exceed $0.6 f_s'$, where f_s' is the specified minimum ultimate tensile strength of the prestressing steel.

Ordinary steel reinforcements, calculated to resist the splitting tensile stresses which occur in the transmission zone of the prestressing forces, have to be provided near the end anchorages of the prestressing cables.

When using post-tensioned cables, the longitudinal prestressing is generally done in steps. In the first step, normally 25–30% of the prestressing force is applied 24 h or less after casting and before any shrinkage cracks may occur.

The remaining prestressing force is applied about 2 weeks later, thus compensating the initial losses in the prestressing force due to shrinkage during this period.

Transverse prestressing is generally done in one step, after completing the longitudinal prestressing, and is then directly followed by grouting all the cables. This transverse prestressing may be 0.60–0.70 the longitudinal prestressing for slabs 700 ft long, and 0.70–0.80 the longitudinal prestressing for slabs 300 ft long.

Measurements have shown a non-uniform movement of a prestressed pavement consisting of 5 bays due to temperature variations, prior to cement grouting. The movement became uniform after grouting. This helps to give a straight expansion joint and reduce the troubles that generally occur at these joints. Cement grouting of post-tensioned cables also increases the ultimate strength of the pavement.

15.5 EXPANSION JOINTS

Expansion joints of prestressed concrete pavements are subject to big variation in their width due to slab length changes resulting from daily temperature cycles, seasonal temperature moisture changes, shrinkage, elastic shortening and creep. The net effect of all these changes must be considered in planning joints between slabs.

The powerful movement at these expansion joints makes it impractical to use the conventional jointing methods in prestressed pavements, even with small slab lengths.

A coefficient of thermal expansion of 6×10^{-6} should be considered in calculating seasonal movement of prestressed concrete pavement slabs.

In comparatively short prestressed concrete slabs, hollow Neoprene sections may be used in the expansion joints. For longer slabs, the type of joint shown in Figs. 15.4 and 15.5 is recommended.

If the length is too big, say more than 600 ft, the joint assembly shown in Fig. 15.6 may be the most convenient.

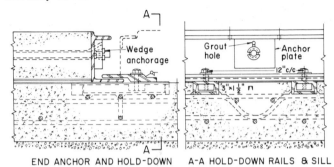

END ANCHOR AND HOLD-DOWN   A-A HOLD-DOWN RAILS & SILL

FIG. 15.4. Details of an expansion joint that prevents upward curling at the corners and edges.

15.6 PERFORMANCE OF PRESTRESSED CONCRETE PAVEMENTS

Both theoretical and practical studies have indicated that prestressed concrete pavements can adequately support interior and edge loads of considerably greater magnitude than those causing bottom surface cracks.

Also, both static and repetitive load tests carried out on highway and aerodrome prestressed pavements have proved that a tremendous increase in load-carrying capacity is induced by prestressing.

The load-carrying advantage obtained by prestressing a concrete pavement is due principally to a change in the criterion of failure from a bottom surface crack to a top surface crack.

Moreover, the records of those prestressed pavements which have been in actual service for some time indicate that their performance from the standpoint of maintenance and repairs needed compares favourably with conventional pavements. Due to the big reduction in the number of transverse joints by using prestressed pavements, a more smooth and comfortable surface for the traffic is obtained.

15.7 ADVANTAGES OF PRESTRESSED CONCRETE PAVEMENTS

A properly designed and constructed prestressed concrete pavement can provide several advantages, such as:

1. absence of cracks from the road surface;
2. a large decrease in the number of transverse joints, thus giving a smooth and comfortable travelling surface; and
3. a big reduction in slab thickness, thus achieving economy in capital cost.

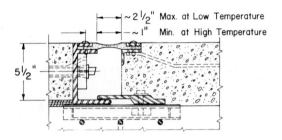

FIG. 15.5. Elastomeric joint closure.

15.8 DISADVANTAGES OF PRESTRESSED CONCRETE PAVEMENTS

1. This type of pavement needs special care in both the design and construction.

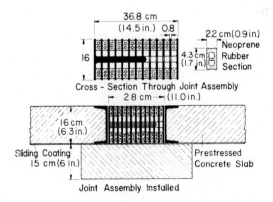

FIG. 15.6. Joint construction adopted on the Dietersheim/Bingem experimental road section. *(Courtesy: Held & Franke.)*

2. The large movement at expansion joints requires special joints that allow horizontal movement but prevent vertical movement, at each end of the slab. Conventional joints cannot be used with this type of pavement.
3. Environmental factors such as weather conditions can have a serious effect on the design and construction.
4. Reaching services beneath the pavement by cutting a trench in the slab can create serious problems unless suitable precautions are taken.

REFERENCES AND BIBLIOGRAPHY

1. ACI Committee 325, Sub-committee VI, 'Proposed Design for Experimental Prestressed Pavement Slab', *Journal of the American Concrete Institute*, April 1968.
2. *Highway Research Record*, No. 44, 'Study of Stresses in Prestressed Concrete Pavements at Maison-Blanche Airport', Washington, D.C., 1963.
3. Kellersmann, G. H., 'A New Sandwich Construction Method for Prestressed Concrete Airfield Pavements', *Proc. 2nd European Symposium on Concrete Roads, Bern, 1973*, 191–203.
4. Leonhardt, F., *Prestressed Concrete Design and Construction*, Second Edition, Wilhelm Ernst, Berlin, 1964, 554–568.
5. Mittelman, G., 'Construction of Prestressed Pavement at an Airport in Portugal', *Journal of the American Concrete Institute*, July 1967.
6. Sargious, M., 'Warping at the Edges of Prestressed and Reinforced Concrete Pavements', *Journal of the American Concrete Institute*, November 1968.
7. Sargious, M. and Bissadah, A., 'An Experimental Investigation on the Behaviour of Prestressed Concrete Pavement under Temperature Variation', *Brücke und Strasse (Berlin)*, November 1968 (in German).
8. Sargious, M. and Bissadah, A., 'Strengthening Prestressed Concrete Pavement', *Journal of the American Concrete Institute*, November 1969.
9. Sargious, M., 'Performance of Joints in Prestressed and Reinforced Concrete Pavements', Research Report CE 72-40, Department of Civil Engineering, The University of Calgary, 1972.

CHAPTER 16

FIBROUS CONCRETE PAVEMENTS

16.1 INTRODUCTION

There are several instances where fibre reinforced concrete may be used to good advantage in pavements. Test sections constructed in the USA indicate that fibrous concrete can function extremely well as a paving material and can be subjected to a large number of load repetitions.

16.2 FIBROUS CONCRETE MIX

Fibrous concrete contains the same materials used for conventional Portland cement concrete plus short steel fibres dispersed in the concrete matrix. Smaller coarse aggregates, aggregate with a smaller ratio of coarse-to-fine, and a higher cement factor, are generally used in the mix. The steel fibres are added in such a way as to achieve, as nearly as possible, a random distribution of fibre throughout the mixture. When a true random distribution is obtained, the fibre reinforces the concrete matrix in all directions and at all points.

The steel fibres generally used are nominally 1 in long and from 10 to 16 mils in diameter. The fibre content generally ranges between 0·5 and 2% by volume. A 1·5% by volume of fibre steel represents about 200 lb for every cubic yard of concrete. The bond between the cement paste and the fibres increases the strength of the mix.

16.3 DESIGN PROCEDURES FOR PAVEMENT SLABS ON GRADE

The procedure recommended for the design of fibrous concrete slab on grade involves three considerations [9].

1. Flexural stress and strength. The slab must be of sufficient thickness to accommodate the flexural stresses imposed by traffic loads. Since the stresses induced by traffic are repetitive and cyclic, a reasonable working stress for the fibrous concrete must be established to ensure performance under fatigue loadings.

2. Elastic deflection. Fibrous concrete pavement slabs are relatively flexible in comparison with conventional concrete slabs because of their reduced thickness and the more ductile nature of fibrous concrete. The value of anticipated elastic deflection must be correctly predicted, since the danger of pumping the supporting materials from beneath the slab increases as the deflection increases.

3. Foundation stress and strength. The anticipated stresses within the supporting material should be determined and examined. These stresses in the underlying layers must be low enough to preclude the possibility of introducing permanent deformation in the supporting materials. Excessive permanent deformations can cause structural failure in the pavement slabs or can result in a rough riding pavement.

16.4 FLEXURAL STRESSES

To resist the same type and volume of loads, the pavement thickness required for fibrous concrete may be in the order of seven-tenths of that required for conventional concrete pavements. The first crack strength of fibrous concrete is approximately twice the flexural strength of conventional concrete. In addition, the cracks that will develop in a fibrous concrete pavement during its life will be held tightly closed and will not spall.

For highways with large numbers of load repetitions, the allowable working stress of fibrous concrete is about 55% of the first crack strength, while for conventional concrete it is 50% of the flexural strength. The ratio of working stress for conventional concrete pavements to that for fibrous concrete pavements can be expressed by:

$$\frac{f_t}{f_F} = \frac{0.50 \, M R_c}{0.55 \, M R_F} \tag{16.1}$$

where f_t = working tensile stress for plain concrete pavements;

f_F = working tensile stress for fibrous concrete pavements;

$M R_c$ = modulus of rupture of plain concrete; and

$M F_F$ = first crack strength of fibrous concrete.

For airfield pavements with 5000 coverages during the design life period, the allowable working stress of fibrous concrete is 80% of the first crack strength; for conventional concrete, it is 75% of the flexural strength. The ratio of working stress for conventional concrete pavements to working stress for fibrous concrete pavements in this case can be expressed by:

$$\frac{f_t}{f_F} = \frac{0.75 \, M R_c}{0.80 \, M R_F} \tag{16.2}$$

Nominal values for $M R_c$ and $M R_F$ yield a working stress ratio of 0.46 for highways and 0.48 for airports.

Westergaard's formulae for computing the stresses show that stress is approximately proportional to the inverse of the square of the pavement thickness. Although the pavement thickness appears at several points in the procedure, the inverse of the square is the main factor in these formulae. As a rule-of-thumb, the ratio of thickness of a fibrous concrete pavement to a plain concrete one can be approximated by:

$$\frac{t_F}{t_c} = \sqrt{\frac{f_t}{f_F}} \tag{16.3}$$

where t_F = thickness of fibrous concrete pavement;
$\quad t_c$ = thickness of plain concrete pavement; and
f_t and f_F are as before.

The ratio of thickness of fibrous to plain concrete is approximately 0.68 for a working stress ratio of 0.46.

It is recommended that the thickness of plain concrete pavement be determined for the given load and for a flexural strength of 600 psi, using the methods described in Chapter 10. Sixty-eight per cent of the plain concrete pavement thickness should then be used to begin flexural stress computations for fibrous concrete pavements. The design process is then a trial-and-error procedure of selecting various pavement thicknesses and computing the corresponding maximum flexural stresses. The design charts presented in Chapter 10 can be used for this purpose.

16.5 ELASTIC DEFLECTIONS

As previously mentioned, the design of a fibrous concrete pavement on grade should include an analysis of the anticipated elastic deflections. For this purpose, the deflections should be calculated at the critical locations, such as at the centre and at the edge of the pavement. The deflection with the largest value is the one that should be considered in the analysis. Although elastic deflections themselves are not harmful to pavement structures, excessive deflections can aggravate pavement pumping or blowing.

A maximum static edge deflection of 0·05 in is recommended for freeways and multi-lane expressways carrying high volumes of truck traffic, while an edge deflection of 0·08 in may be allowed for highways and streets carrying small volumes of truck traffic.

For airfield pavements with less than 5000 coverages during the design life period, a maximum static edge deflection of 0·12 in is recommended, to guard against pavement pumping. If the number of coverages is between 5000 and 25 000, the static edge deflection should not exceed 0·08 in.

The elastic deflection in a concrete pavement, due to an interior load acting on an elliptical area, can be calculated from the following formula developed by Westergaard:

$$y = \frac{P}{8Kl^2} \left[1 - \frac{a^2 + b^2 + 4x^2 + 4y^2}{16\pi l^2} \log_{10} \frac{E_c h^3}{K\left(\frac{a+b}{2}\right)^4} - \frac{a^2 + 4ab + b^2}{16\pi l^2} + \frac{(a-b)(x^2 - y^2)}{2\pi l^2 (a+b)} \right] \quad (16.4)$$

where y = the deflection at point (x, y) from the centre of the load;
P = the load;
K = the modulus of subgrade reaction;
l = the radius of relative stiffness;
E_c = the modulus of elasticity of the concrete; and
a and b = semi-axes of the ellipse representing the imprint area of the tyre.

The value of the maximum deflection at the centre of the load can be obtained from eqn. (16.4) by putting zero values for x and y.

Also, the influence charts developed by Pickett and Ray in 1951 [7] can be used to compute the elastic deflections for loads acting at the centre or along the edge of the pavement.

16.6 FOUNDATION STRESSES

The use of fibrous concrete in pavements will result in thinner slabs. These thin slabs have greater flexibility and do not distribute the load over as large an area as thicker, stiffer conventional concrete pavements. For this reason, the use of fibrous concrete will increase the stresses transmitted to the foundation, and precautions should be taken to prevent the foundation stresses from becoming so great as to introduce permanent deformation into the foundation materials.

Foundation stresses can be determined by using the elastic layered theory and the computer programs available for the purpose [4, 6]. An inherent problem in using this theory is the assignment of a value for the modulus of elasticity and Poisson's ratio to a soil layer. Reasonable results are generally obtained by using 1500 times the California Bearing Ratio (CBR) value of the layer for the modulus of elasticity of the layer. The effect of different values for Poisson's ratio on the vertical subgrade stress is relatively small. A value of 0·30–0·40 for Poisson's ratio may be used.

The allowable vertical stress that can be imposed on a subgrade soil is a function of the strength of the subgrade soil. Peattie [5] has developed the relationship shown in Fig. 16.1, between subgrade soil CBR and permissible vertical stress. The curve of Fig. 16.1 was developed by computing the required pavement thickness for soils with various CBRs, and then determining the theoretical vertical stress developed under them.

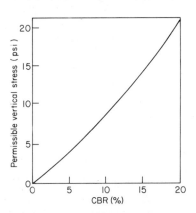

FIG. 16.1. California bearing ratio vs. permissible vertical stress. (From [5].)

This curve represents the approximate endurance limit of the pavement, and multiple applications of vertical stresses of these magnitudes have very little cumulative effect.

Peattie's curve was derived from flexible-highway-pavement data. One major difference between flexible and rigid pavements is that rigid pavements are capable of bridging over weak supporting soils, while the materials used in flexible pavements have little or no capacity for bridging. For this reason, the allowable foundation stresses shown in Fig. 16.1 can be increased by at least 50%, in the case of highways, for CBR values of 10% or below. In addition, highway pavements are subjected to several times as many traffic loadings as airfield pavements. Accordingly, the allowable foundation stresses shown in Fig. 16.1 can be at least doubled, in the case of airfields, for CBR values of 10% or below. This recommendation has been verified by field tests on fibrous concrete pavements, in which vertical stresses on a subgrade with a 4% CBR were about 10 psi, over 6000 coverages.

It should be noticed, however, that Peattie's curve and the above recommendations are applicable only if the materials are compacted to reasonably high densities. Caution should be exercised in the case of widely spread heavy loads or poorly compacted soil layers when considering foundation stresses. In addition, great care is needed in assigning modulus of elasticity values to soil layers. For example, an increase in the modulus of elasticity of more than double the modulus of the underlying layer, in a layered pavement system, is extremely difficult to obtain without a stabilising agent.

Example

The following example is presented to illustrate the recommended design procedure for fibrous concrete pavements on grade.

A fibrous concrete airfield pavement is to be designed to receive aircraft with tricycle landing gears. Each main gear consists of dual wheels spaced 37 in centre to centre, and the contact area for each tyre is 267 in^2. The design is required for C traffic areas with light traffic that produces 3000 coverages at a 100 000 lb gear load during the design life period. The subgrade is composed of a clay material having a CBR of 5% and is covered with a 4 in thick filter course, with a CBR of 10%. The modulus of subgrade reaction for the foundation system is 150 lb/in^3. The allowable working stress for the fibrous concrete in flexure is 1200 psi.

Solution. Computation of flexural stress. The thickness of a conventional concrete pavement necessary to support the given loading conditions can be determined from the chart presented in Fig. 16.2, for the aircraft considered in this problem [1]. The design chart is entered on the left ordinate with a flexural strength of 600 psi, since the factor of safety is included in the chart. A horizontal projection is made to a subgrade modulus of 150 psi, by interpolation; a vertical projection is then made to the C traffic area line and a horizontal projection is carried to the right to give the required thickness of plain concrete. This thickness is 14 in, as indicated by the broken line in Fig. 16.2. A tentative flexural stress computation is

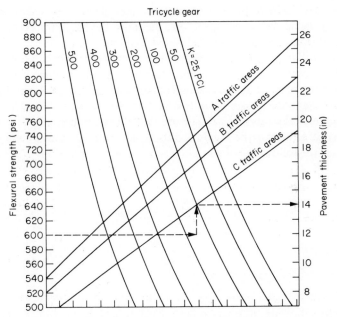

FIG. 16.2. Design curves for concrete airfield pavements. 100 000 lb gear load, twin wheels spaced 37 in c.c., 267 in².

(Courtesy: USA Department of Army and Department of the Air Force.)

performed for a fibrous concrete pavement in the range of 8–9 in thickness, using Westergaard analysis for the edge loading case and with an allowance for 25% load transfer at the joints. The results are:

Thickness (in)	Stress (psi)
8·0	1 330
8·5	1 225
9·0	1 130

Since the stress corresponding to a thickness of 8·0 in exceeds the given allowable flexural stress of 1220 psi, only the 8·5 or 9·0 in thickness should be used for further analysis.

Determination of elastic deflection. Using the influence charts of Pickett and Ray to compute the elastic deflection for the 8·5 in thick pavement under the dual-wheel gear, along a jointed edge with 25% load transfer, a value of 0·085 in for the deflection is obtained. This value is smaller than the permissible deflection of 0·12 in; therefore the 8·5 in thick pavement provides an acceptable level of elastic deflection.

Computation of foundation stress. The permissible vertical stresses for the filter course and subgrade are double the values obtained from Fig. 16.1, for the reasons already discussed. Accordingly, in this example, the filter course material can tolerate a vertical stress of 17 psi and the subgrade a vertical stress of 7 psi. The use of the layered theory in computing the vertical stresses in the 4 in thick filter layer and the subgrade underneath provides the following values:

Pavement thickness (in)	Vertical stress in filter course (psi)	Vertical stress in subgrade (psi)
8·5	10·4	9·1
9·0	9·5	8·3
9·5	8·7	7·7
10·0	8·1	7·0

The above values show that the subgrade stresses control the design. They also indicate that a pavement thickness of 10 in over a 4 in filter course would be recommended. It may also be found that a 9 in thick fibrous concrete pavement over a 12 in thick filter course is an acceptable alternative.

16.7 JOINTS

As with conventional concrete pavement construction, construction joints should be placed between adjacent lanes, and at other locations where concrete placement will be interrupted for 1 h or more. Load transfer devices or thickened edges are recommended to avoid high stresses at the construction joints or at the free edges of the pavement. Shear transfer can be accomplished by using dowelled or key joints. However, keyed joints are acceptable only for slabs greater than 9 in in thickness and if the design is for medium or light loading.

Contraction joints, perpendicular to the direction of paving, are required in fibrous concrete pavement to control cracking from concrete shrinkage.

These joints should be of the dummy-groove type, in which a weakened plane is formed either by sawing a groove or by placing an insert in the plastic concrete. The spacing of contraction joints is a function of the coefficient of subgrade friction between the slab and foundation, and of the percentage of fibre steel. Experience with fibrous concrete pavement indicates that slab lengths of up to 100 ft are acceptable for pavements using conventional or high-early-strength cements.

16.8 SUBSURFACE DRAINAGE

As mentioned before, fibrous concrete pavements will deflect more than conventional concrete pavements under the same load and are thus more susceptible to pumping or blowing. For this reason, it is essential to have a free-draining filter course of 4 in thickness or more beneath a fibrous concrete pavement, to provide positive drainage across the entire pavement cross-section. In addition, the allowable elastic deflections discussed before are still applicable, even though a free-draining filter course is required. This is essential, since the possibility of either blowing the finer fraction of the filter course or pumping subgrade materials up into the filter course still exists. Careful attention is also needed where there are abrupt changes in grade, such as at intersections with conventional concrete pavements, to prevent ponding conditions beneath the pavement.

16.9 FIBROUS CONCRETE OVERLAYS

The use of a thin fibrous concrete overlay for strengthening or improving the condition of an existing pavement is particularly beneficial for urban expressways where vertical clearance at bridges is limited. Also, the use of fibrous concrete could be advantageous in resurfacing of airfield pavements by increasing their load-bearing capacity. Two fibrous reinforced concrete overlay test sections were constructed recently at Tampa International Airport in Florida, and have demonstrated that fibrous concrete can be satisfactorily produced in central mix plants and can be placed with slipform paving equipment.

Fibrous concrete overlays on existing rigid pavements may be of either the non-bonded, the partially bonded or the fully bonded type. Fibrous concrete overlays on existing flexible pavements should be treated as slabs on grade.

(a) *Non-bonded overlay.* The use of a non-bonded overlay is recommended for application where the base concrete pavement is in poor structural condition and needs a large increase in its structural capacity. In this case, a bond-breaking course between the overlay and the concrete is necessary.

The following formula can be used to determine the required thickness of a plain concrete overlay:

$$h_c = \sqrt{(h^2 - Ch_e^2)} \qquad (16.5)$$

where h_c = thickness of plain concrete overlay;
$\quad h_e$ = thickness of existing pavement;
$\quad h$ = design thickness of an equivalent strength single plain concrete slab; and
$\quad C$ = condition factor.

The following values of C can be used:

$\quad C = 1.00$ (base pavement is in good condition);
$\quad C = 0.75$ (base pavement has initial non-progressive cracks due to load);
$\quad C = 0.35$ (base pavement is badly cracked or crushed).

Once the thickness, h_c, of a plain concrete overlay is determined, it should be multiplied by a factor of the order of 0·55–0·65 to obtain the required thickness of fibrous concrete overlay.

(b) *Partially bonded overlay.* This type of overlay is recommended for situations where an increase in load-carrying capacity is required and the base pavement is in fair condition.

The following formula can be used for determining the required thickness of a plain concrete overlay for this particular case:

$$h_c = 1.4\sqrt{(h^{1.4} - Ch_e^{1.4})} \qquad (16.6)$$

where h_c, h, h_e and C are as before.

The thickness, h_c, obtained from eqn. (16.6) for a plain concrete overlay should be multiplied by a factor in the range of 0·55–0·65 to obtain the thickness of the fibrous concrete overlay.

(c) *Bonded overlay.* A bonded overlay can only be used if the structural condition of the base pavement is very good. Even if this condition is satisfied, several disadvantages are apparent when using fully bonded overlays; for example, the elaborate and high cost of surface preparation, the positioning of the fibrous concrete in the compression zone of the slab and the positioning of load transfer devices across the joints.

The equation recommended for determining the thickness of fibrous concrete overlay, in this case, is as follows:

$$h_F = 0.9\,(h - h_e) \tag{16.7}$$

where h_F = thickness of fibrous concrete overlay;

$\quad h$ = thickness of plain concrete pavement desired, as determined from rigid-pavement design charts; and

$\quad h_e$ = thickness of base pavement.

Due to the high crack-resistant properties of fibrous concrete, it may not be necessary to match base pavement jointing schemes if partially bonded or non-bonded fibrous concrete overlays are used. Also, load transfer devices may only be required for non-bonded overlays.

16.10 RESEARCH NEEDS

The use of fibrous concrete in pavement slabs on grade or in overlays is still at an early stage. Items that need further research are: thickness requirements, types of sub-base needed, optimum fibre content, cost considerations, spacing between transverse joints and maximum size of coarse aggregate best suited for fibrous concrete paving mixtures. For example, aggregate size of $\frac{3}{4}$–1 in may be suitable for thicker pavement sections, while aggregate size of $\frac{3}{8}$–$\frac{1}{2}$ in may be suitable for thicknesses less than 3 in.

In addition, studies are needed to evaluate the performance of fibrous concrete pavements and overlays under repeated loads, and to examine the crack patterns that may develop when this material is used in pavements.

REFERENCES AND BIBLIOGRAPHY

1. Department of the Army and Department of the Air Force, 'Rigid Pavements for Airfields Other than Army', TM 5-824-3, AFM 88-6, Chapter 3, December 1970.
2. *Fibrous Concrete Construction Material for the Seventies, Conference Proceedings,* M-28, December 1972. Construction Engineering Laboratory, P.O. Box 4005, Champaign, Illinois, 61820.
3. Gray, B. H., 'Fiber Reinforced Concrete: A General Discussion of Field Problems and Applications', Construction Engineering Research Laboratory, US Army Corps of Engineers, Technical Manuscript M-12, April 1972.
4. Michelon, J., 'Analysis of Stresses and Displacement in an N-Layered Elastic System Under a Load Uniformly Distributed on a Circular Area', Computer Program, California, September 1963.

5. Peattie, K. R., 'A Fundamental Approach to the Design of Flexible Pavements', *Proc. 1st International Conference on the Structural Design of Asphalt Pavements, University of Michigan, Ann Arbor, August 1962.*

6. Pentz, M., Jones, A. and van Kempton, H., 'Layered Systems Under Normal Surface Loads', Computer Program, Koninklijke/Shell Laboratorium, Amsterdam, Holland.

7. Pickett, G. and Ray, G., 'Influence Charts for Rigid Pavements', *Trans. ASCE*, 1951.

8. Westergaard, H. M., 'New Formulae for Stresses in Concrete Pavements', *Trans. ASCE*, 1948.

9. Yrjanson, W. A., 'Pavement Application of Fibrous Concrete', *Proc. Conference on Fibrous Concrete Construction Material for the Seventies*, M-28, Illinois, 1972.

CHAPTER 17

MATERIALS, METHODS OF CONSTRUCTION AND PERFORMANCE EVALUATION OF RIGID PAVEMENTS

17.1 INTRODUCTION

Pavement performance depends to a large extent upon the quality of the materials used and the techniques applied in construction. Quality control of materials together with high standards of supervision and operator skill are vital to the production of a good pavement.

17.2 CONCRETE MATERIALS

17.2.1 Cements

The types of cement generally used in concrete pavements are:

1. Ordinary Portland cement. It has a medium rate of setting and hardening and is used for most types of highway work.

2. Rapid hardening Portland cement. It hardens more rapidly than ordinary Portland cement and is used when concrete of higher early strength is required. Special care should be taken during the first days after casting to protect the concrete pavement from the sun, otherwise wide cracks may develop. The use of this cement in pavements is not recommended except in special conditions.

3. Portland blast-furnace cement. Its properties are very similar to those of ordinary Portland cement except that it evolves less heat, is slightly more resistant to some forms of chemical reaction and gains strength more slowly at early stages.

4. Sulphate-resisting Portland cement. It is used in pavements where soils or groundwater containing sulphates are in contact with concrete.

17.2.2 Aggregates

The maximum size of aggregate in general use for highway work varies between $\frac{3}{4}$ and $1\frac{1}{2}$ in, with the smaller size preferred for pavements carrying heavy traffic. Coarse aggregate is the material retained on a $\frac{3}{16}$ in sieve. Fine aggregate is the natural sand, crushed rock, crushed gravel sand or other material passing a $\frac{3}{16}$ in sieve.

The two important properties of aggregates for concrete in pavements are durability and cleanliness. The aggregates should be hard and should not contain materials that are likely to decompose or change in volume when exposed to the weather.

Aggregates with the full range of grading formed by adding together the required proportion of single-size materials are better than materials that are supplied containing all sizes of aggregate. Also, the ratio to the total of fine aggregate, passing sieve $\frac{3}{16}$ in, is important.

17.2.3 Water

The essential requirement is that the water should be reasonably free from impurities such as dissolved salts, organic matter and suspended solids which may adversely affect the properties of the concrete. Sea-water does not normally affect the strength or durability of concrete. However, it may sometimes have adverse effects, which can mar the appearance of the work.

17.2.4 Concrete

The workability of the concrete is the property which determines the amount of work necessary to produce full compaction, and is the most important property in the plastic state. Workability can be measured by either the compacting factor test or the slump test.

Although the workability of concrete increases as water content and maximum size of the aggregate increase, concrete strength decreases with increase in water content and maximum size of aggregate. In addition, concrete with high water content will be more permeable and will shrink more than a concrete with low water content.

Depending upon weather conditions, concrete with workability between 0·85 and 0·90 compacting factor ($\frac{1}{2}$–$1\frac{1}{2}$ in slump) generally gives good results. The higher compacting factor or slump is used in the case of warm dry weather.

Thorough mixing of the concrete is important to get a uniform mix in which no segregation takes place. Segregation in a concrete mix occurs when some of the constituents tend to separate from the main mass. Segregation indicates, in general, poor aggregate grading, incorrect water content or faulty handling techniques, and can usually be overcome if sufficient care is given to quality control of the concrete.

The concrete mix must be designed to suit the compaction means available on site, in order to produce the required density. Compaction is generally defined as the removal of the air entrapped in mixing and handling and is reached by working the concrete.

The abrasive effect on high-quality dense concrete caused by rubber-tyred vehicles is negligible. Concrete of 4000 lb/in^2 minimum cube crushing strength after 28 days is generally satisfactory in this respect. However, tracked vehicles or similar do have an abrasive effect, and it is recommended that a concrete having a minimum 28 day crushing strength of 5000–6000 lb/in^2 and made with a rough rock aggregate be used in such cases.

To reduce the possibility of frost damage, a fully compacted concrete with a low water/cement ratio should be used. In recent years, it has been a common practice to remove snow and ice from paved areas by sprinkling calcium chloride or salt onto the surface. This procedure aggravates the damage produced by frost. A practical solution for this case is to use an air entraining agent in the concrete as a safeguard against scaling.

17.2.5 Admixtures

No admixtures should be used in concrete pavements unless it can be clearly shown that it will be advantageous. There are several types of admixtures that can be used in improving the concrete properties, and before using them it is desirable to carry out preliminary tests under the actual site conditions to ensure that the improvement desired is in fact achieved and that no detrimental side effects result.

Accelerators such as calcium chloride are used to speed the progress of concrete work or to allow of working in cold weather. Their effect is to accelerate the hydration of the cement, thus producing more heat; as such, they should be used in a concentration not exceeding 2% by weight of the cement.

Air entraining agents increase the resistance of the concrete to frost by incorporating non-connected minute bubbles of air into the concrete mix. For this reason, all concrete which is likely to be subjected to freezing and thawing cycles and to have neat salt applied to its surface, should contain an air entraining agent. The recommended amount of entrained air necessary to prevent frost damage is $4\frac{1}{2} \pm 1\frac{1}{2}\%$. Entrained air also increases workability, reduces segregation, reduces strength and reduces bleeding at the surface.

Water reducing agents increase the workability of the concrete mix without an increase in the water content. Their use is sometimes accompanied by a loss of strength.

Retarders are used when placing conditions are difficult and when the mortar must remain plastic for a considerable time, such as in hot dry weather.

17.3 CONCRETING

17.3.1 Hot-Weather Conditions

When concrete is cast in hot-weather conditions, it tends to dry out as a result of evaporation of the mixing water. The practical difficulties of the reduced time for transporting and placing the concrete due to earlier setting in hot weather must be taken into consideration. Also, special attention should be paid to protection and curing. Curing should be started as soon as possible, preferably immediately after the final finishing.

Retarders may be used where the concrete is subject to a good degree of control.

Tests have shown that concrete strength decreases as the temperature of the freshly mixed concrete and the initial curing temperature increases.

Plastic cracks are those that occur in the surface of fresh concrete soon after it has been placed and before it has fully hardened. They appear mostly on horizontal surfaces and their principal cause is rapid drying of the concrete at the surface. If the rate of evaporation exceeds the rate at which water rises to the surface, it is possible that plastic cracking will occur on the concrete immediately after compaction. These cracks can be avoided by improved curing methods and by protecting the concrete from the sun by tents supported by frameworks mounted on wheels.

 In general, it is not advisable to cast concrete pavements when the temperature is higher than 107°F in the shade.

17.3.2 Cold-Weather Conditions

Concrete can be laid in freezing conditions provided the concrete itself is prevented from freezing before it has hardened sufficiently. If concrete reaches a crushing strength in excess of 800 lb/in², it is unlikely to be damaged by frost. The precautions for casting concrete in cold weather are designed to prevent the water in the concrete from freezing and to obtain 800 psi crushing strength as soon as possible. If the water contained in the fresh concrete freezes, the mass expands so that normal strength can never be reached and disintegration usually takes place.

The precautionary measures in cold-weather concreting are mainly the extension of the curing time and the insulation of the concrete after placing to prevent loss of heat. If light frost occurs at night, it is important also to make sure that the aggregates are not frozen. If there is severe frost day and night, then it is essential to heat the water and aggregate, mix the concrete near the job, enclose the placed concrete and provide continuous heating, in addition to the previous precautions.

It should be noted, however, that the strength gain of concrete is very low at the freezing temperature, and at temperatures below freezing there is almost no increase in strength.

17.3.3 Concrete Curing

Water is essential for the chemical action of setting and hardening of concrete. Also, there is normally an adequate quantity of water available in the mix for full hydration, and it is necessary to ensure that this water is either retained or replenished to enable the chemical reaction to continue until the concrete is fully hardened. Concrete made with ordinary Portland cement, blast-furnace or sulphate-resisting Portland cement, needs curing for at least 7 days under average conditions. If rapid hardening Portland cement is used, this period can be reduced to 3 days. When concreting takes place in cold-weather conditions, the normal curing period should be doubled.

It is especially important to provide adequate curing as early as possible. Strength lost by lack of warmth or moisture during the first few days cannot be regained by subsequent curing.

The methods commonly used in curing are:

1. Spraying with water.
2. Covering the surface and edges of the slab with clean and wet burlap or cotton mats, which should be overlapped at least 4 in. Sometimes sand that is kept continuously wet can be used.
3. Covering with plastic sheeting with sufficient strength to prevent tearing and prevent evaporation of the water. The sheets should cover the entire surface and edges, and should have weights on them to prevent displacement. They should be overlapped at least 4 in. Sheets which have a white coating on the top surface are more effective in reducing temperature variation in the slab. This is particularly important in hot weather.

4. Spraying with curing membrane compounds shortly after casting. These compounds should be tested to make sure that they can retain at least 95% of the moisture in the concrete. White-pigmented membranes are particularly useful because their high light reflectance reduces the temperature variations in the concrete.

17.4 CONSTRUCTION OF CONCRETE PAVEMENTS

After compacting the subgrade or sub-base, the top surface underlying the pavement should be graded to the required surface level. In pavement projects that are expected to receive heavy traffic, a sub-base layer is usually desirable unless the subgrade is of high quality. Well-graded sands or gravels, lean concrete, cement stabilised soil or artificial materials can be used as sub-base materials to provide a hard, smooth layer upon which to work.

17.4.1 Form Work

The next step is to lay the side-forms for casting the concrete. Steel forms supported on a continuous bed of cement mortar or concrete (at least 3 in thick) are generally used for this purpose. Forms should be aligned over long lengths in order to ensure accuracy. They should be well supported with an adequate number of steel pins, which are driven into the ground through holes in the form. Support may be given to the vertical face of the form by means of back stiffeners. Also, each form is provided with a sliding locking plate at one end and housing at the other to allow of proper matching-up of the formwork both in line and in level at each end.

Forms often serve not only as shaping for the sides of the concrete slab and its top surface, but also as support rails, which are generally needed for running the machines used in pavement construction. If such machines are used, particular care should be taken to ensure that the forms are rigidly supported, so that they will not be displaced during the spreading and compaction of the concrete.

It should be emphasised that a large percentage of the major irregularities in a concrete pavement surface can be attributed to the use of defective forms which are inadequately placed. For this reason, it is important to use good-quality forms and employ suitable labour to lay them accurately.

After completion of form laying and aligning, the subgrade or sub-base is further prepared to the exact elevations and shape of the surface (often within $\pm\frac{1}{2}$ in) to ensure uniform slab thickness. A subgrade planer which moves along the side forms can be used for this purpose.

In large projects, especially where slipform pavers are used, automatically controlled subgrade trimmers can be utilised to advantage. These machines appeared a few years ago and have greatly improved the accuracy of preparing the subgrade or sub-base to the required elevations and shape of surface. In these machines, electronic sensors follow the string lines for alignment and grade. The crown is set into the cutting edges of the trimmer.

When the final preparation of the subgrade or sub-base is completed, the surface should be covered with reinforced waterproof paper or polyethylene sheeting. Sometimes the surface is instead sealed with a bituminous material blended with a thin layer of sand. This prevents the escape of cement paste into the subgrade or sub-base layer and will reduce the friction between the slab and sub-base. Another alternative is to wet the surface just before the concrete is cast in order to prevent water from being absorbed from the concrete when it is placed. This alternative should only be used where there is a hard smooth surface to the sub-base.

17.4.2 Concrete Mixing

For constructing highway and airfield pavements, batching and mixing of the concrete can be done using one of the following three methods:

1. Batching is carried out at a central plant. The dry batched material is then transported to mixing and spreading machines at the site. The site mixer moves parallel to the spreading and compacting machines, so that the freshly mixed concrete is always near to the location where it has to be placed.

   This method allows of excellent control over the mixing time. Since water is added at the site just prior to spreading, there is no chance for the concrete to set prior to casting. Even if there is a breakdown in equipment at the site, the material can be kept unmixed in the mixer for later use. In addition, any defect that may be found in the concrete during construction can be rectified immediately before starting another batch. However, the use of this method becomes difficult when continuous placing of large quantities of concrete is required throughout the day.

2. A central plant is used to batch and wet mix the concrete before it is transported to the spreading machines at the site. Both the batching equipment and the mixing machines are located together at the central plant, thus obviating the necessity to move the mixer

continuously during construction. In this method, more sophisticated storage facilities and batching and mixing equipment are generally used, which permit of a high degree of quality control.

Some problems may result in using this method if the wet concrete has to be transported long distances to the site. If the travelling time exceeds, say 20–30 min, the concrete will start to stiffen to a degree which may detrimentally affect its consistency and workability on arrival at the site. In addition, hauling on bumpy roads causes segregation in the wet mixture, which in turn causes defects in the compacted concrete. Also, in case of failure of a major piece of equipment at the site, all concrete in transit between the mixer and the site may have to be disposed of if it cannot be diverted for other purposes.

3. This method is a compromise between the previous ones. Batching is made at a central plant and transit-mixer vehicles are used to transport and mix the materials at the same time on their way to the site.

 If the construction site is far from the central plant, the addition of water and mixing of the materials can be started on the way just before the transit-mixer reaches the site. By doing this, the problems of segregation and stiffening of the concrete can be practically eliminated.

 However, the size of the drums of most transit-mixers is smaller than the on-site or central plant mixers. In addition, the operating cost of the transit-mixing equipment is generally high. These two factors limit the use of this method to situations where there is a number of small projects distributed over a large area.

17.4.3 Concrete Spreading

It is important to spread the concrete in a way which ensures uniform compaction. Unless spreading is made by hand, the following two categories of spreading machines are generally used.

1. Box-hopper machines, which can move longitudinally on the side rails and transversely across the subgrade or sub-base in order to deposit the concrete. The concrete is transmitted from the mixing machines to the hopper by lorries, transit-mix vehicles or similar. The hopper, which can be opened at the bottom, consists of a V-shaped power-operated box mounted on a wheeled chassis. Once filled with the concrete to a uniform height, it can travel transversely between the forms and deposit the wet concrete to the desired height.

The machine on which the hopper is mounted can move longitudinally on the side-forms to allow spreading of a new layer.

To allow for subsidence occurring during compaction, the concrete must be spread to a thickness which is larger than finally required. The amount of extra thickness, usually called surcharge, depends on the workability of the concrete mix. It generally ranges between 25 and 18% of the compacted thickness for concrete having workabilities within the usual range, i.e. with compaction factors between 0·80 and 0·85, respectively.

If steel reinforcement is used in the concrete pavement, then the concrete is usually spread in two layers, with the reinforcement mat placed on top of the lower layer. To ensure complete bond between the two layers, two spreading machines are sometimes used. This will allow the second layer to be placed before the first layer starts to set.

2. Spreading machines with a moving blade or rotating screw to distribute the concrete heaped on the subgrade or sub-base. The freshly mixed concrete, which is dumped in small well-distributed heaps, is distributed uniformly across the pavement width by transversely moving blade or screw spreaders. These blade or screw spreaders are part of wheel-mounted machines which can move longitudinally along the side-forms.

17.4.4 Concrete Compaction and Finishing

The next stage after spreading the concrete to the desired depth is to compact it in order to get a dense homogeneous slab of concrete which is free from voids, honeycombs and surface irregularities. This compaction can be accomplished by either surface vibration, internal vibration or tamping of the wet concrete. The most widely used process is compaction with power-propelled surface vibrators supported by the side-forms and moved longitudinally as compaction goes on.

The finishing operation should then start immediately after completing the compaction. The function of finishing is to obtain a pleasant appearance for the surface, free from depressions and irregularities, highly skid-resistant and with good light reflection properties. A heavy beam which oscillates transversely across the width of the slab, immediately behind the compactor, is normally used to remove irregularities. The transverse ridges left by this finishing can be removed by a longitudinal beam which is set parallel to the centreline of the road and then moved from one side of the casted slab to the other with a wiping motion which eliminates irregularities along the direction of travel.

Finally, an adequate skid-resistant surface texture should be built into the concrete pavements in the transverse direction, as soon as possible and certainly before the disappearance of the surface water sheen. One or more of the following texturing methods can be used: (a) burlap dragging, (b) brooming, (c) use of plastic combs.

The pavement surface should include both fine and coarse texture. The fine texture, which is formed by the sand in the cement–mortar layer, imparts the adhesion component in the tyre–pavement interaction. The coarse texture, which is formed by the ridges of mortar left by the method of texturing the surface, has the dual role of imparting the hysteresis component and providing drainage under the tyre.

A grooving machine recently developed by Britain's Cement and Concrete Association can produce transverse channels of $\frac{1}{4} \times \frac{1}{4}$ in section in concrete while it is still plastic. Such a texture is recommended by the US Federal Aviation Administration.

The transverse channels are created by vibrating ribbed plate across the surface of the slab. This grooving head can be supplied either with a regular pattern for airfield work or a 1–2 in pitch random spacing for roads to eliminate the high-frequency whine normally created by vehicles travelling at speed over grooved surfaces.

The grooving head has an operating width of 4 ft at each pass and is suspended from a transversing unit mounted on the main carriage. It incorporates a hydraulically driven rotary vibrator with variable frequency and amplitude control and is raised and lowered by double-acting hydraulic rams.

Through this groove-forming technique, the concrete remains in good condition and the finish is such that an excellent road surface riding quality can be achieved. The life of the texture is 10–20 years.

The machine is designed to form part of the concreting train and runs on rails behind the finisher.

Concrete curing, discussed previously, should start immediately after the disappearance of the water sheen following texturing. Durability and wear resistance of the concrete surface are influenced to a large extent by the effectiveness of curing, which should be carried out carefully to avoid damaging the surface texture.

The time allowed to elapse between mixing the concrete and final finishing should not exceed $1\frac{1}{2}$ h. In hot-weather conditions, less time may be necessary. The irregularities in the finished surface should be within the usual specified tolerance of $\frac{1}{8}$ in/10 ft.

17.5 SLIP-FORM PAVER

This machine, which was originated in the USA in 1949, can replace all the machines in the conventional paving train except the mixer [3]. It is able to lay concrete in a single pass without using side-forms.

When slip-form pavers are used, the subgrade and sub-base are prepared in a way similar to that used for the conventional methods of construction. Instead of using side-forms, the paver is guided by a sensing device which feels its way along stretched wires fixed to correct line and level on both sides of the machine.

After spreading the mixed concrete uniformly on the subgrade or sub-base ahead of the machine, the paver moves forward, carrying with it its own side-forms and an adjustable strike-off plate. This plate, which is set slightly above the height of the finished slab, is moved forward and backward by hydraulic rams, to spread the concrete to the correct amount between the side-forms of the paver.

Two L-shaped vibrating plates with tamping bars behind them follow the strike-off plate and compact the concrete surface. A wide extrusion plate, extending the full width between the forms, follows the tamping bars. This plate, which forms the most important part of the paver, is set to the required thickness and crown before operation. With its front slightly higher than its rear, this plate acts as an orifice through which the vibrated concrete is forced as the paver moves forward. This extrusion of the concrete gives it sufficient stability to stand with vertical sides after the paver has gone by.

A transverse reciprocating belt follows the extrusion plate and gives a preliminary finish to the concrete surface. Initial support is given to the finished concrete, as the paver moves forward, by a pair of trailing side-forms held in position by transverse trusses. The last truss can also be used to drag burlap or similar material behind it to provide the skid-resistant surface texture. A little final surface and edge finishing is generally required, which is done by experienced workmen using straight edges.

In multi-lane construction, dowel bars at transverse joints and tie bars in longitudinal joints can be incorporated during casting. Also, if a slip-form paver is to be used in constructing a slab adjacent to an existing one, one track can be run in a raised position on the edge already laid.

The advantages of slip-form pavers are:
1. If large quantities of materials, coupled with large-capacity batching and mixing plants are available, the output of slip-form pavers can

be as much as 400 linear feet per hour. This output is much higher than can usually be achieved with conventional machinery.

2. The machine lays the concrete in a single pass without the need to use side-forms.

3. In large pavement projects, slip-form pavers produce a most economically constructed concrete pavement that has the same smooth riding qualities as a conventionally constructed rigid pavement.

4. The riding quality of pavements constructed by this machine appears to be largely independent of variations in the loose density of the spread concrete.

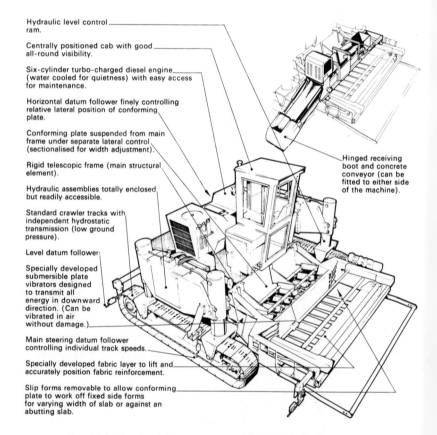

Hydraulic level control ram.

Centrally positioned cab with good all-round visibility.

Six-cylinder turbo-charged diesel engine (water cooled for quietness) with easy access for maintenance.

Horizontal datum follower finely controlling relative lateral position of conforming plate.

Conforming plate suspended from main frame under separate lateral control (sectionalised for width adjustment).

Rigid telescopic frame (main structural element).

Hydraulic assemblies totally enclosed but readily accessible.

Standard crawler tracks with independent hydrostatic transmission (low ground pressure).

Level datum follower.

Specially developed submersible plate vibrators designed to transmit all energy in downward direction. (Can be vibrated in air without damage.)

Main steering datum follower controlling individual track speeds.

Specially developed fabric layer to lift and accurately position fabric reinforcement.

Slip forms removable to allow conforming plate to work off fixed side forms for varying width of slab or against an abutting slab.

Hinged receiving boot and concrete conveyor (can be fitted to either side of the machine).

Fig. 17.1. Details of the components of the CPP60 slip-form paver.
(Courtesy: Stothert & Pitt Ltd.)

The disadvantages of slip-form pavers are:

1. Reinforcement is more difficult to introduce in a slab formed by a slip-form paver.
2. Heavy rain, if it occurs soon after placing the concrete, can cause slumping of the pavement edge, because the edge is not protected with side-forms.

Because of the economy that can be achieved by using slip-form pavers in large projects, its potential is high and its use is increasing in several countries. The different components of a slip-form paver recently developed in the United Kingdom for medium-sized projects are shown in Fig. 17.1.

17.6 PERFORMANCE EVALUATION OF RIGID PAVEMENTS

The performance of a pavement is a measure of its accumulated service or the adequacy with which it serves its purpose. The evaluation of pavement performance involves a study of the functional behaviour of an entire stretch of pavement. For this purpose, information is needed on the trend of the effect of load application on the ability of the pavement to serve traffic.

The performance of a pavement can be predicted by studying the pavement safety, serviceability and durability.

17.6.1 Safety

The design and construction of highway and airfield pavements is undertaken to provide safe carrying platforms for the loads planned to operate on them at given tyre pressures. Such pavements are built up of materials specified to have a minimum strength and thickness. Because of variations in cement strength, the properties of concrete, the compaction of bases and sub-bases as well as the thickness of construction, the final pavement strengths must themselves vary. Generally, the variations are supposed to be such as to give strength above the minima specified. However, in many situations, unseen effects such as changes in soil moisture content or subsoil water level after some years of service, as well as general overloading, may reduce the strength of the pavement. Thus, the actual load-bearing capacity of a pavement at any time may vary from the design figure and can only be determined by full-scale load testing.

17.6.2 Serviceability

There are five fundamental assumptions associated with the pavement serviceability concept. These may be summarised as follows.

1. Highways and airports are for the comfort and convenience of the travelling public. A good highway or airport is one that is not only safe but also smooth.
2. The user's opinion as to how he is being served by highways or airports is on the whole subjective.
3. There are, however, characteristics of highways and airports that can be measured objectively which, when properly weighed and combined, are in fact related to the user's subjective evaluation of the ability of the highway or airport to serve him.
4. The mean evaluation of all users should be a good measure of highway serviceability.
5. The performance of a pavement can be described if one can observe its serviceability from the time it was built until the time its performance evaluation is desired.

A major development resulting from the AASHO Road Test was the introduction of the term 'Present Serviceability Index'. This index represents the degree to which the public considers itself to be served by the pavement system. In the USA a PSI of zero represents a very poor pavement, while a PSI of 5 represents an excellent pavement.

The following formula can be used in computing the Present Serviceability Index:

$$PSI = 12 \cdot 54 - 4 \cdot 49 \log_{10} R - 0 \cdot 01 \sqrt{C+P} - 1 \cdot 38 D^2 \qquad (17.1)$$

where R = roughness (in/mile), as measured by a roughometer or profilometer;

C = pronounced cracking ($ft^2/1000 \ ft^2$);

P = patching ($ft^2/1000 \ ft^2$); and

D = average rut depth in both wheel paths (in), as measured by a 4 ft long straight edge.

In an extensive study made recently in the USA, a complete stress analysis was performed for all rigid pavement slabs in the AASHO ROAD TEST for which serviceability data was available. The critical stresses for each loading case and slab were plotted against the number of load repetitions needed to reduce the present serviceability index to 2·5. Each tandem-axle load was considered as two single-axle load repetitions. It was found that for the conditions that existed in the Road

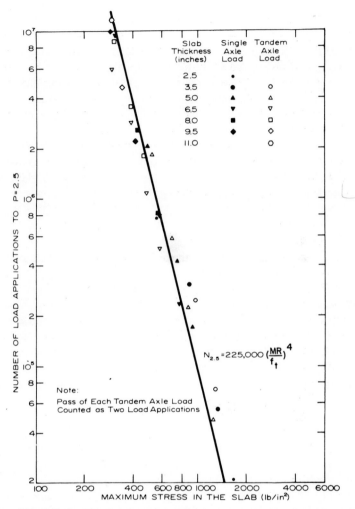

FIG. 17.2. Relationship of performance to maximum combined tensile stress in slab.
(From [20].)

Test, a unique relationship results regardless of slab thickness and axle loading.

It can be shown that the data in Fig. 17.2 can be fitted by the expression [19]:

$$N_{2\cdot5} = 225\,000 \left(\frac{MR}{f_t}\right)^4 \tag{17.2}$$

where $N_{2.5}$ = number of load repetitions needed to reduce the Present Serviceability Index to 2·5;

MR = the modulus of rupture of the concrete used in the pavement slab (for AASHO Road Test slabs MR = 790 psi); and

f_t = absolute maximum tensile stress in the pavement slab caused by traffic load, P, placed in some extreme position such as the slab edge or the slab corner.

Using eqn. (17.2) and considering the fact that the pavement stress, f_t, for AASHO Road Test conditions is found to be proportional to wheel load, P, and inversely proportional approximately to the 1·25 power of the slab thickness, h, has led to the following general relationship between principal variables in the AASHO Road Test:

$$N_{2.5} = C(MR)^4 h^5 P^{-4} \tag{17.3}$$

where C is a constant.

Equation (17.3) suggests that slab thickness should be increased as the fifth root of the anticipated number of load applications.

17.6.3 Durability

The term 'pavement life' cannot be given an exact definition. Some engineers and highway agencies consider the life of a concrete pavement ended when the first overlay is placed. On this basis, concrete pavement life may be less than 20 years on some projects with design, material or construction defects, and more than 50 years on other projects without these defects. Rigid pavements should be designed for a life period of not less than 20 years, and 25 years or more is recommended.

The number of load applications has a direct effect on the pavement life. If the number of load applications is high, then a lower allowable tensile stress for the concrete should be considered in the design. This will in turn produce a larger thickness.

Equation (17.3) indicates that under otherwise equal circumstances the pavement life may be increased 1·8 times by adopting a 9 in instead of an 8 in slab thickness. At the same time, the pavement life may be reduced to one-half by adopting a 7 in instead of an 8 in slab thickness.

It also follows from the equation that the pavement life varies as the fourth power of the concrete strength. This points out the importance of quality materials in pavement construction. A 10% increase in strength may mean a 50% increase in pavement life; a 20% increase in strength may mean doubling the pavement life. At the same time, a 10% reduction in concrete strength may mean reducing the pavement life to 65% of the normal expectation; a 20% reduction in strength may mean reducing the life to 40% of the normal expectation.

Equation (17.3) may provide a rational basis for evaluation of effects of overload and mixed traffic on pavement life. A consistent 20% overload may reduce the pavement life to one-half that normally expected. One application of double load is equivalent to 16 applications of normal load; at the same time, 16 applications of the half-load in mixed traffic should be equivalent to one application of normal load. About 6500 applications of a 2 kip axle load should be equivalent to one application of an 18 kip load.

17.7 FACTORS AFFECTING RIGID PAVEMENT PERFORMANCE

Climate, including frost action and precipitation. A sub-base course consisting of gravel, sand, crushed stone or a stabilised material will improve the performance of the pavement with respect to climate. Performance surveys have shown that each type of sub-base course listed above will give satisfactory performance, as long as the gradation is properly controlled and the base is compacted to a high relative density. The choice of a particular type of sub-base will depend upon economical factors in the area. In considering gravel, crushed stone and sand bases, however, sand bases have shown, in some cases, better performance than the other materials. The reasons for this are that sand base course materials will result in less pumping action than gravel, and also more fines can be permitted in a sand base than in the larger-sized materials.

However, recent research work conducted in Germany [18] has indicated that sub-base layers of sand and gravel underneath concrete pavements in service for a long period of time show a very low compaction (< 90% of the simple Proctor density test). This demonstrates that sand and gravel, provided as frost protection layers, become loosened by the effects of traffic. The average deformation modulus for these materials is only 8100 psi (570 kg/cm^2). The explanation given for this is that under the load the sand in the sub-base layer is slightly compressed. When the load is subsequently removed, a vacuum occurs under the pavement. Accordingly, the sand, from which the air cannot escape quickly enough, is dragged along with the air and loosened in consequence. The process is similar to the pumping effect, except that there is air instead of water in this case. This process, although it does not lead to any loss of material at the centre of the slab, causes the loosened sand to be displaced sideways at the edges, and joints and voids are formed.

This research indicates that traffic does not compact the material under rigid pavements; rather it loosens unbound sand and gravel layers, contrary to what was previously believed. For this reason, stabilised sub-bases are being used now on a large scale in the USA and Europe, under the rigid pavements of major highways and airfields.

A non-uniform sub-base can cause pavement distress and should be avoided if possible.

De-icing agents used during the frost periods. Different types of de-icing agents are generally used in Winter during the process of snow and ice removal from the pavement surface. When the concrete comes in contact with the de-icing salts or similar materials, it has to possess a high resistance to frost/salt action if significant freezing/de-icing damage is to be avoided.

In spite of the great amount of research done in this field, the cause of frost/salt damage to rigid pavements when the concrete is not sufficiently resistant has not yet been completely clarified. Research in the USA indicates that when salt is placed on the concrete surface, its concentration diminishes in the deeper layers and the moisture in these deeper concrete layers freezes first. Thus, when the upper layer freezes, there is a lack of space for expansion, and the resulting hydraulic pressure forces off the top layers.

Concrete pavements constructed in locations where there is snow in Winter must have a high resistance to frost/salt action. In addition to the standard requirements for high-quality concrete, the following items should be given special attention [4].

1. The aggregates used must be sufficiently hard, frost-resistant and free of harmful materials such as loam and clay.

2. The content of fines—cement and fine materials up to 0·01 in (0·25 mm)—and the amount of fine mortar should be restricted to a minimum. The content of fines, e.g. with concrete having $1\frac{1}{4}$ in (32 mm) maximum size of aggregate should not exceed 680 lb/yd^3 (400 kg/m^3); with concrete having $\frac{5}{8}$ in (16 mm) maximum size of aggregate the fine content should be less than 760 lb/yd^3 (450 kg/m^3).

3. The water/cement ratio should be as low as possible. A water/cement ratio of the order of 0·42–0·45 should be aimed at.

4. The concrete must contain a certain amount of small air pores. These air pores are produced by the addition of a suitable air entraining agent. The required air porosity content for high frost/salt resistance depends on the composition of the concrete and the climatic con-

ditions. In general, it is necessary to obtain an air porosity of at least 4%.

5. Young concrete cast in late autumn, which will not quite have reached the required high strength and will not have been able to evaporate part of the excess water needed for mixing, should be properly protected. By impregnating the concrete with an impregnation fluid, such as linseed oil or epoxy resin, it can be protected from frost/de-icing salt effects in the first winter. Impregnation can also increase the frost/salt resistance of concrete pavements and expand the life of old pavements which do not contain sufficient air porosity.

Load repetition of heavy vehicles or aircraft. An increase in the value of the factor of safety against flexural failure will produce an increase in slab thickness. This will in turn increase the pavement life under these load repetitions. Also, a control on the quality of concrete used and the quality of construction will improve the pavement performance.

17.8 PERFORMANCE OF CONTINUOUSLY REINFORCED CONCRETE PAVEMENTS

The performance of continuously reinforced concrete pavement test sections with varying sub-base, subgrade and slab thickness characteristics was evaluated by the Texas Highway Department [19]. The evaluation was made in terms of steel strain, deflection, crack pattern, pumping and traffic. Results indicated that each of the parameters mentioned affects the performance of the pavement in varying degrees. Pavement thickness, sub-base type and subgrade were all found to affect deflection.

Pavements with 0·5% longitudinal steel perform satisfactorily as true continuous pavements with good load transfer. Increasing the amount of longitudinal steel beyond 0·5% will cause a slight decrease in pavement deflection under load.

The crack pattern development was found to be related to pavement age, sub-base friction, concrete flexural strength and curing temperature. Most of the cracks develop during the first 4 months after casting, and crack spacings are larger for concretes with higher flexural strength.

Crack distribution in a certain length of the pavement indicates performance. Generally, a normal distribution reflects satisfactory performance while a skewed distribution indicates unsatisfactory performance.

In the above study, and based on percentage evaluation, twice as many jointed pavements as continuous pavements were found to be pumping.

However, the use of lime stabilised sub-bases directly under continuously reinforced concrete pavements has been unsatisfactory. Lime treated soil loses some of its integrity when it becomes wet, and erodes and pumps like a fine-grained material.

Also, the study indicated that continuous pavements are performing with a significantly higher Present Serviceability Index than are jointed concrete pavements with the same number of equivalent 18 kip axle applications.

17.9 PERFORMANCE OF PRESTRESSED CONCRETE PAVEMENTS

Both theoretical and practical studies have indicated that prestressed concrete pavements can adequately support interior and edge loads of considerably greater magnitude than those causing bottom surface cracks.

Also, tests carried out on prestressed concrete pavements for highways and airfields, using static and repetitive loads, have indicated that there is a tremendous increase in load-carrying capacity induced by prestressing. The load-carrying advantage obtained by prestressing is principally due to a change in the criterion for failure from a bottom surface crack to a top surface crack.

The records of prestressed concrete pavements which have been in actual service for some time indicate that their performance compares favourably with conventional pavements, from the standpoint of maintenance and repairs needed. Due to the big reduction in the number of transverse joints by using prestressed pavements, a more smooth and comfortable surface for the traffic is obtained.

REFERENCES AND BIBLIOGRAPHY

1. ACI Committee 325, Sub-committee VI, 'Prestressed Pavement—A World View of its Status', *ACI Manual of Concrete Practice*, Part 1, 1968, 325-37 to 325-46.
2. ACI Committee 325, Sub-committee VI, 'Proposed Design for Experimental Prestressed Pavement Slab', *Journal of the American Concrete Institute*, April 1968.
3. American Road Builders Association, 'Slip-Form Paving in the United States', *Technical Bulletin 263*, 1967.
4. Bonzel, J., 'Concrete with High Frost and Salt Resistance', *Proc. 2nd European Symposium on Concrete Roads, Bern, 1973*, 251–264.

5. Chastain, W. E. Sr, Beanblossom, J. A. and Chastain, W. E. Jr, 'AASHO Road Test Equations Applied to the Design of Portland Cement Concrete Pavements in Illinois', *Highway Research Record*, No. 90, 1965.

6. Clayton, E. G., 'Slip Form Pavement Shows Excellent Performance on Iowa Country Roads', *Better Roads Magazine*, Portland Cement Association, April 1969.

7. *Highway Research Board Special Report 95*, 'Rigid Pavement Design', Washington, D.C., 1968.

8. Leyder, J. P., 'Ten Years Experience with Deep Transverse Grooves Formed in Fresh Concrete Roads', *Proc. 2nd European Symposium on Concrete Roads, Bern, 1973*, 311–319.

9. Majidzadeh, K. and Talbert, L. O., 'Performance Study of Continuously Reinforced Concrete Pavements', Final Report, Ohio Dept. of Highways, September 1971.

10. Marcc, M., 'Recent Airfield Runway Construction Techniques', *Proc. 2nd European Symposium on Concrete Roads, Bern, 1973*, 157–181.

11. O'Flaherty, C. A., *Highways*, Arnold, 1967.

12. Osawa, Y., 'Strength of Prestressed Concrete Pavements', *Journal of the Structural Division, Proc. ASCE*, October 1962.

13. Portland Cement Association, 'Concrete Pavement Manual—Suggested Practices for Office and Field', Fifth Edition, Illinois, 1955.

14. Portland Cement Association, 'Interim Recommendation for the Construction of Skid-Resistant Concrete Pavement', Technical Bulletin No. 6, August 1969.

15. Schwartz, D. R., 'Continuously Reinforced Concrete Pavement Performance in Illinois', *Proc. 24th Annual Ohio Highway Engineering Conference, Ohio State University, Columbus, April, 1970*.

16. Sebastayan, G. Y., 'Portland Cement Concrete Airport Pavement Performance in Canada', *Highway Research Record*, No. 46, 1963, 101–125.

17. Sharp, D. R., *Concrete in Highway Engineering*, Pergamon Press, 1970.

18. Siedek, P., 'Sub-Bases for Concrete Pavements', *Proc. 2nd European Symposium on Concrete Roads, Bern, 1973*, 11–19.

19. Treybig, H. J., 'Performance of Continuously Reinforced Concrete Pavement in Texas', *Highway Research Record*, No. 291, 1969, 32–47.

20. Vesic, A. S. and Saxena, S. K., 'Analysis of Structural Behaviour of Road Test Rigid Pavements', *Highway Research Record*, No. 291, 1969, 156–158.

21. Yoder, E. J., *Principles of Pavement Design*, Wiley, 1959.

CHAPTER 18

ECONOMICS OF RIGID PAVEMENT CONSTRUCTION AND MAINTENANCE

18.1 INTRODUCTION

Among the main requirements in choosing the type of pavement to be used, are the following: (a) economy in construction and maintenance costs; (b) least interruption to traffic due to maintenance and repairs; (c) safety and durability; and (d) performance and comfort.

The first item will be discussed in this chapter. It should be noted that the use of design charts in determining the required pavement thickness, for the purpose of calculating costs and comparing alternative solutions, can simplify the procedure to a large extent.

The initial costs of pavement construction form only one element in the cost analysis of a pavement system. Resurfacing and maintenance costs can also be a governing factor in determining the type of pavement to be used. For accurate calculations, the delays in vehicle movement and the inconvenience to traffic should be evaluated and considered as a part of the maintenance costs.

Since the design of prestressed concrete pavements is different from that of jointed concrete or continuously reinforced concrete pavements, its cost analysis will be considered separately.

18.2 JOINTED CONCRETE AND CONTINUOUSLY REINFORCED CONCRETE PAVEMENTS

The main elements affecting the costs of these pavements are: (a) strength of subgrade or sub-base; (b) thickness of pavement slab; (c) amount of steel reinforcement; and (d) maintenance.

420

The strength of subgrade or sub-base depends, to a large extent, upon the quality and degree of compaction. The pavement thickness required to carry a volume of traffic depends upon the quality of the concrete mix and amount of cement and, to a lesser degree, upon the strength of the subgrade. Steel reinforcement, if used in jointed concrete pavements, is dependent upon slab thickness, subgrade friction, spacing between joints and the steel strength. In continuous pavements, the percentage of longitudinal steel depends mainly upon the tensile strength of the concrete, yield strength of the steel used and the subgrade friction.

The pavement cost is a major item in the over-all cost of a highway or an airfield. Accordingly, it is important to make a proper economic analysis of the pavement and use the results in the decision making process. The following fundamental steps are proposed for such an analysis:

1. Identify the feasible methods for compacting the subgrade or sub-base of the pavement under study after specifying a minimum foundation strength for the design. The strength that is expected to be reached by using each method and the corresponding cost per unit area can then be estimated.

2. Set a minimum strength for the concrete to be used in construction of the pavement, and use this value to determine the required thickness of a jointed concrete pavement corresponding to each subgrade or sub-base strength, estimated in step 1.

3. Calculate the cost of the concrete per unit area of the pavement for each alternative design and add to this cost the corresponding cost of compacting the subgrade or sub-base. The minimum cost value defines the type of compaction and the corresponding subgrade or sub-base strength that should be used in the design.

4. Use the value of the subgrade or sub-base strength determined in step 3 to design the same jointed concrete pavement but for concrete of higher quality. Three to five variable concrete strengths, within the range of the concrete quality that can be reached for the project, can be selected for this purpose. Based on the unit cost for each type of *in situ* concrete, the corresponding cost per unit area of concrete pavement can be calculated according to the thickness obtained from the design. The minimum cost will define the type of concrete to be used.

5. Add to the value of minimum cost obtained in longitudinal and transverse joints and cost of do any, calculated for a unit area of pavement. Th forcement per unit area, if any, should also be i

the minimum cost of construction for this type of rigid pavement, and corresponds to a concrete of known quality. The alternative design which dominates other possible solutions within one pavement type is now defined and is the one to be used in steps 7 and 8.

6. Repeat steps 4 and 5 for other types of rigid pavements, such as continuously reinforced concrete pavements, to obtain the minimum cost of construction for this type of pavement and the corresponding concrete quality to be used.

7. To consider the effect of maintenance costs in comparing alternative types of concrete pavements, eqn. (18.1) can be applied to calculate the annual cost per mile of travelled way for each alternative type:

$$C_1 = \mathrm{CRF}_n\, A_1 + M_1 \qquad (18.1)$$

where C_1 = annual cost of travelled way per mile;

A_1 = initial construction cost of travelled way per mile
 = cost per unit area obtained in step 5 or 6 multiplied by area of travelled way per mile;

M_1 = annual maintenance cost of travelled way per mile;

CRF_n = the capital recovery factor during the analysis period n

$$= \frac{r(1+r)^n}{(1+r)^n - 1};$$

r = the interest rate (5–10% and generally taken as 6%); and

n = the analysis period (generally taken as 40 years).

The annual construction cost of the pavement, obtained by multiplying the initial total construction cost by the capital recovery factor, is based upon amortising the entire investment over the analysis period. In other words, the salvage value of the project is considered zero at the end of this period and is not entered into the computation.

8. The pavement type with the least annual cost of travelled way per mile is supposed to be the best from the point of view of economy. The concrete mix which will provide the minimum cost for this type of pavement is also defined.

This proposed procedure does not include the effect of the presence or absence of a sub-base layer on pavement economy. The decision as to whether a sub-base layer should be used or not, is mainly dependent upon the environmental conditions in the area where the project will be constructed.

The least annual cost of a concrete pavement can also be compared with that of a flexible pavement, designed for the same purpose, to determine which one is better for this particular situation.

18.3 PRESTRESSED CONCRETE PAVEMENTS

For approximate estimates, the basic cost of a prestressed concrete pavement can be divided into five main items. In each item, the cost includes that of material, workmanship, equipment, overhead, profits, etc. The five items are:

1. Cost of concrete including mixing, shuttering, casting, curing, etc.
2. Cost of prestressing steel in longitudinal and transverse directions, including cost of cable ducts, if any, placing in position and prestressing operations.
3. Cost of anchor heads.
4. Cost of expansion and construction joints.
5. Cost of abutments, sleeper slabs, anchor beams and infill concrete.

18.3.1 Formulation of a Cost Formula

If for a certain prestressed pavement, values x_1, x_2, x_3 and x_4 are assigned for the variables, and cost coefficients $a_1, a_2, \ldots, a_7$ are assigned for the costs of the different items, then an equation can be formulated to give the total cost of the pavement per unit area as follows:

$$a_1 x_1 + (x_2 - 4000)a_2 x_1 + a_3 w x_3 + 2a_4 \times \frac{x_3}{x_4} + a_5 + \frac{a_6 + a_7}{x_4} = Z \qquad (18.2)$$

where Z = cost per ft^2 of pavement;

x_1 = thickness of pavement slab (ft);

x_2 = concrete quality expressed in terms of its ultimate compressive strength f_c' (psi):

x_3 = required cross-sectional area of prestressing steel in in^2/foot width of pavement

$$= \frac{f_p}{f_s'} \cdot x_1 \cdot 144 ;$$

f_s' = allowable stress in the prestressing steel after losses $<0.6 \times$ the specified minimum ultimate strength of the prestressing steel (psi);

$w x_3$ = required weight of prestressing steel in lb/ft^2 horizontal area of pavement = $12 \times 0.284 \times x_3 = 3.408\, x_3$;

x_4 = length of pavement slab between expansion joints = L (ft);

a_1 = *in situ* cost per ft^3 of concrete of quality 4000 psi;

a_2 = additional cost per ft^3 of concrete for each 1000 psi increase in compressive strength;

a_3 = cost per lb of prestressing steel placed in position, including cost of cable ducts, if any, and cost of prestressing operations;

a_4 = cost of one anchor head per in^2 of cross-sectional area of prestressing tendon;

a_5 = cost of expansion and construction joints per ft^2 horizontal area of pavement;

a_6 = cost of abutments, if any, per ft width of pavement per slab between expansion joints; and

a_7 = cost of sleeper slab, anchor beams and infill concrete per ft width of pavement per slab between expansion joints.

The effect of each independent variable (x_1, x_2, and x_4) on the final cost in eqn. (18.2) can be determined in steps by choosing different values for one variable, and determining the corresponding values of the dependent variable while keeping the other independent variables constant. Once the value of that particular variable which gives the minimum cost in the first step is determined, this value is considered for the remaining steps while different values are chosen for the next variable, and so on.

In order to get practical results, certain constraint conditions, such as shown below, may be imposed on eqn. (18.2):

$$f_F + 150 < f_p < 800 \text{ psi} \tag{18.3}$$

where $f_F = \dfrac{F\gamma}{2} \cdot x_4 = \dfrac{145}{2 \times 144} F \cdot x_4 = 0.503 \, F \cdot x_4$

(F = coefficient of subgrade friction)

or

$$0.503 \, F \cdot x_4 + 150 < f_p < 800 \text{ psi} \tag{18.4}$$

$$4000 < x_2 < 6000 \tag{18.5}$$

$$300 < x_3 < 700 \tag{18.6}$$

Here an upper limit of 800 psi is imposed in eqn. (18.4) for the prestressing stress, to avoid the occurrence of excessive compressive stresses at the top surface of the pavement slab.

Example

The pavement for the runway, taxiways and terminal apron of a new airport can be made either of prestressed concrete of the post-tensioned type or as a continuously reinforced concrete. The single-wheel load that should be considered in the design is 75 kips with a tyre pressure of 100 psi. No differentiation is needed between the design for the runway and that for the taxiways and apron. The modulus of subgrade reaction established from tests is 300 psi and the coefficient of subgrade friction is 0·60.

Given: Allowable stress in the prestressing steel, f_{se} = 150 000 psi
Longitudinal reinforcement in the continuously reinforced concrete pavement = 0·7%
Transverse reinforcement = 0·18%

Allowable concrete flexural stress $f_t = \dfrac{MR}{FS}$

Modulus of rupture $MR = 9\sqrt{f_c'}$
Factor of safety in prestressed concrete pavement, $FS = 1·1$
Factor of safety in continuously reinforced concrete pavement, $FS = 1·80$
$a_1 = \$1·20/\text{ft}^3$
$a_2 = \$0·12/\text{ft}^3$
$a_3 = \$0·40/\text{lb}$
$a_4 = \$20.00$
$a_5 = \$0.10/\text{ft}^2$
$a_6 = 0$
$a_7 = \$60.00$

Required: (a) Minimum feasible cost per ft^2 if a prestressed concrete pavement is considered.
(b) Thickness, prestressing stress, concrete quality and length of pavement that can lead to this cost.

TABLE 18.1 ALLOWABLE TENSILE STRESSES FOR CONCRETE OF DIFFERENT QUALITIES

f_c' (psi)	MR (psi)	f_t (psi) Prestressed concrete	f_t (psi) Continuously reinforced concrete
4 000	570	519	316
5 000	635	576	353
6 000	695	630	386

TABLE 18.2 DETERMINATION OF THE COST PER UNIT AREA FOR THE PRESTRESSED CONCRETE PAVEMENT SLAB

x_1	x_2	$f_p + f_t$	f_p	$f_F + 150$	x_3 (in^2)		x_4	$a_1 . x_1$	$\dfrac{(x_2 - 4\,000)}{a_2 . x_1}$	$a_3 w x_3$ ($)		$2a_4 . \dfrac{x_3}{x_4}$	a_5	$\dfrac{a_6 + a_7}{x_4}$	Z	Remarks
(ft)	(psi)	(psi)	(psi)	(psi)	long.	tran.	(ft)	($)	($)	long.	tran.	($)	($)	($)	(%)	
7/12		1 425	906													f_p exceeds 800 psi
8/12	4 000	1 320	800	361	0·51	0·31	700	0·8	0	0·70	0·42	0·047	0·10	0·086	2·153	Use x_1 = 8/12 ft for next step
9/12		1 240	721		0·52	0·31	700	0·9	0	0·71	0·42	0·048	0·10	0·086	2·264	
8/12	4 000	1 320	800	361	0·51	0·31	700	0·8	0	0·70	0·42	0·047	0·10	0·086	2·153	
	5 000		744		0·475	0·285		0·8	0·08	0·65	0·39	0·043	0·10	0·086	2·149	Use x_2 = 6 000 psi for next step
	6 000		690		0·440	0·265		0·8	0·16	0·60	0·36	0·040	0·10	0·086	2·146	
8/12	6 000	1 220	590	241	0·38	0·266	300	0·80	0·16	0·52	0·36	0·086	0·10	0·200	2·226	
		1 270	640	301	0·41	0·266	500	0·80	0·16	0·56	0·36	0·054	0·10	0·12	2·154	This case gives the min. cost
		1 320	690	361	0·44	0·264	700	0·80	0·16	0·60	0·36	0·040	0·10	0·086	2·146	

The value of x_3 in the transverse direction = F times the value of x_3 in the longitudinal direction, where $F = 0.70$ for $L = 300$ ft; $F = 0.65$ for $L = 500$ ft; $F = 0.60$ for $L = 700$ ft.

(c) Minimum feasible cost per ft^2 if a continuously reinforced concrete pavement is used.

(d) Comparison between the costs of the two alternative types of pavements.

Solution. The values of MR for different concrete strength and the corresponding values of f_t to be used in the design of prestressed concrete pavement and the continuously reinforced concrete pavement are shown in Table 18.1.

Case A. Prestressed concrete pavement. In this case four variables x_1, x_2, x_3 and x_4 are involved. Equation (18.2) together with the constraint conditions given in eqn. (18.4)–(18.6) are applied and the results are shown in Table 18.1. The chart shown in Fig. 14.8 is used in determining the thickness and stress $f_t + f_p$ corresponding to different slab lengths. In this case, x_3 can be considered as a dependent variable since it is directly proportional to f_p, which depends upon x_1, x_2 and x_4:

$$x_3 = \frac{f_p}{f_{se}} \cdot x_1 \cdot 144 = \frac{144}{150\,000} f_p \cdot x_1 = 0{\cdot}96 \cdot 10^{-3} \cdot f_p \cdot x_1$$

The minimum cost of $2·146/ft^2, between the studied cases is obtained if a slab of 8 in thickness and 700 ft length is used together with a concrete of quality 6000 psi.

Although the solution obtained here is the most economical of the cases considered in Table 18.2, it is not the global optimal solution.

There is no single curve that can relate load to stresses. Accordingly, the known optimisation techniques may not be applicable in this situation.

TABLE 18.3 DETERMINATION OF THE COST PER UNIT AREA FOR THE CONTINUOUSLY REINFORCED CONCRETE PAVEMENT SLAB

f'_c (psi)	f_t (psi)	x_1 (ft)	a_1x_1 ($)	$(x_2 -4\,000) \cdot a_2 \cdot x_1$ ($)	x_3 (in^2) long.	x_3 (in^2) tran.	a_3wx_3 ($) long.	a_3wx_3 ($) tran.	a_5 ($)	Z ($)
4 000	316	$\dfrac{15{\cdot}5}{12}$	1·55	0	1·30	0·33	0·89	0·23	0·10	2·77
5 000	353	$\dfrac{14{\cdot}5}{12}$	1·45	0·145	1·22	0·31	0·84	0·21	0·10	2·745
6 000	386	$\dfrac{13{\cdot}5}{12}$	1·35	0·27	1·13	0·29	0·78	0·20	0·10	2·70

Case B. Continuously reinforced concrete pavement. The chart given in Fig. 12.11 is used to determine the required thickness for different concrete qualities. The results for the thickness, area of steel, weight of steel and costs corresponding to different concrete qualities are given in Table 18.3, where an *in situ* value of $0·20 per lb of reinforcing steel is considered.

This shows that the use of a concrete of quality 6000 psi will give the most economical cost of 2·70 $/ft² for the three cases considered. In this example, the use of prestressed concrete for the pavement results in a cost saving of about 20% over that of a continuously reinforced concrete pavement.

REFERENCES AND BIBLIOGRAPHY

1. Baldock, R. H., 'Determination of the Annual Cost of Highways', *Highway Research Record*, No. 12, 1963.
2. Lake, J. R., 'Factors Affecting the Choice of Road Pavements', *Proc. 2nd European Symposium on Concrete Roads, Bern, 1973*, 355–364.
3. Lowenberg, H., 'The Economics of Concrete Motorway Pavements', *Proc. 2nd European Symposium on Concrete Roads, Bern, 1973*, 327–341.
4. O'Flaherty, C. A., *Highways*, Arnold, 1967.
5. Ray, G. K., 'Concrete Pavement for Economic Light Traffic Rural Roads', *Proc. 2nd European Symposium on Concrete Roads, Bern, 1973*, 381–395.
6. Sargious, M. and Wang, S. K., 'Economical Design of Prestressed Concrete Pavements', *Journal of the Prestressed Concrete Institute*, July–August 1971.
7. Vauthier, Ph., 'Rural Roads—Techniques and Economics', *Proc. 2nd European Symposium on Concrete Roads, Bern, 1973*, 365–379.

PART IV

PAVEMENT REHABILITATION AND SYSTEMS ANALYSIS

CHAPTER 19

PAVEMENT MAINTENANCE

19.1 INTRODUCTION

Pavement maintenance can be defined as the routine work performed to
keep a pavement, under normal conditions of traffic and environmental
situations, as nearly as possible in its as-constructed condition. In other
words, it is the preservation of the surfaces in good condition or to the
standard to which they were originally built.

When not enough funds are available, pavement rehabilitation should
be done according to a priority scheme. A rational approach to the
establishment of priorities can be developed by employing, for example,
the Present Serviceability Index (PSI) used in the AASHO Road Test
to evaluate performance of the sections. Pavement ratings on the scale
0–5, as used on the AASHO Road Test, can be determined visually and
independently by each member of an engineering team, which should be
set up to review all pavements which need resurfacing. On the basis of
these surveys, priorities for resurfacing projects can then be established,
with low PSI-rating projects receiving first consideration for resurfacing.

19.2 REPAIR OF PAVEMENT DEFECTS

The early detection and repair of minor defects in pavements is the most
important work done by the maintenance crew. Cracks and other surface
breaks, which in their first stages are almost unnoticeable, may develop
into serious defects if not soon repaired. Maintenance work requires
proper supervision, skilled workmen and good workmanship. Repairs
should preferably be made during warm (50°F and above) and dry weather.

431

Repairs that must be done during cold or damp weather require much greater care and have less chance of being satisfactory.

19.3 MAINTENANCE OF FLEXIBLE PAVEMENTS

Saturation of the granular bases underneath asphalt-surfaced pavements is the cause of many maintenance problems. Cracks of the familiar alligator or chicken-wire pattern generally develop in the surface of the asphalt pavement when the granular bases become soft. The cause of the distress should be eliminated in these cases, since these problems are not solved by filling cracks or placing skin patches.

When investigating surface failures which appear to be related to excessive deflection, the base should be checked for plastic fines or trapped water. If these are present, repair may call for digging out the broken area to sound material, improving drainage and patching with asphalt patching mixture. High-quality hot-mixed patching mixtures, properly compacted, result in longer-lasting patches. Many patches bleed, become unstable and are subject to pushing after placement, if an excess of asphalt is used in the mixture or if the patch is not allowed to cure before subjecting it to traffic.

If the base of a deep patch is made with untreated material it should be primed with 0·2–0·3 gal liquid asphalt/yd^2. The prepared edges of the surface surrounding the area being patched should be tack coated, with a very light application of liquid asphalt, to ensure a bond between them and the patch material.

Defects in flexible pavements and their methods of repair are summarised in the following paragraphs.

Cracking. Cracking takes many forms such as alligator cracks, edge- or lane-joint cracks, reflection cracks, slippage cracks and widening cracks. Figure 19.1 shows the different forms of cracks.

The alligator cracks, which form a series of small blocks resembling an alligator's skin or chicken-wire, are generally caused by excessive deflection of the surface over unstable subgrade or lower courses of the pavement. The unstable support is usually the result of saturated granular bases or subgrade and the affected areas in most cases are not large. However, repeated loads exceeding the load-carrying capacity of the pavement may cause cracks over an entire section of pavement.

Permanent repair for such a situation can be effected by removing the surface and base as deep as necessary, draining accumulated water, if any, and patching with asphalt patching mixture as described before. The cut should be rectangular with faces straight and vertical and should extend at least a foot into good pavement outside the cracked area.

If a temporary repair is needed for areas with cracks narrower than $\frac{1}{8}$ in, an aggregate seal coat patch can be used. After cleaning the cracked area, liquid asphalt is sprayed on the surface, usually at a rate of 0·15–0·25 gal/yd^2. The cover aggregate is then immediately applied and the seal coat is rolled with rubber-tyred equipment. A good aggregate size for this type of patch is $\frac{1}{4}$ in to No. 10 screenings.

Edge cracks, which are longitudinal cracks a foot or so from the edge of the pavement, usually develop when there is a lack of lateral (shoulder) support. They may also be caused by settlement of the material underlying the cracked area, due to poor drainage, frost heave or shrinkage from drying out of the surrounding earth.

Edge-joint cracks are caused by the separation of the joint between the pavement and the shoulder due to alternate wetting and drying beneath the shoulder surface, shoulder settlement or mix shrinkage or due to trucks straddling the joint. Lane-joint cracks are longitudinal separations along the seam between two paving lanes and are usually caused by a weak seam between adjoining spreads in the courses of the pavement.

Reflection cracks are cracks in asphalt overlays which reflect the crack pattern and movements in the pavement structure underneath. They occur most frequently in asphalt overlays on Portland cement concrete and on cement treated bases. They may also occur in overlays on asphalt pavements whenever cracks in the old pavement have not been properly repaired. Longitudinal reflection cracks which show up in the asphalt overlay above the joint between the old and new sections of a pavement widening are usually called widening cracks.

Edge cracks, edge- or lane-joint cracks, reflection cracks and widening cracks can be repaired by cleaning out the crack and filling with emulsion slurry or liquid asphalt mixed with sand. When cured, the top of the crack is sealed with liquid asphalt and the surface of crack filler is then sprinkled with dry sand to prevent pickup by traffic.

In the case of edge cracks, if the edge of the pavement has settled, it can be brought up to grade by applying a tack coat to the cleaned surface, and then spreading and compacting hot asphalt plant-mixed material in the settled area.

Shrinkage cracks are interconnected cracks forming a series of large blocks. Frequently, they are caused by volume change of fine aggregate asphalt mixes that have a high content of low-penetration asphalt. These

Alligator cracks

Edge crack

FIG. 19.1. Different forms of cracks. *(Courtesy: The Asphalt Institute.)*

Edge joint crack

Lane joint crack

FIG. 19.1—*contd.*

Reflection crack

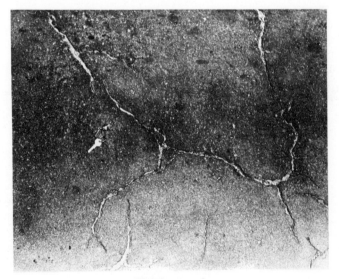

Shrinkage cracks

FIG. 19.1—*contd.*

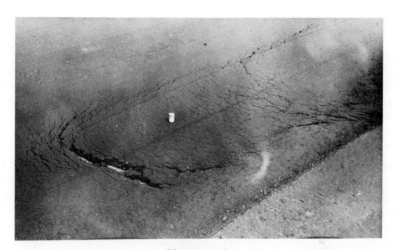

Slippage cracks

Widening crack

FIG. 19.1—*contd.*

cracks can be repaired by removing all loose matter from the pavement surface, wetting the surface and all crack faces with water and applying a tack coat of asphalt emulsion diluted with equal parts of water. The tack coat should be applied when all surfaces are uniformly damp with no free water. An asphalt emulsion slurry mixture is then poured into the cracks, and when it is cured until firm, the whole surface is slurry sealed. The surface seal should be allowed to cure until firm enough to prevent pickup by traffic.

Slippage cracks, which are crescent-shaped cracks pointing in the direction of the thrust of wheels on the pavement surface, are usually caused by the lack of a good bond between the surface layer and the course beneath. These cracks are repaired by removing the surface layer from around the crack to the point where good bond between the layers is found, cleaning and tack coating the exposed surface, patching with plant-mixed asphalt material and then compacting thoroughly.

Distortion. Pavement distortion is any change of the pavement surface from its original shape and is usually caused by such things as too little compaction of the pavement courses, too many fines in surface mixtures, too much asphalt, swelling of underlying courses or settlement. Distortion takes a number of different forms such as grooves or ruts, shoving, corrugations, depressions or upheaval. Figure 19.2 shows the different forms of pavement distortion. As with any other defect, the type of distortion and its cause must be determined before the correct remedy can be applied. Methods of repair can range from levelling the surface by filling with new material to complete removal of the affected area and replacement with new material.

Ruts are channelised depressions which may develop in the wheel tracks of the pavement due to consolidation or lateral movement under traffic in one or more of the underlying courses, or by displacement in the asphalt surface layer itself. They can be repaired by applying 0·05–0·15 gal/yd^2 of a light tack coat to the cleaned depressed area, spreading and compacting dense-graded asphalt concrete in the channels and then placing a thin overlay of hot plant-mixed material. If the pavement is not to be overlaid, a sand seal is placed over the patched areas to prevent entrance of water.

Corrugation is a form of plastic movement typified by ripples across the asphalt pavement surface, while shoving is a form of plastic movement resulting in localised bulging of the surface. They usually occur at points where traffic starts and stops, on hills where vehicles brake on the downgrade or on sharp curves. Asphalt layers lacking stability due to the use of a mixture too rich in asphalt, or with too soft an asphalt cement, are subject to corrugation and shoving; these are also caused if too high a proportion of fine aggregate is used, or if coarse or fine aggregate is used which is too round or smooth-textured.

If the corrugated pavement has an aggregate base with a thin surface treatment, the repair can be made by scarifying and breaking up the surface. The broken-up surface material is then mixed with the base material to a depth of 4 in. After compacting, reshaping and priming the base, a new surface treatment can be applied.

In the case of a thick asphalt surface, the corrugations are planed to a smooth surface with a heater–planer. The planed surface is then covered with a hot plant-mixed asphalt seal coat or asphalt emulsion slurry seal.

Shoved areas are repaired by deep patching, as in the case of alligator cracks discussed before.

Grade depressions are localised low areas of limited size which may or may not be accompanied by cracking. They may be caused by traffic heavier than that for which the pavement was designed, by settlement of the lower pavement layers or by poor construction methods. These depressions are not only a source of pavement deterioration, but also a hazard to motorists, especially in freezing weather. They can be repaired by filling with hot plant-mixed asphalt material, after cleaning the entire area and applying a light tack coat of $0.05–0.15$ gal/yd^2 asphalt emulsion, diluted with equal parts of water. The tack coat should be allowed to cure before filling and the edges of the patch should be feather-edged by careful raking and manipulation of the material, without allowing segregation of the mixture. The patch should then be thoroughly compacted, and a sand seal placed on the patched area to prevent the entrance of water.

Upheaval is the localised upward displacement of a pavement due to swelling of the subgrade or a part of the pavement structure. It is most commonly caused by expansion of ice in the lower courses of the pavement or subgrade. The swelling effect of moisture on expansive soils may also cause upheaval in the pavement. To repair such areas, deep patching can be used, as in the case of alligator cracks discussed above.

Disintegration. Disintegration is the breaking up of a pavement into small loose fragments, which, if not stopped in its early stages, can spread until the pavement requires complete rebuilding. Figure 19.3 shows the different forms of pavement disintegration.

Potholes, which are bowl-shaped holes of various sizes, are usually caused by weakness in the pavement resulting from too little asphalt, too

Channels (ruts)

Corrugations

FIG. 19.2. Different forms of pavement distortion. *(Courtesy: The Asphalt Institute.)*

Shoving

Upheaval

FIG. 19.2—*contd.*

thin asphalt surface, too many or few fines, or poor drainage. Temporary repair usually involves cleaning the hole of loose material and water, softening asphalt surfacing surrounding it and filling the hole with asphalt emulsion mixture. After raking smooth, the mixture is compacted, then dried with an infra-red heater. Deep patching can be used in permanent repairs, as in the case of alligator cracks. The procedure for such a repair is explained above.

The second common type of disintegration is ravelling, which is the progressive separation of aggregate particles in a pavement from the surface downward or from the edges inward. Usually the fine aggregate comes off first, and then larger and larger particles are broken free as the erosion continues. Causes of ravelling are lack of compaction, construction during wet or cold weather, dirty or disintegrating aggregate, too little asphalt in the mix or overheating of the asphalt mix. The emergency repair of ravelling surfaces involves cleaning the surface, applying a fog seal at the rate of 0·1–0·2 gal/yd^2 and closing to traffic until the seal has cured. The fog seal is a light application of slow-setting asphalt emulsion diluted with an equal amount of water.

In addition to the above, a surface treatment in the form of slurry seal, sand seal, aggregate seal or plant-mixed surface treatment should be applied if a permanent repair is required.

Skid hazard. One of the most frequent causes of slippery asphalt pavements is a thin film of water on a smooth surface. Also, a thick film of water can cause a vehicle moving at high speeds to leave the pavement surface and skim over the water. A film of asphalt on the surface, or polished aggregate in the surface course, generally produces a smooth pavement condition.

Bleeding or flushing is the upward movement of asphalt in an asphalt pavement, which results in the formation of a film of asphalt on the surface. This defect usually occurs in hot weather and is caused by too much asphalt in one or more of the pavement courses. Also, heavy traffic may cause added compression of a pavement that contains too much asphalt, and force the asphalt to the surface.

Bleeding can be corrected, in many cases, by applying $\frac{3}{8}$ in maximum size slag screenings, sand or rock screening to the affected area. The aggregate, which should be heated to at least 300°F, is spread at the rate of 10–15 lb/yd^2 and is rolled with a rubber-tyred roller. When the aggregate has cooled, loose particles are broomed off and the process is repeated if necessary.

When bleeding is light, a plant-mixed surface treatment or an aggregate

FIG. 19.3. Different forms of pavement disintegration. Top: pothole. Bottom: ravelling.
(Courtesy: The Asphalt Institute.)

FIG. 19.4. Skid hazard due to defects in pavement surface. Top: bleeding asphalt. Bottom: polished aggregate in pavement surface. *(Courtesy: The Asphalt Institute.)*

seal coat, using absorptive aggregate, is sometimes the only treatment needed.

When bleeding is heavy, the asphalt film is removed with a heater–planer and the surface is left as planed, or a seal coat is applied.

Polished aggregate, which includes both naturally smooth uncrushed gravels and crushed rock that wears down quickly under the action of traffic, is quite slippery when wet. The only way of repair in this case is to cover the surface with a skid-resistant treatment and compact with rubber-tyred and steel-wheel rollers. This treatment can be in the form of a hot plant-mixed surface treatment, a sand seal or an aggregate seal. The aggregate must be hard, angular and non-polished, such as slag silica sand. A light tack coat should precede this process.

Figure 19.4 shows the defects in pavement surface that may cause skid hazard.

19.4 MAINTENANCE OF RIGID PAVEMENTS

Bituminous sealing compounds and asphalt have several usages in maintenance of rigid pavements. Joints and cracks in rigid pavements are sealed to prevent surface water seepage, to protect joint fillers and to keep out foreign matter. A few years ago, penetration-grade asphalt, alone or with added filler, was the sealing material most commonly used. It is still used in many places. However, rubber-asphalt compounds of the cold-applied mastic or hot-applied types, have gained favour in recent years. Among other advantages, they have less of a tendency to become brittle in cold weather and to soften and track under traffic in hot weather.

Sealing of joints can be done conveniently and rapidly by trailer-mounted pressure applicators which will handle either hot or cold applied sealing materials. They have control valves which help to avoid excessive spilling of the sealing material on the pavement surface, and are equipped with nozzles of the size and shape needed to convey the sealing material into the opening.

After cleaning out, just enough sealing material should be placed in the joint or crack to fill it. Because hot sealer may shrink upon cooling, when used to fill deep crevices, enough additional material should be added to fill the opening flush with the surface. On airfield pavements serving jet planes, fuel-resistant compounds should be used.

Also, when cavities occur beneath a rigid pavement, due to slab pumping or settlement of the subgrade, they should be filled with undersealing

Corner crack

Restraint crack

FIG. 19.5. Different forms of cracks in rigid pavements. *(Courtesy: The Asphalt Institute.)*

Transverse cracks

Longitudinal crac.

FIG. 19.5—*contd.*

Diagonal crack

FIG. 19.5—*contd.*

asphalt to restore the pavement's support and prevent further erosion. This undersealing fills the cavities, raises sunken slabs and forms a water-proof layer which prevents fine material from pumping out again.

Asphalt overlays can be used to restore smoothness in distorted parts of rigid pavements, to make slippery parts skid-resistant or to salvage and strengthen disintegrating portions of these pavements.

The following paragraphs summarise the maintenance requirements for rigid pavements. Defects in Portland cement concrete slabs, their causes and methods of repair, are also included.

Joints and cracks. Conventional concrete pavements are provided with transverse joints at intervals to allow the volume changes due to shrinkage and temperature variations to take place without producing serious intermediate cracks.

In addition to the transverse joints, there are longitudinal construction joints between the lanes as well as between the outer lanes and the shoulders.

All joints and cracks must be kept sealed with some adhesive material, such as asphalt with or without an additive, to prevent damage to the pavement from water and foreign matter. The method of sealing joints and cracks is essentially the same.

Joints need periodic maintenance by ploughing out old defective seal to a depth of 1 in, sand-blasting the vertical faces of the joint, cleaning the joint, inserting sponge rubber, plastic, or tape to the bottom of the groove and finally sealing in one pour.

Cracks in rigid pavements can take several forms, these being transverse, longitudinal, diagonal, corner and restraint cracks. Figure 19.5 shows the different forms of cracks.

Corner cracks are diagonal cracks forming a triangle with a longitudinal edge or joint and transverse joint or crack. They can be caused by traffic loads on unsupported corners or curled or warped slabs, as well as by loads over weak spots in the subgrade under the slabs.

The common and most economic procedure of repair is to remove the broken corner and level and prime the sub-base. A tack coat is then applied to the sides of the slab and dense-graded asphalt concrete is placed in layers not exceeding 4 in each in thickness. Finally, the asphalt concrete is compacted and levelled flush with the surrounding pavement.

Diagonal cracks are generally caused by traffic loads on unsupported slab ends. When the foundation settles or the slab curls, the subgrade soil pumps out and a diagonal crack takes place. Repairing this type of crack consists of sand-blasting the vertical faces of the crack to a depth of at least 1 in, cleaning the crack, half filling with a rubber-asphalt compound, undersealing the slab and then filling the crack again with the same compound up to the surface.

Longitudinal cracks can be caused by concrete shrinkage, expansive subgrade or sub-base, combined warping and load stresses or loss of support due to edge pumping. They are approximately parallel to the centre line of the pavement.

Transverse cracks are approximately at right angles to the pavement's centreline and can be caused by overloading, repeated bending of pumping slabs, failure of soft foundations, frozen joints and shrinkage of the concrete.

Both longitudinal and transverse cracks can be repaired by sand-blasting the vertical faces of the crack to a depth of at least 1 in, cleaning the crack and then filling with a rubber-asphalt compound. If the crack is caused by pumping, the void beneath the pavement must be undersealed.

Restraint cracks are caused by foreign matter, such as hard gravel, becoming lodged deep in a transverse joint and restraining the slabs from expanding. They develop near the outside edges of the pavement and progress in an irregular path towards the longitudinal joint.

The repair in this case consists of ploughing out the old sealer and foreign matter in the transverse joint to the depth required to remove all contamination, and sand-blasting the vertical faces of the joint and crack;

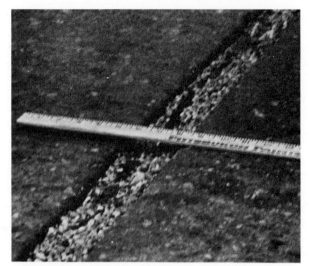

Fault

Pumping

FIG. 19.6. Different forms of distortion in rigid pavements.

both should then be cleaned and sealed in two pours with a rubber-asphalt compound.

Distortion. Distortion can be defined as any change of the pavement surface from its original shape. This can be caused by foundation settlement or by the presence of expansive or frost susceptible soils. Different types of distortion are shown in Fig. 19.6.

One of the most predominant types of rigid-pavement distortion is faulting. Because pumping results in faulting or sunken slabs, it is included in this category.

A fault is a difference in elevation of two slabs at a joint or a crack caused by inadequate load transfer between slabs along with consolidation or shrinkage of underlying layers; or it may be due to pumping out of the foundation material.

Faulting can be repaired by ploughing out old sealer from the joint or crack to a depth of 1 in, sand-blasting the vertical faces of joint or crack, cleaning and then half filling with a rubber-asphalt compound. The next step is to raise the slab to the original grade with a high-softening-point undermealing asphalt cement, and finish filling the joint or crack with the rubber-asphalt compound.

Pumping is slab movement under passing loads resulting in the ejection of mixtures of water, sand, clay and/or silt, and can occur along transverse or longitudinal joints and cracks and along pavement edges. It is caused by the presence of free water on or in the subgrade or sub-base along with heavy loads passing over the pavement surface and deflecting the slab.

Pumping can be repaired by filling the voids beneath the pavement with a high-softening-point undersealing asphalt and resealing the joints, following the procedure explained before for joint maintenance.

Disintegration. Disintegration is the breaking up of a pavement into small loose fragments, which must be arrested in its early stages; otherwise the pavement will deteriorate until it requires complete rebuilding.

Blowups, spalling and scaling are common types of disintegration in rigid pavements and are shown in Fig. 19.7.

A blowup can be defined as the localised buckling or shattering of a rigid pavement, occurring usually at a transverse crack or joint. It is mainly caused by excessive expansion and building up of pressure in the slabs until they either buckle or shatter, crumbling along the transverse joint or crack.

To repair a blowup, the damaged portion of the slab should first be removed by sawing a straight neat cut; the sub-base should be levelled, if required. If asphalt concrete is used in the repair, the sub-base should

Blowup (buckling)

Blowup (shattering)

FIG. 19.7. Different forms of disintegration in rigid pavements.
(Courtesy: The Asphalt Institute.)

Scaling

Spalling

FIG. 19.7—*contd.*

be primed and a tack coat applied to the sides of the slab. Dense-graded asphalt concrete is then placed and compacted in layers not exceeding 4 in each and the surface is finished flush with the surrounding pavement.

If Portland cement concrete is used in the repair, then either one slab or two should be completely removed, depending upon whether the blow-up is along a transverse crack or along a transverse joint. The sub-base is then levelled if necessary and new Portland cement concrete slabs cast. In this case, it is important to provide longitudinal and transverse joints that are similar to the corresponding construction joints in the original pavement.

Scaling is the peeling away of the surface of Portland cement concrete due to overfinishing, improper mixing, unsuitable aggregates, improper curing or the chemical action of de-icing salts.

The scale resistance of Portland cement concrete pavements can be increased by treating the surface with linseed oil antispalling compound. In this case, attention should be given to the loss in skid resistance due to the presence of the linseed oil. While a dry pavement may recover its skid resistance in less than 6 h after treatment, a wet pavement may need more than 2 days for that recovery.

A temporary repair can be made to scaled areas that are $\frac{3}{8}$ in or less in depth. The procedure here is to remove all loose particles from the surface, clean the area and apply an asphalt-emulsion slurry seal to restore the surface to the original grade.

Extensive and deep scaling needs an asphalt concrete or Portland cement concrete overlay.

Spalling is the breaking or chipping of the pavement at the joints, cracks or edges. Some of the major causes of spalling are hard pieces of gravel becoming lodged in a joint or crack, improperly installed load transfer devices, improper forming and sawing of joints and weak mortar.

Asphalt concrete may be used in repairing spalls. The procedure starts with chipping out the spalled area to sound material, squaring the edges and making the sides as vertical as possible. After blowing out the area and applying a light tack of asphalt emulsion diluted with equal parts of water to the sides of the concrete, the hole is filled with a dense-graded asphalt concrete, compacted and levelled flush with the surrounding pavement.

Epoxy or polyester resins or Portland cement concrete can also be used in repairing spalls. The hole to be filled in these cases should be prepared in the form of a trapezoidal notch with the larger dimension at the bottom. In the case of epoxy or polyester resins, mortar mixes leaner than 1 : 4

should not be used. It should also be noticed that both resin and mortars are susceptible to alternate heating and cooling, alternate freezing and thawing and continued contact with moisture.

If Portland cement concrete is used in the repair, the concrete mix should be properly designed and the surfaces of the old concrete around the hole treated with a thin layer of cement mortar or other binding agent, before casting.

Skid hazard. A major cause of slipperiness in rigid pavements is polished aggregates in the surface, as shown in Fig. 19.8. It may also develop from surface contamination.

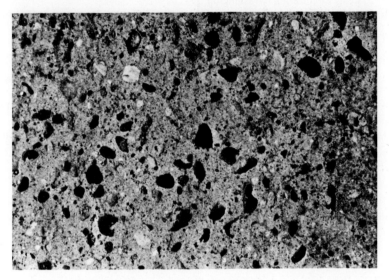

FIG. 19.8. Skid hazard due to polished aggregate in pavement surface.
(Courtesy: The Asphalt Institute.)

Polished aggregates include both naturally smooth uncrushed gravels and also crushed rock that wears down quickly under the action of traffic. These are extremely slippery when wet.

The most effective and economical way to repair a pavement with polished aggregates is to apply a light tack to the surface, spread a course of asphalt mix using sharp angular non-polishing aggregate, and roll with rubber-tyred and steel-wheeled rollers. The thickness of the overlay should be at least 3 in to reduce the probability of reflection cracking.

REFERENCES AND BIBLIOGRAPHY

1. Fickes, L. A. and Rhodes, C. C., 'A Field Study of Joint and Crack Resealing Methods and Materials', *Highway Research Board Bulletin 166*, 1957.
2. Ghosh, R. K., 'Concrete Repairs with Epoxy and Polyester Resins', *Highway Research Record*, No. 327, 1970, 12–17.
3. Hennes, R. G. and Eckse, M., *Fundamentals of Transportation Engineering*, Second Edition, McGraw-Hill, 1969.
4. *Highway Research Record 146*, 'Pavement and Bridge Maintenance', Washington, D.C., 1966.
5. *Highway Research Record*, No. 359, 'Maintenance Operations and Applied Systems Engineering', Washington, D.C., 1971.
6. Runkle, S. R., 'Skid Resistance of Linseed Oil Treated Pavements', *Highway Research Record*, No. 327, 1970, 1–11.
7. Sharp, D. R., *Concrete in Highway Engineering*, Pergamon Press, 1970.
8. Swanberg, J. H., 'Pavement Rehabilitation: Background and Introduction', *Highway Research Record*, No. 300, 1969, 1–3.
9. The Asphalt Institute, 'Asphalt in Pavement Maintenance', Manual Series No. 16, First Edition, December 1967.
10. The Asphalt Institute, 'Asphalt Surface Treatments', Manual Series No. 13, Second Edition, November 1969.
11. Winnitoy, W. E., 'Rating Flexible Pavement Surface Condition', *Highway Research Record*, No. 300, 1969, 16–26.
12. Yoder, E. J., *Principles of Pavement Design*, Wiley, 1959.

CHAPTER 20

PAVEMENT OVERLAYS

20.1 INTRODUCTION

Due to the rapid increase in traffic volumes and loads using highways and airfields, more maintenance is required than that necessary to merely preserve the pavement's original load-carrying capability. Pavements are exposed to loads and traffic volumes beyond those for which they were originally designed.

Highway and airfield pavements can be strengthened by application of overlays to correct defective areas or to increase the load-carrying capacity of the pavement. These overlays may be flexible or bituminous, or of cement concrete. While a flexible overlay consists of a base course and a wearing course, a bituminous overlay is one made up of hot-mix asphalt without a base course.

The base pavement, which is the existing pavement needing strengthening, can also be either a flexible or rigid pavement. The various combinations of base pavement and overlay construction are illustrated in Fig. 20.1.

Overlay design procedure requires investigating the present properties and strength of the subgrade and sub-base layers underlying the existing pavement. Both laboratory and site tests may be needed for this purpose. Changes in the properties of the materials forming the pavement foundation, due to water or frost effects after constructing the base pavement, should be carefully studied.

In addition, it is important to evaluate the base pavement so that a proper design of the overlay can be made. Modern equipment has been developed in recent years to facilitate and expedite the evaluation procedures.

457

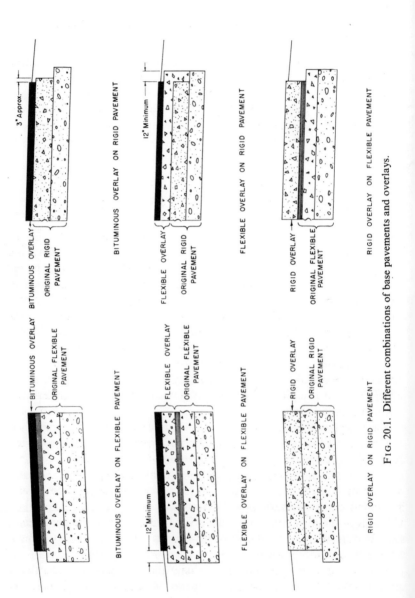

FIG. 20.1. Different combinations of base pavements and overlays.

20.2 EQUIPMENT FOR PAVEMENT EVALUATION

Some methods of overlay design are based on determining the thickness of the overlay layer which will reduce the deflection of the base pavement to a certain designated value. Other methods are based upon estimating the PSI of the base pavement and determining the overlay thickness which will bring the PSI up to a specified value. For this purpose, measurements of deflection and roughness of the base pavement are important for overlay design.

20.3 MEASUREMENT OF PAVEMENT DEFLECTION

Benkelman beam. This device has a narrow beam, which can be slipped between the dual tyres of the rear axle of a loaded truck to measure the total pavement rebound deflection. This rebound deflection is the amount of vertical rebound of a surface occurring when a load is removed from the surface.

The US Federal Highway Administration-type Benkelman Beam, shown in Fig. 20.2, has the following dimensions:

Length of probe arm from pivot to probe point = 8 ft 0 in
Length of measurement arm from pivot to dial = 4 ft 0 in
Distance from pivot to front legs = 10 in
Distance from pivot to rear legs = 5 ft $5\frac{1}{2}$ in
Lateral spacing of front support legs = 13 in

The test procedure can be summarised as follows:

1. Select and mark a point to be tested in the outer wheel path, and centre the dual tyres of the right rear wheel of a 5 ton truck above that point. The truck should have an 18 000 lb rear-axle load equally distributed on two wheels, each with dual tyres. The tyres should be $10 \cdot 0 \times 20$, 12-ply inflated to a pressure of 80 psi.

2. Insert the Benkelman beam's probe between the dual tyres and place the probe point over the selected test point.

3. Remove the locking pin from the beam, drop the probe point to the pavement and adjust the rear legs so that the plunger of the beam is in contact with the dial gauge stem. Adjust the dial gauge to read 0·4 in.

4. Start buzzer and record initial reading of dial when dial needle stops moving.

Benkelman beam in position

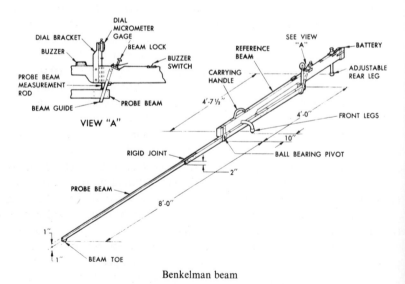

Benkelman beam

Fig. 20.2. Measurement of rebound deflection by Benkelman beam.
(Courtesy: The Asphalt Institute.)

5. Move the truck forward about 30 ft and record final reading of dial when the dial needle stops moving.

6. Subtract the final dial reading from the initial dial reading. The difference multiplied by 2 is the actual pavement rebound deflection.

7. Record air and pavement temperatures, so that measured rebound deflection can be adjusted for temperature.

Dynaflect. Some US states use a special Dynaflect for measuring deflections in a rapid and inexpensive way. The Dynaflect used in the State of Utah consists of a small trailer towed behind a vehicle with the control unit located in the vehicle beside the operator. It measures the dynamic deflection between two rigid steel wheels (force wheels) that are subjected to an oscillatory load of 1000 lb peak-to-peak varying sinusoidally at the rate of 8 c/s. The magnitude of the deflections are sensed by means of five geophones spaced at 12 in intervals, with the first geophone placed between the force wheels. The deflections are read on the meter in the control unit in the tow vehicle. The deflection value is shown by the vertical scale on a semi-logarithmic plot, while the distance of the geophones from the force wheels is shown on the horizontal scale. The highest point represents the deflection under the geophone located between the Dynaflect force wheels. The various deflection lines or basins at each test site on each project are connected at the high points to indicate trends.

In a study done in Texas, the relationship between the Dynaflect deflections and those obtained with the Benkelman Beam for 18 000 lb single-axle loads, was found to be approximately 1 to 28.

The performance–deflection relationships for flexible pavements, which were established as part of the AASHO Road Test, were modified for the Dynaflect and used to develop two groups of curves. These curves are based on a PSI of 2·5 and relate Dynaflect deflections, daily equivalent 18 kip loads and pavement life. The first group is for fall deflections and the second one is for spring deflections under the environmental conditions that prevail in the State of Utah. Figure 20.3 shows the group of curves for the autumn deflections and Fig. 20.4 is for the spring deflections. The horizontal scale represents the Dynaflect deflections in mils (one mil equals one-thousandth of an inch) and the vertical scale is years of service. Each curve represents daily equivalent 18 kip loads. For a given deflection on a pavement of known traffic, the anticipated remaining years of service can be determined from these curves. The reverse procedure can also be used to determine allowable deflections when a certain service life is desired. For example, if a service life of 20 years is desired on a pavement of 100 daily equivalent 18 kip loads, the allowable deflection for autumn would be about 0·85 mils and for Spring about 1·4 mils. The autumn deflections are generally more reliable in determining overlay requirements. The surfacing required for rehabilitation is then the amount required to reduce the actual deflection to the allowable deflection.

To determine the thickness of a new overlay course required to reduce

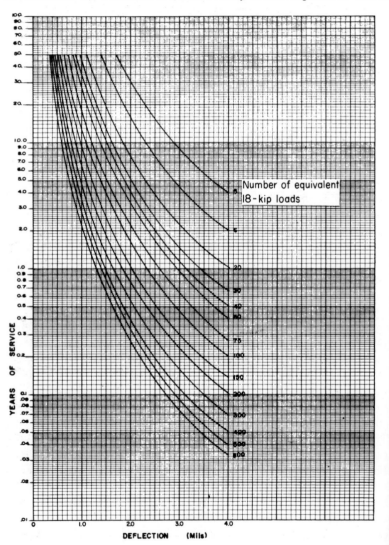

FIG. 20.3. Autumn deflection curves. (From [8].)

the deflection to a point at which it will meet the design-life criteria, the curve developed by the California Department of Highways, shown in Fig. 20.5, can be used. The reduction in deflection (%) is shown in the

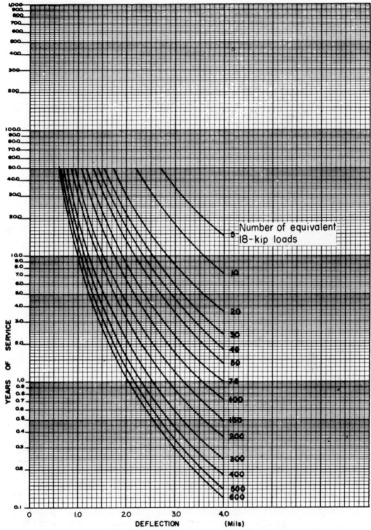

FIG. 20.4. Spring deflection curves. (From [8].)

vertical scale, and the required thickness of a gravel overlay is shown on the horizontal scale. An equivalency of 3·0 to 1 for bituminous surface to gravel can be used. This relationship is shown by an additional horizontal scale on the graph. For any given reduction in deflection, the surfacing requirements can be rapidly determined.

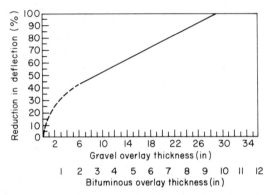

FIG. 20.5. Thickness of overlay required to reduce deflection. (From [8].)

Deflection measurements using the Dynaflect in the State of Utah have indicated that maximum deflections on the good-stability projects are all below 0·75 mils, whereas the poor-stability projects have maximum deflections generally above 1·5 mils.

20.4 MEASUREMENT OF PAVEMENT ROUGHNESS

Roadmeter. The Portland Cement Association (PCA) roadmeter, developed in 1965, provides a rapid, simple and inexpensive way of measuring pavement roughness. This roughness is a principal ingredient of the PSI established as a result of the AASHO Road Test. In general, present serviceability is a function of transverse and longitudinal profile. However, patching, cracking, faulting and spalling no doubt contribute to a certain extent.

The PCA roadmeter is an electromechanical device installed in a standard passenger car, to measure the number and magnitude of vertical deviations between the body of the automobile and the centre of the rear-axle housing. Any vertical movement between the rear-axle housing and a package deck placed just behind the rear seat is translated into a horizontal movement of a steel cable and a corresponding movement to a recorder. The steel cable is connected to the top centre of the rear-axle housing in the car and is brought vertically through the floor to the package deck.

At this point the cable is passed over a transverse-mounted pulley and restrained by a tension spring attached to a small post on the package deck. Halfway between the pulley and the tension spring, a microswitch mounted on a rectangular plate is attached to the steel cable. The switch is forced by its own internal compression spring on to a copper switch plate and is therefore always in a partially compressed state. The switch plate is divided into 23 segments, each $\frac{1}{8}$ in long, so that any transverse roller movements derived through the action of the steel cable are measured in $\frac{1}{8}$ in increments of vertical motion. Output from the switch-plate circuit is fed to the visual indicators of the road-car deviations and is also connected to eight high-speed electric counters capable of recording electrical impulses. The counters sum the impulses received from the segments of the switch plate, according to the magnitude of the impulses relative to the road-car deviations. Individual counters are connected to switch-plate segments which correspond to road-car deviations of $\pm\frac{1}{8}$, $\frac{1}{4}$, $\frac{3}{8}$, $\frac{1}{2}$, $\frac{5}{8}$, $\frac{3}{4}$, $\frac{7}{8}$ and 1 in. The centre segment, which is used for the initial zero reference, has no electrical connection. The summation (Σ counts) is obtained by reading off the number accumulated on each counter for a pre-established length of pavement (usually 1 mile), and then multiplying these numbers by factors of 1, 2, 3, 4, 5, 6, 7 and 8.

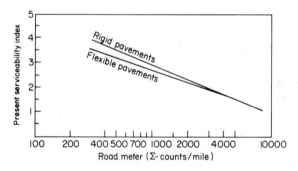

FIG. 20.6. Correlation of PSI and road meter data.

The roadmeter must then be calibrated to a universal standard, in order to be able to derive PSI values from the Σ counts. This calibration can be done by finding the PSI of different pavement sections using a rating panel or other devices such as a roughometer or a profilometer. The Σ counts/mile are determined for the same sections using the PCA roadmeter. PSI and roadmeter data can then be correlated for the cases of rigid and flexible pavements, as shown in Fig. 20.6. In order to obtain reliable results, the automobile in which the roadmeter is used must be in excellent condition, particularly the suspension components. Tyre pressures must be maintained at 24–26 psi and the operating speed should be maintained at 48–52 mph.

The roadmeter should be operated only when the air temperature is above 10°F, because the characteristics of the automobile's suspension system will probably change at low temperatures.

Roughometer. The US Bureau of Public Road roughometer is a single-wheeled trailer with a centrally located recording wheel. An integrator connected to the axle of the recording wheel by a steel cable is coupled to an electric counter calibrated to record inches of vertical movement. The recording system thus measures the inches of vertical movement of

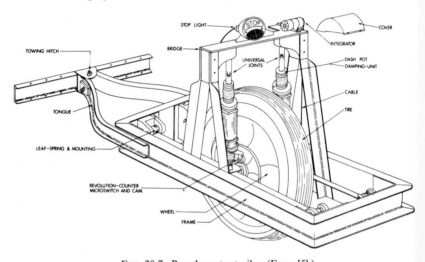

FIG. 20.7. Roughometer trailer. (From [5].)

the axle relative to the top of the suspension system. A second counter records the revolutions of the recording wheel so that the roughness of any length of pavement can be recorded. A standard operational speed of 20 mph is specified to ensure that the conditions of the test are exactly repeatable. Figure 20.7 shows the different components of the roughometer trailer.

Profilometer. The profilometer is basically a 16-wheeled articulated carriage which supports a detecting and recording device at a constant height above the road surface. The unit is designed so that only $\frac{1}{16}$ of the vertical movement of any single wheel is transmitted to the mounting of the detector wheels. The detector wheel, which tracks the line of travel, is located in the centre of the chassis.

The profilometer plots a profile of the road surface on a natural vertical scale, and measures the number of different sizes in intervals of 0·1 in by means of a classifier composed of electrical counters. The roughness value in in/mile is determined by the sum of all downward vertical motions in each interval.

Another type of profilometer is the CHLOE profilometer developed for use in the AASHO Road Test to measure the slope variance. In this device the change in angle between two reference lines is the measure of the pavement profile roughness. One reference line is the line connecting the centre of two slope wheels which follow the road and are relatively close together. The second reference line is determined by a 20 ft long member which is supported by a trailer hitch on the back of a towing vehicle, and a wheel supporting the rear end of the member.

20.5 BITUMINOUS AND FLEXIBLE OVERLAYS

There are several possible methods for determining the thickness of an asphalt concrete overlay, with or without a base course, needed to strengthen an existing pavement. Some of these methods require an evaluation of the existing pavement, while others need an analysis of pavement deflection.

Where the pavement is sufficiently strong and needs only correction of surface defects, a thin overlay of $\frac{1}{2}-\frac{3}{4}$ in may be all that is required. However, where strengthening is needed, a more substantial thickness of overlay should be used. Three of the most popular methods of design of asphalt overlays are explained below.

20.5.1 Asphalt Institute Method of Overlay Design

The Asphalt Institute has prepared an asphalt overlay design manual which describes two methods for determining the thickness requirements of an asphalt concrete overlay [7]. The first method, referred to as the component analysis method, requires an evaluation of the existing pavement to compare with a new full-depth asphalt pavement design. The second method, called the pavement deflection method, needs an analysis of pavement deflection and is partially based on elastic-layered theory. Both methods, which are based on field observations and engineering experience, provide thicknesses of asphalt concrete overlays directly.

TABLE 20.1 FACTORS FOR CONVERTING THICKNESS OF
EXISTING PAVEMENTS COMPONENTS TO EFFECTIVE
THICKNESS (T_E)

Classi- fication	Description of material	Conversion factors[a]
I	Native subgrade in all cases.	0·0
II	(a) Improved subgrade constructed with predominantly granular materials that may contain some silt and clay but have PI of 10 or less. (b) Lime-modified subgrade constructed from high-plasticity soils having a PI greater than 10.	0·0–0·2
III	(a) Granular sub-base or base constructed with reasonably well graded, hard aggregates having some plastic fines and CBR not less than 20. Upper part of range is used if PI is 6 or less; lower part of range is used if PI is more than 6. (b) Cement modified sub-bases and bases constructed from low-plasticity soils that have a PI of 10 or less.	0·2–0·3
IV	(a) Granular base constructed with a non-plastic granular material complying with established standards for high-quality aggregate base. Upper part of range is used. (b) Asphalt surface mixtures that have large, well-defined crack patterns, spalling along the cracks and appreciable deformation in the wheel paths, showing some evidence of instability. (c) Portland cement concrete pavement that has been broken into small pieces, 2 ft or less in maximum dimension, prior to overlay construction. Upper part of range is used when sub-base is present; lower part of range if used when slab is on subgrade. (d) Soil–cement bases that have developed extensive crack patterns as shown by reflected surface cracks, and may exhibit pumping; pavement shows minor evidence of instability.	0·3–0·5
V	(a) Asphalt surfaces and underlying asphalt bases[b] that exhibit appreciable cracking and crack patterns, have little or no spalling along the cracks and remain essentially stable even though exhibiting some wheel path deformation. (b) Appreciably cracked and faulted Portland cement concrete pavement that cannot be effectively undersealed. Slab fragments, ranging in size from approximately 1 to 4 yd^2, are well seated on the subgrade by heavy pneumatic rolling. (c) Soil–cement bases that exhibit little cracking, as shown by reflected surface crack patterns, and that are under stable surfaces.	0·5–0·7

Table 20.1 *(continued)*

Classification	Description of material	Conversion factors[a]
VI	(a) Asphalt concrete surfaces that exhibit some fine cracking, small intermittent cracking patterns and slight deformation in the wheel paths though they remain stable.	0·7–0·9
	(b) Liquid asphalt mixtures that are stable and generally uncracked, show no bleeding and exhibit little deformation in the wheel paths.	
	(c) Asphalt treated base, other than asphalt concrete.[b]	
	(d) Portland cement concrete pavement that is stable and undersealed, has some cracking, but contains no pieces smaller than about 1 yd².	
VII	(a) Asphalt concrete, including asphalt concrete base, that is generally uncracked and has little deformation in the wheel paths.	0·9–1·0
	(b) Portland cement concrete pavement that is stable, undersealed and generally uncracked.	
	(c) Portland cement concrete base, under asphalt surface, that is stable and non-pumping, and exhibits little reflected surface cracking.	

[a] Values and ranges of conversion factors are multiplying factors for conversion of thickness of existing structural layers to equivalent thickness of asphalt concrete. These conversion factors apply only to pavement evaluation for overlay design. In no case can they be applied to original thickness design.

[b] Asphalt concrete base, asphalt macadam base, plant mixed base and asphalt mixed *in situ* base.

The component-analysis method is applicable to either rigid or flexible base pavements, while the deflection method is applicable only where the original pavement is flexible.

Component-Analysis Method

This method is applicable to overlay design for both highways and airfields. The first step is to evaluate the existing pavement and convert it to an effective thickness of asphalt concrete. The conversion factors that can be used for this purpose are given in Table 20.1 for different materials (including Portland cement concrete) and different pavement conditions. The equivalent thickness of asphalt concrete can be determined, for the existing pavement, by selecting the appropriate conversion factor for each pavement layer and multiplying the thickness of that layer by the conversion factor. The effective full-depth asphalt pavement thickness, T_E, can then be obtained by adding the results for each layer.

The second step is to determine the required thickness, T_A, of a new full-depth asphalt pavement for the existing condition of the subgrade and the predicted traffic loads and volumes. Methods covered in Chapters 6 and 7 for the design of flexible pavements, especially the Asphalt Institute design method, can be used for this purpose.

The overlay thickness in inches of asphalt is then determined as follows:

$$T = T_A - T_E \tag{20.1}$$

where T = required thickness of asphalt concrete overlay (in);
$\quad T_A$ = thickness of a new full-depth asphalt pavement (in); and
$\quad T_E$ = the total effective thickness of the existing pavement in inches of asphalt concrete.

For Portland cement concrete pavements, a minimum thickness of $4\frac{1}{2}$ in is recommended to minimise reflection cracking, especially above the joints. In some situations, special treatment may be needed at the expansion and contraction joints of these pavements before putting down the asphalt concrete overlay.

The minimum thickness of overlay recommended for existing flexible pavements is 3 in.

A flexible overlay can sometimes be used, if the required thickness of asphalt concrete overlay is larger than the recommended minimum thicknesses of pavement surface, as given in Chapter 6. In this case, the overlay may consist of an asphalt concrete surface layer and a base layer of asphalt treated or untreated granular material. A thin soft base layer of sand treated with bitumen is sometimes useful in minimising reflection cracking, and reducing the transmission of joint movements in the base pavement to the overlay. The methods explained in Chapter 6 to determine the thickness of the different layers of the flexible pavement, for each particular type of base, can be used in calculating the thickness of the base layer of a flexible overlay.

Example
An existing pavement, in good condition, consists of 3 in of asphalt concrete surface, 5 in of crushed stone base and 6 in of granular sub-base.

The pavement needs strengthening to accommodate heavier traffic. The design of a new full-depth asphalt pavement indicates the need for a thickness of 10 in. What is the required thickness of asphalt concrete overlay?

Solution

Thickness of a new full-depth asphalt pavement, T_A = 10 in

Conversion factor for the existing asphalt surface layer = 0·80

Conversion factor for the existing base layer = 0·40

Conversion factor for the existing sub-base layer = 0·25

Effective thickness of existing pavement, T_E =

$$3·0 \times 0·80 + 5·0 \times 0·40 + 6 \times 0·25 = 2·40 + 2·0 + 1·5 = 5·9 \text{ in}$$

Thickness of asphalt concrete overlay =

$$T_A - T_E = 10 - 5·9 = 4·0 \text{ in}$$

Pavement-deflection method. This method is intended only in design for highways. In the technique, the strength of the existing pavement is evaluated by measuring the rebound deflection using the Benkelman beam and an 18 kip single-axle load. The pavement under consideration is divided into sections, using engineering judgement and experience, and the measurement locations determined accordingly. Special locations such as poorly drained or broken-up areas should be tested and dealt with separately, as it is likely that a change in overlay thickness or other special treatment will be required in these areas.

Pavement deflections are obtained by testing the outer wheel path at a minimum of 20 locations per mile, the locations being selected by a random sampling procedure. The measured rebound deflections are reduced to a representative rebound deflection value, which is the mean of adjusted measured rebound deflections plus two standard deviations $(\bar{x} + 2S)$. This represents a deflection level that would be exceeded in only 2% of the length of the pavement.

Measured rebound deflections should be adjusted for temperature and for the most critical period of the year, as regards pavement performance. Temperature adjustment factors, derived from correlation between the rebound deflection and the mean temperature of the asphalt layers of a pavement during the test, are shown in Fig. 20.8. Curve A should be used for most of the cases, since it has the greatest data support. Curve B should only be used in special situations, such as for those pavements with 4 in or more of total asphalt thickness, on a weak foundation.

In some climatic environments, especially where freezing conditions prevail during the Winter, certain periods of the year are more critical for pavement performance than others. The ratio of the rebound deflection in Spring to that in Autumn can reach 2·5 in areas where the freezing depth underneath the pavement is large. For this reason it is extremely important to determine the rebound deflection that reflects the most critical period. One of the following two methods can be used for this purpose: (1) Take the rebound measurements during the most critical period. (2) Multiply the rebound measurements, taken at any time, by a factor representing the ratio of critical period deflection to the deflection for the date of test.

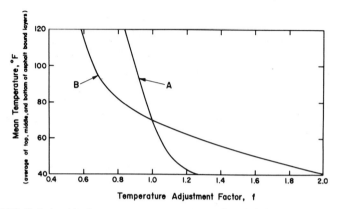

FIG. 20.8. Relationship between mean temperature of asphalt layers and temperature adjustment factor. (From [7].)

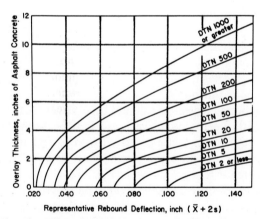

FIG. 20.9. Chart for design of asphalt overlays on flexible pavements. (From [7].)

Once the critical rebound deflection is determined, together with the design traffic number (DTN) described in Chapter 6, the thickness of the asphalt concrete overlay can be determined using Fig. 20.9. One first reads the deflection on the abscissa, proceeds vertically to the desired DTN curve and then horizontally to read the overlay thickness on the ordinate.

If for any reason a flexible overlay consisting of an asphalt concrete surface layer and a base is recommended, the procedure explained in the component-analysis method for its design can also be followed here.

Example
The critical rebound deflection for an existing pavement is 0·08 in, and it is required to strengthen the pavement to receive traffic with a DTN of 200. What will be the thickness of the asphalt concrete overlay?

Solution
From Fig. 20.9, the required thickness of the asphalt concrete overlay is 4·7, say 5 in.

To compare this method with other deflection methods, the design curves for the Asphalt Institute, for a state and for five countries, are

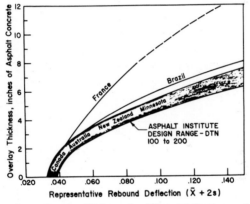

FIG. 20.10. Comparison of overlay thickness for different deflection methods. (From [7].)

shown in Fig. 20.10. The shaded area in Fig. 20.10 is the area between the Asphalt Institute design curves for DTN 100 and DTN 200. The design curves for Canada, Australia and New Zealand are for moderately heavy traffic with an annual average daily traffic (AADT) approximately equal to 1000, equivalent to 100 18 kip axle loads per day in one lane. The design curve for Brazil is for moderately heavy traffic and that for France is for an AADT of 3700 vehicles. A seasonal factor of 1·8 is used for the Canadian curve.

The curves in Fig. 20.10 show good agreement except for France, which has a 29 000 lb maximum allowable single-axle load.

20.5.2 AASHO Interim Guide Method

The AASHO method for the design of flexible pavements for highways, explained in Chapter 6, can also be used in the design of flexible overlays in either rigid or flexible pavements. The structural number of the existing pavement can be estimated, in terms of the flexible-pavement guides, using eqn. (20.2):

$$SN = a_1 D_1 + a_2 D_2 + a_3 D_3 \qquad (20.2)$$

where D_1, D_2 and D_3 are the thicknesses of the surface, base and sub-base courses, respectively;

$a_1 = 0.40$ for asphalt concrete surface layers that are generally uncracked and have little deformation in the wheel paths, or for Portland cement concrete pavements that are stable, non-pumping, undersealed and generally uncracked;

$a_1 = 0.30–0.39$ for asphalt concrete surface layers that exhibit some fine cracking and slight deformation in the wheel paths, or for Portland cement concrete pavements that are stable and undersealed, have come cracking, but contain no pieces smaller than about 1 yd^2;

$a_1 = 0.22–0.29$ for asphalt concrete surface layers that exhibit appreciable cracking and some wheel path deformation, or appreciably cracked and faulted Portland cement concrete pavements that cannot be effectively undersealed;

$a_2 = 0.14$ for a base layer consisting of crushed stone; and

$a_3 = 0.11$ for a sandy-gravel sub-base layer.

If the base pavement is of the rigid type, the structural number is that of the concrete slab alone and is determined by multiplying the pavement thickness D_1 by the factor a_1 corresponding to the pavement condition. Experience and engineering judgement should be used to accurately estimate the values of the coefficients a_1, a_2 and a_3. The structural number of the existing pavement is then subtracted from that required for a new design. This Δ-value is then used to determine the required thickness of a flexible overlay, by applying eqn. (20.2) and considering a value of 0.44 for the coefficient a_1.

If a bituminous overlay is recommended, then the thickness of this overlay can be obtained by dividing the Δ-value by 0.44.

20.5.3 The FAA Method

The Federal Aviation Administration recommends the following formulae for determining the thicknesses of flexible and bituminous overlays required for strengthening rigid airfield pavements [2 and 3]:

for flexible overlays, $$t_f = 2 \cdot 5 (Fh - h_e) \tag{20.3}$$

for bituminous overlays,

$$t_b = \frac{(Fh - h_e)}{0 \cdot 6} + \frac{t_s}{3} \tag{20.4}$$

where t_f = thickness of flexible overlay;
$\quad t_b$ = thickness of bituminous overlay;
$\quad t_s$ = required thickness of surface course for the flexible overlay;
$\quad h_e$ = thickness of the original Portland cement concrete pavement;
$\quad h$ = required thickness of an equivalent single concrete slab placed directly on the subgrade or sub-base;
$\quad F$ = factor which depends upon subgrade class
$\qquad$ = 0·80 for subgrade class R_a;
$\qquad$ = 0·90 for subgrade class R_b;
$\qquad$ = 0·94 for subgrade class R_c;
$\qquad$ = 0·98 for subgrade class R_d;
$\qquad$ = 1·00 for subgrade class R_e.

20.6 PORTLAND CEMENT CONCRETE OVERLAYS

The three common cases of rigid overlays are: (1) a concrete overlay on existing flexible pavements, (2) a concrete overlay on existing rigid pavements without separation, and (3) a concrete overlay on rigid pavements with separating course.

20.6.1 Concrete Overlay on Existing Flexible Pavements

In this case, the original flexible pavement is considered as sub-base for the concrete overlay. Field tests or experience should be used to determine the modulus of subgrade reaction K. The value of K will depend upon the thickness and condition of the existing flexible pavement, as well as on the K value of the subgrade.

For an approximate estimation of the value of K, the thicknesses of the sub-base and base layers of the existing flexible pavement should be added

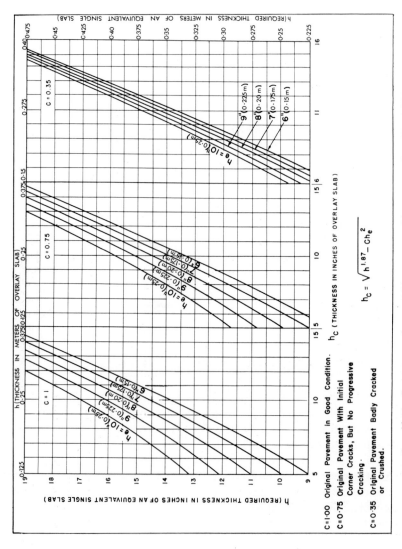

FIG. 20.11. Concrete overlay on rigid pavement without separation. (From [6].)

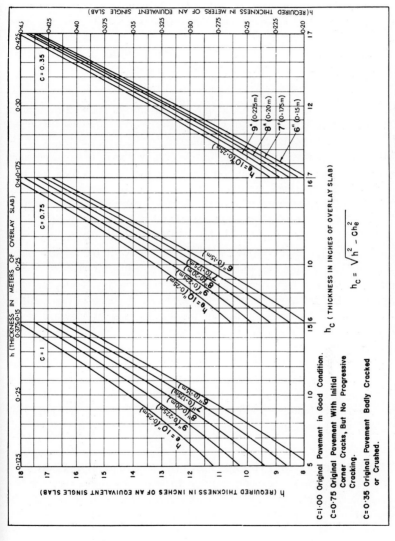

FIG. 20.12. Concrete overlay on rigid pavement with separating course. (From [6].)

together. The value of K at the top of the base layer of the flexible pavement can be determined from Fig. 10.7, if the value of K for the subgrade is known. Figure 10.8 can then be used to determine the K value on top of the existing asphalt layer, in spite of the fact that this figure is for the case of cement treated bases. An asphalt course is generally of higher value than a cement treated layer.

Once the modified value of K is known, the methods of design of rigid pavements can be used to design the concrete overlay.

20.6.2 Concrete Overlay on Existing Rigid Pavements Without Separation

For this case, it is assumed in designing the overlay that the two pavements are acting together, to some degree, in carrying the traffic loads. For this reason, it is important to have a certain degree of bond between the existing and the new concrete pavements. The bond stresses, together with the inherent shear resistance that is generally developed through friction between the top surface of the existing pavement and the bottom surface of the new one, help in getting the two slabs to act together.

Bond can be developed by several methods. One of the practical and economical ways to develop it is to clean the surface of the existing pavement, wet the surface and then spread a cement mortar layer of 0·08 in thickness, 15 min before placing the concrete overlay. However, it is important to carry out tests to investigate the effectiveness of this method.

It is important to make sure that the joints in the overlay are of the same type and are placed at the same locations as those in the existing pavements.

Figure 20.11 can be used to determine the concrete overlay thickness, once the required thickness of an equivalent single concrete slab, placed directly on the subgrade or sub-base, is calculated. Three groups of curves are shown in the figure, for the cases of good, moderately cracked and badly cracked base pavements.

20.6.3 Concrete Overlay on Rigid Pavement With Separating Course

If a separating course is recommended, for any reason, between the two slabs, the effect of a given thickness of rigid overlay is reduced, and the curves shown in Fig. 20.12 should be applied.

A separation course is generally applied if the type of concrete overlay used is different from the type of the base pavement. An example of this case is a continuously reinforced concrete overlay on a jointed concrete pavement.

REFERENCES AND BIBLIOGRAPHY

1. Cornelius, D. F., 'The Assessment of the Riding Quality of Concrete Roads', *Proc. 2nd European Symposium on Concrete Roads, Bern, 1973*, 209–223.
2. Federal Aviation Agency, 'Airport Paving', US Government Printing Office, 1962.
3. Hennes, R. G. and Eckse, M., *Fundamentals of Transportation Engineering*, McGraw-Hill, Second Edition, 1969.
4. *Highway Research Record*, No. 359, 'Maintenance Operations and Applied Systems Engineering', Washington, D.C., 1971.
5. Chong, G. J. and Phang, W. A., *Highway Research Board Special Report 133*, 'Pavement Evaluation Using Road Meters', Washington, D.C., 1973.
6. International Civil Aviation Organisation, 'Aerodrome Physical Characteristics —Part 2', *Aerodrome Manual*, Second Edition, 1965.
7. Kingham, R. I., 'Asphalt Overlay Design,' *Highway Research Record*, No. 300, 1969, 37–42.
8. Liddle, W. J. and Peterson, D. E., 'Utah's Use of Dynaflect Data for Pavement Rehabilitation', *Highway Research Record*, No. 300, 1969, 10–15.
9. McCullough, B. F., 'What an Overlay Design Procedure Should Encompass', *Highway Research Record*, No. 300, 1969, 43–49.
10. McCullough, B. F., 'Overlay Design: What are the States Presently Doing?' *Highway Research Record*, No. 300, 1969, 4–9.
11. McCullough, B. F. and Monismith, C. L., 'A Pavement Overlay System Considering Wheel Loads, Temperature Changes, and Performance', *Highway Research Record*, No. 327, 1970, 64–82.
12. National Cooperative Highway Research Program 9, 'Pavement Rehabilitation —Materials and Techniques', Highway Research Board, Washington, D.C., 1972.
13. Vaswani, N. K., 'Method for Separately Evaluating Structural Performance of Subgrades and Overlaying Flexible Pavements', *Highway Research Record*, No. 326, 1971, 48–62.
14. Yoder, E. J., *Principles of Pavement Design*, Wiley, 1959.

CHAPTER 21

PAVEMENT SYSTEMS ANALYSIS

21.1 INTRODUCTION

A system can be described as a device or scheme which behaves according to some description to accomplish an operational process. Thus, a pavement can be defined as a system obeying physical laws that transforms the effects of input variables into various responses characterised by an output function. Design of a system like a pavement needs a co-ordinated set of procedures, to detail the use of money and materials in the most economical combinations.

Systems analysis can be defined as 'a systematic approach that provides decision makers with a complete, accurate and meaningful summary of the information necessary to clearly define issues and alternatives'. It deviates from conventional common-sense, being an organised procedure that considers all factors together in an integrated form, to achieve the best results.

Piecemeal solutions that have been adopted in existing design procedures are oriented towards emphasising certain important features of design and neglecting certain others. For example, traffic loads and volumes have been the main inputs to the design, and the guiding decision criteria have been to limit the level of stress or deflection. However, technological developments have brought the realisation that stress or deflection analysis is but one criterion of several that are important for making decisions.

21.2 FLEXIBLE PAVEMENTS

21.2.1 Pavement System

Figure 21.1 shows a conceptual pavement system developed by F. N. Finn and others [6, 7]. The block diagram indicated in this figure is intended to

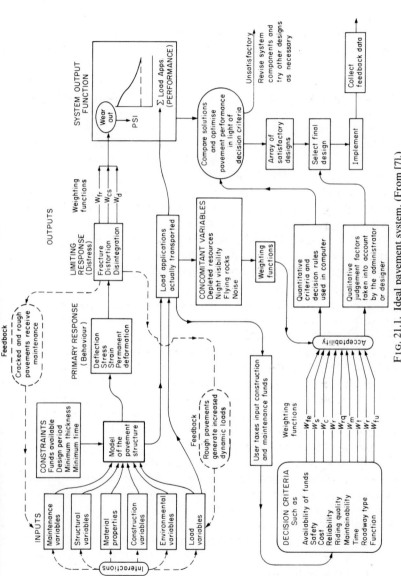

Fig. 21.1. Ideal pavement system. (From [7].)

systematise the many variables included in the structural design, operation and management of a total pavement system. It also emphasises the interaction of many of the variables involved. The figure shows that the inputs interact to produce output responses in the pavement system. It also shows that the expected output of a structural systems model is a behavioural characteristic, such as deflection or strain, that results at a certain stage in distress. The terms 'rupture, distortion, and disintegration', are used to describe all types of distress. When they are combined with appropriate weighting functions, they yield a wear-out curve or system output function for the pavement. This system output function may then be used as the objective function in the systems engineering process for asphalt concrete pavement design. However, to develop a method for relating to these factors is rather a difficult task.

A research team in Texas [9, 16] has simplified the problem and developed a working pavement design system that accomplishes this purpose, as shown in Fig. 21.2. For relating inputs to outputs, the team used only a deflection model, in this case surface deflection under the load. The tie between expected deflection and expected performance was made empirically from equations developed at the AASHO Road Test relating deflections to performance.

21.2.2 Pavement Performance
Because the output function is defined in terms of performance, it may be useful to avoid confusion by defining performance as well as other terms. The following definitions are used in this chapter.

1. Performance is a measure of the accumulated service provided by a facility, i.e. the adequacy with which a pavement system fulfils its purpose. It is often specified by a performance index, and as such is a direct function of the serviceability history of the pavement.
2. Present serviceability is the ability of a specific section of pavement to serve current traffic in its existing condition. The existing condition means the condition on the date of rating and not that at any future or past date.
3. Behaviour is the response of a pavement to load, environment and other inputs. This response is usually a function of the mechanical state (i.e. stress, strain or deflection) which occurs as a primary response to the input.
4. Distress modes are those responses which lead to some form of distress when reaching a certain limit, e.g. deflection under load is a

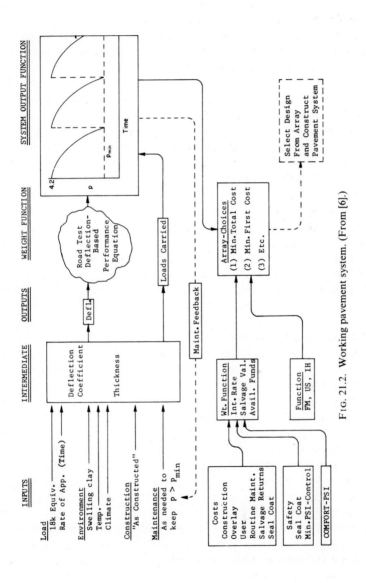

Fig. 21.2. Working pavement system. (From [6].)

mechanism that can lead to fracture. Some behavioural responses may not provide distress mechanisms.

5. Distress manifestations are the visible consequences of various distress mechanisms that usually lead to a reduction in serviceability.
6. Fracture is the state of being broken apart and includes all the different types of cracking, spalling, and slippage.
7. Distortion is the state of change of the pavement from its original shape condition. This change is permanent or semi-permanent, as opposed to transient changes such as deflections.
8. Disintegration is the state of being decomposed into constitutive elements, and includes stripping, ravelling and scaling.

21.2.3 Serviceability Concept

The function or primary operating characteristic of a pavement at any particular time is the level of service it provides to the users. In turn, the change of serviceability with time is some measure of pavement performance. This performance, together with the cost and benefit implications, is the primary consideration of design and over-all management system of pavements.

In general, there are two types of pavement evaluation. The first type is a functional evaluation, identified by the serviceability–performance concept, and indicates how well a pavement is currently serving its function. The second type is a mechanistic evaluation of the pavement, associated with determining the pavement's mechanical condition with the purpose of improving future performance. It is mainly an indicator of action needed to maintain serviceability, and in that sense may be a precursor to the serviceability evaluation.

One of the best methods of defining and obtaining serviceability was established at the AASHO Road Test and was based on subjective evaluation of the riding quality provided by a pavement at a given time. Correlations with physical measurements of the surface characteristics were performed for a large set of test pavements, and the result was termed the Present Serviceability Index. This PSI has been extensively used to predict pavement serviceability. Integrating the PSI over time or over the summation of load applications produced what was termed performance.

21.2.4 Performance as a System Output

The pavements management process consists of a variety of planning, design, construction, operation and research activities. Recent attempts to define this process in terms of a formal systems framework have explicitly

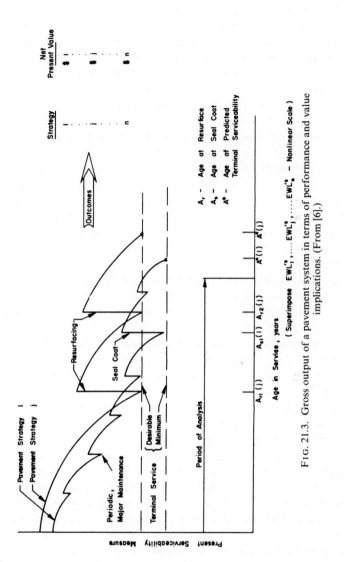

FIG. 21.3. Gross output of a pavement system in terms of performance and value implications. (From [6].)

recognised that one of the major activities involved is that of performance evaluation. This in turn necessitates that the system outputs to be evaluated, must first be defined.

To illustrate the idea, Fig. 21.3 shows the gross output of two alternative pavement strategies in terms of their serviceability–age histories (performance) and the associated value implication. A pavement strategy, in this figure, includes the structural design, the materials used, the construction process, the resurfacing, and the maintenance procedures. For each pavement strategy, a large number of variables, such as traffic, materials, climate, construction, maintenance and others combine to produce a corresponding performance profile. Age rather than sum of equivalent load applications is taken as the abscissa in Fig. 21.3, so that value implications can be taken into account. Furthermore, it is not sufficient to predict only the initial serviceability or the terminal age, for the intermediate portion is also important in checking the design strategies and their programmes for maintenance and resurfacing. In addition, this intermediate part is essential for an adequate exploration of the implications of raising the terminal serviceability level.

The measurement of the outputs of a pavement system during its service life, i.e. the evaluation of its performance, constitutes a major activity in the management system. In Fig. 21.4 the principal elements of this activity are shown as a portion of the over-all pavement management system.

21.2.5 Pavement Distress and System Failure

Distress can normally be evaluated by two basic approaches. The first approach is a functional evaluation of the effect of distress on the pavement's ability to serve traffic today. The second approach is a mechanistic evaluation of distress with an eye to finding out the current physical condition of the pavement, the causes for this condition and its effect on the future performance of the pavement.

The difference between the two approaches mentioned above is the key to the problem of relating pavement behaviour to pavement performance. For example, a crack in the pavement surface may have a minor effect or no effect on how well the pavement is serving traffic today (present serviceability). On the other hand, the engineer who looks at this crack in terms of mechanistic evaluation may think of it as a local failure, since a crack will cause intrusion of water, increased deflection and rapid deterioration of the pavement.

It is recommended that adequate attention be given to combining various pavement behaviour and distress factors into an over-all perform-

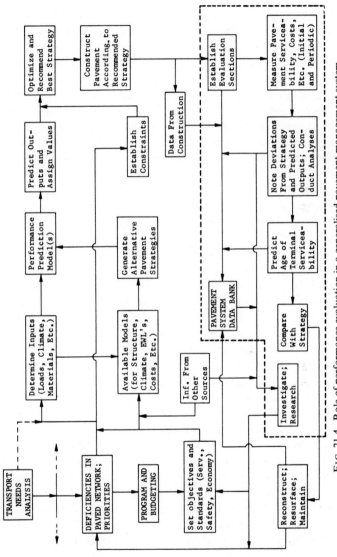

Fig. 21.4. Role of performance evaluation in a generalised pavement management system. (From [6].)

ance function. A proper identification of this performance function is essential for solving the pavement problem in the context of systems analysis.

A definition of failure should also be fully specified, and a differentiation made between failure of the pavement system and a material failure. While failure of a pavement material is generally not a catastrophic occurrence, pavement failure is a condition that usually develops gradually over a long period of time. In this framework, the system output exceeds some limiting value formulated by the decision criteria. As shown in Fig. 21.2, the pavement system output and the decision criteria should be considered together, since the decision criteria are used to evaluate the system output and to make a judgement of pavement performance. Thus, failure can be defined by the decision criteria as some limiting value of the system output.

As shown in Fig. 21.2, the behaviour of a pavement structure may be quantified in terms of its primary response. The limiting response, designated as distress, may be expressed conceptually as:

$$\mathbf{D I}(\mathbf{x}, t) = \mathbf{F}_{s=o}^{s=t} [\mathbf{R}(\mathbf{x}, s), \mathbf{S}(\mathbf{x}, s), \mathbf{D}(\mathbf{x}, s), \mathbf{x}, t] \qquad (21.1)$$

where t = time;

$\mathbf{x}$ = a space variable;

$\mathbf{D I}(\mathbf{x}, t)$ = distress index, a function of space and time;

$\mathbf{R}(\mathbf{x}, t)$ = measure of rupture;

$\mathbf{S}(\mathbf{x}, t)$ = measure of distortion; and

$\mathbf{D}(\mathbf{x}, t)$ = measure of disintegration.

Since distress is spatial, it can best be considered on a unit volume basis. The distress index, as described in eqn. (21.1), is a function of the history of fracture, distortion and disintegration from time zero to current time, t. The three parameters, rupture, distortion and disintegration, in addition to being functions of space and time, are functions of five input classes of variables. The distress index is presented schematically in graph 7 of Fig. 21.5, if subtracted from some constant, C, that depends on the scaling factors involved, it will give a serviceability curve as shown in graph 8 of the figure.

Since failure of the pavement may be expressed as a condition where distress from the system output exceeds an acceptable level based on decision criteria, a definition of an index for these criteria is in order. This decision criteria index, DCI $(\mathbf{x}, t)$, is a function of riding quality, economics, safety, maintainability and other factors, in addition to space and time. However, riding quality is supposed to be independent of time, since there is a minimum allowable rideability for any given type of pavement regardless of time. Also, safety is time-invariant, because it should have some minimum acceptable level, for given conditions, that should not be exceeded during the life of a pavement. Graphs 4, 5 and 6 of Fig. 21.5 are illustrations of possible acceptable levels for each decision

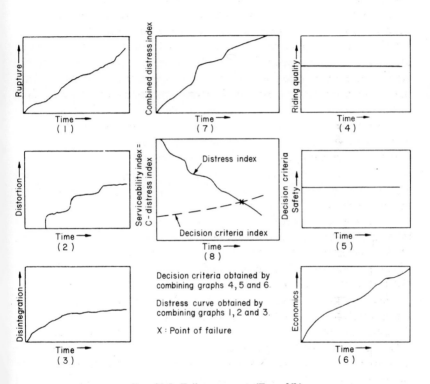

FIG. 21.5. Failure concept. (From [6].)

criterion. The design criteria index, which combines riding quality, safety and economics, is presented by the dashed curve in graph 8. The point of intersection of this curve with the curve representing the distress index might then represent failure for the system, i.e. when the pavement is performing at less than the desired level.

21.3 RIGID PAVEMENTS

21.3.1 Formulation of Conceptual Design System

According to the definition of a system, a comprehensive formulation of the design process characterising various technical and economic aspects is needed, before a more realistic working design system can be formulated for immediate use. Figure 21.6 shows many of the factors that may be involved in the conceptual pavement design system.

Physically, the pavement system is characterised by the properties, amount and arrangement of the materials used, as well as the quality of construction. It may be considered as an operator that reacts in the form of responses, when acted on by the excitation functions; the latter are often termed systems inputs. These responses are generally characterised by an immediate mechanical state, defined by stresses, strains and deflections, which is eventually followed by time-dependent accumulated effects in the forms of rupture, distortion, disintegration and decreased skid-resistance. The response defined by the mechanical state of the system is termed a primary response. The progressive effects resulting from the repetitive or continued existence of the state of primary response produce a limiting response which is the actual criterion of pavement failure. For example a limiting response, such as the roughness of a pavement, may influence the dynamic effect of loads to an extent that may subjectively be considered as failure, unless maintenance is provided. Discomfort to the rider, which is determined by the vibrations of vehicles moving on the pavement, is the ultimate measure of pavement deterioration. The average of these human responses characterises the extent to which the travelling public is served, i.e. serviceability.

Performance, which represents the serviceability–age history of the pavement, is very important in evaluating the cost implications of the pavement system. The AASHO Road Test, in a simplified version, helped to develop a concept for directly predicting the pavement serviceability by correlating the subjective ratings of pavements to their objective characteristics.

21.3.2 Generation and Evaluation of Alternatives

Part of the system shown in Fig. 21.6 involves generation and evaluation of potential alternative strategies so as to select the best. In this context, a strategy is defined as a set of resource allocations for a design; this design should last the required life, according to the specifications laid

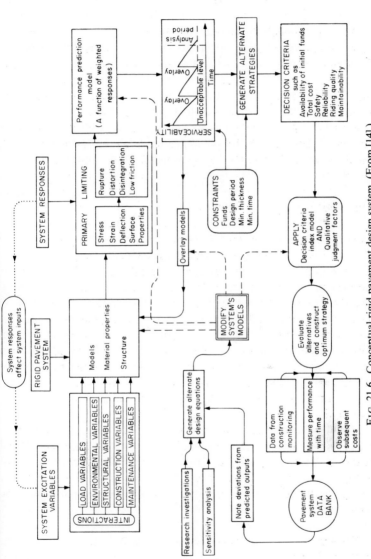

Fig. 21.6. Conceptual rigid pavement design system. (From [14].)

down for it. Possible strategies are evaluated with regard to a set of decision criteria, such as those shown in Fig. 21.6, to obtain the optimum. Each decision criterion should be quantified and weighted. These criteria can then be combined to define a function, referred to as a decision criteria index. The simplest form of this function is the subjective evaluation of various factors such as riding quality, safety and availability of funds.

A further step in the long-range planned objectives of the pavement management system is implementation and feedback. The pavement system data bank is an important part of the feedback subsystem. Among other things, it consists of performance evaluation of the optimal strategies implemented in the past; the subsequent expenditures on these strategies are also observed.

21.3.3 Working System for Rigid Pavements

The development of an operational rigid pavement system can be described by objectives, inputs, constraints and options, decision criteria and output. The following is a brief discussion of each of these items.

Objectives

A large payoff can be obtained from past research while the ultimate design system and its models are being developed. For this reason, it is important to gather the available research literature, analyse the significant models and try to formulate the first version of a rigid-pavement design system. Such models, which are important links in the system and for which research is not available, may be developed mathematically.

Inputs

System inputs can be described by the following general groupings.

1. *Load variables.* This input provides information concerning initial traffic weights, volume and distribution, as well as traffic growth with time.
2. *Traffic delay variables.* This set of variables can be used to analyse the consequent cost of overlay construction incurred due to the inconvenience to people delayed in traffic. This inconvenience can be considered by calculating the costs of traffic delays and operating time losses.
3. *Performance variables.* These variables define the levels of initial and terminal serviceability for an initial or overlaid pavement structure. They can be used to determine the life of such a pavement structure

when its serviceability index is allowed to drop from its initial value to a specified minimum level.

4. *Material variables.* The engineering properties of all materials used are determined experimentally or defined theoretically, and as inputs to the system are used in predicting the life of an initial or overlaid pavement structure.

The properties of non-homogeneous materials used in pavements change from point to point and are also functions of time and environment. These changes can be taken into account by using the dispersion data of the tests conducted to determine these properties. A design value can then be determined by specifying a certain level of confidence desired for the design with respect to any particular property.

5. *Cost variables.* The criterion of present value of total over-all cost can be used for this working system to indicate the preference for one design over another. A number of cost inputs, including costs of materials, construction, maintenance and other operations, are essential for the working system during the process of its evaluation of different design strategies.

For rational economic analysis, an interest rate should be included, so that the future investments may be properly evaluated; a salvage value of the pavement at the end of the analysis period should also be estimated.

6. *Miscellaneous variables.* Inputs such as the number and width of lanes of the facility to be designed fall in this category, as do parameters used in other inputs and in the computer program.

Constraints and Options

A primary set of constraints is generally specified for the working system; examples of these are the minimum and maximum thicknesses of the materials to be provided. These constraints allow of only a number of possible designs; these can be analysed and checked against a number of other constraints located at later stages in the working system. This second set of constraints rejects certain designs, designating them as infeasible, and selects the feasible designs from all the generated possible designs. Different options can be incorporated in the system, with respect to the types of designs desired. For example, the system can be restrained to design one or all of the following:

1. Pavement type: jointed and continuously reinforced concrete pavements.

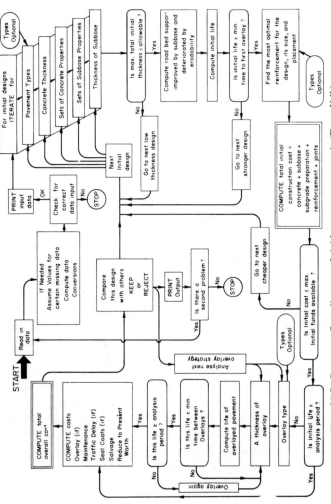

Fig. 21.7. Summary flow diagram of rigid-pavement system 1. (From [14].)

2. Overlay type: asphalt concrete and Portland cement concrete overlays.
3. Reinforcement type: deformed bars and wire mesh.

The system constraints and options are in many situations the designer's decisions to generate a reasonable type and number of solutions. In some cases, however, they may be the actual physical limitations advocated by special conditions of design and construction.

Decision Criteria

One of the prime decision criteria for the selection of optimal pavement strategy is the total over-all cost. In cases where there are budget constraints, availability of initial funds may be another decision criterion. which also acts as a restraint. Other criteria such as safety, riding quality and maintainability are controlled by a minimum specified serviceability level.

Output

The ordered set of alternative design strategies, with pertinent information for each strategy, are summarised in a table based on increasing order of the present worth of total over-all costs: judgement should then be exercised in making a decision. The relative importance to the decision maker of the various economic, social and experience factors will play a role in his decision.

21.3.4 Description of a Working System

A working systems model called Rigid Pavement System 1 (RPS1), has been recently developed at the Center for Highway Research of the University of Texas at Austin. An RPS1 computer program has been written to solve various performance, structural and cost models. The solution results in arrays of design strategies that are stored and scanned for optimisation by various techniques built into the program.

Figure 21.7 shows a summary flow chart for the working system. The design process can be broadly divided into the following parts:

(a) read data, check against invalid inputs, and print input data;
(b) generate possible initial designs;
(c) select feasible initial designs;
(d) design sub-bases, reinforcements, and joints;
(e) develop overlay strategies (Portland cement concrete or asphalt concrete overlays) for those feasible initial designs that reach the minimum specified serviceability level in times less than the analysis period;

(f) analyse cost of all strategies;
(g) store, optimise and scan; and
(h) print output.

Figure 21.8, which shows the general overlay performance patterns built into the program, illustrates the relative difference in the designed

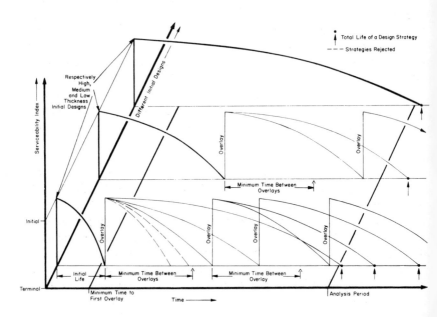

FIG. 21.8. Illustrative performance patterns of overlays in rigid-pavement system 1.
(From [14].)

performance patterns of structurally strong, medium and weak initial designs. The structural model used in RPS1 to represent the serviceability loss due to traffic is an extension of the one of the AASHO Road Test equations. In addition, the stochastic variations of the material properties are developed by using statistics.

Example
A rural highway with two lanes in each direction is to be designed to carry about 10 000 initial average daily traffic, with a 5% yearly growth. The total number of 18 kip equivalent axle loads expected to be carried during a lifetime of 20 years is 5 million.

The highway passes through an area of moderate swelling clays; the modulus of subgrade reaction has a mean value of 100 pci with a standard deviation of 15 pci. The flexural strengths of the concrete that can be used are 450 and 650 psi with standard deviations of 40 and 60 psi, respectively. Two alternative sub-bases, granular and cement treated, are available with modulus values of 20 000 and 900 000 psi, respectively. Two alternative types of deformed bars and one type of wire mesh are available for use as reinforcing steel. The modulus of elasticity of the asphalt concrete that may be used for overlays is 200 000 psi. In the cases of Portland cement concrete overlays, the same concrete as used in the initial construction should be provided.

Initial serviceability index values of 4·2 and 4·0 are specified for initial and overlay construction, respectively, while a minimum serviceability index of 2·5 must be maintained at all times.

Initial funds are limited and restrict the possibility of using large thicknesses in the initial state. In addition, the pavement will not be overlaid for at least 5 years after the initial construction and for 6 years after any overlay construction. Each overlay construction will cause traffic disturbance over $\frac{1}{2}$ mile of road length; 5% of the vehicles will be stopped because of this construction. During the 8 h daily working period, it is assumed that 48% of the average daily traffic will arrive, and that one lane will be overlaid while traffic is diverted to the second lane.

Any acceptable design should have a confidence level of 95% with respect to each of the material properties; adequate maintenance should also be provided.

21.3.5 Discussion of Systems Analysis and Results for the Example

The specified combinations of concrete and sub-base thicknesses provided 196 initial possible designs. Because of the imposed constraints such as non-availability of sufficient initial funds, several designs are rejected by the program. The remaining 79 feasible initial designs provide 751 design strategies, out of which 652 are feasible within the program limitations.

Problem 1, in Fig. 21.9, is the optimal design for this example. The effect of two important input parameters is introduced by solving for the two alternative conditions identified in parts (b) and (c) of Fig. 21.9. Most of the optimal designs in each of the three problems shown in this figure will require either one or two overlays during their service lives. The interesting result that the program favours concrete pavement designs with overlays contradicts the traditional practice, which has never been in favour of stage construction for rigid pavements. The inadequate

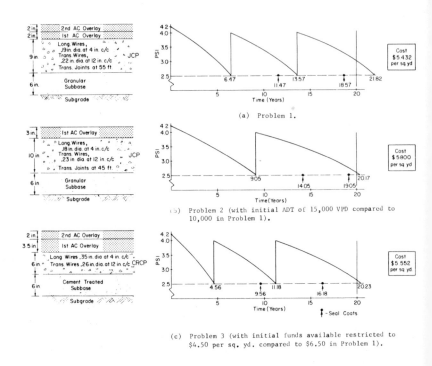

(a)   Problem 1.

(b)   Problem 2 (with initial ADT of 15,000 VPD compared to 10,000 in Problem 1).

(c)   Problem 3 (with initial funds available restricted to $4.50 per sq. yd. compared to $6.50 in Problem 1).

FIG. 21.9. Systems analysis of example problems. (From [14].)

analytical techniques available in the past for designing composite pavements, and the limited knowledge of the various aspects of these pavements, have encouraged this traditional trend.

21.3.6 Summary
A successful pavement design requires the use of economy, performance and other decision criteria in the analysis of various possible design strategies; a search must be made for the optimum design. Such a solution process can be evolved in a systems framework, which should be handled in an integrated form.

The example given by Kher *et al.* [14] to illustrate the use of the R P S1 program has been provided here, merely to indicate the type of data required when such an approach is used.

REFERENCES AND BIBLIOGRAPHY

1. Canadian Good Roads Association, 'Field Performance Studies of Flexible Pavements in Canada', *Proc. 2nd International Conference on Structural Design of Asphalt Pavements, University of Michigan, Ann Arbor, 1967*, 1087–1098.
2. Carey, W. N. and Irick, P. E., 'The Pavement Serviceability–Performance Concept', *Highway Research Bulletin 250*, 1960.
3. Haas, R. C. G. and Hudson, W. R., 'The Importance of Rational and Compatible Pavement Performance Evaluation', *Highway Research Board Special Report 116*, 1971, 92–111.
4. Hall, A. D., *A Methodology for Systems Engineering*, Chapter 8, Van Nostrand, 1962.
5. *Highway Research Board Special Report 61*, 'The AASHO Road Test', 7 reports, 1961–1962, Washington, D.C.
6. Hudson, W. R., 'Serviceability Performance and Design Consideration', Highway Research Board 126, 1971, 140–149.
7. Hudson, W. R., Finn, F. N., McCullough, B. F., Nair, K. and Vallerga, B. A., 'Systems Approach to Pavement Design, Systems Formulation, Performance Definitions and Materials Characterization', Materials Research and Development, Inc., Oakland, Calif., March 1968.
8. Hudson, W. R. and Kennedy, T. W., 'Parameters of Rational Airfield Pavement Design System', *Transportation Engineering Journal of the American Society of Civil Engineers*, **99**, No. TE 2, May 1973, 235–253.
9. Hudson, W. R., McCullough, B. G., Scrivner, F. H. and Brown, J. L., 'A Systems Approach Applied to Pavement Design and Research', Texas Transportation Institute, Texas A & M University, Research Report 123–1, March 1970.
10. Hudson, W. R., Teske, W. E., Dunn, K. H. and Spangler, E. B., 'State of the Art of Pavement Condition Evaluation', *Highway Research Board Special Report 95*, 1968.
11. Hutchinson, B. G., 'Principles of Subjective Rating Scale Construction', *Highway Research Record*, No. 46, 1964, 60–70.
12. Hutchinson, B. G. and Haas, R. C. G., 'A Systems Analysis of the Highway Pavement Design Process', *Highway Research Record*, No. 239, 1968, 1–24.
13. Kher, R. K., 'A Systems Analysis for Rigid Pavement Design', Ph.D. dissertation, University of Texas at Austin, 1970.
14. Kher, R. K., Hudson, W. R. and McCullough, B. F., 'Comprehensive Systems Analysis for Rigid Pavements', *Highway Research Record*, No. 362, 1971, 9–20.
15. Phillips, M. B. and Swift, G., 'A Comparison of Four Roughness Measuring Systems', *Highway Research Record*, No. 291, 1969, 227–235.
16. Scrivner, F. H., McFarland, W. F. and Carey, G. R., 'A Systems Approach to the Flexible Pavement Design Problem', Texas Transportation Institute, Texas A & M University, Research Report 32-11, 1968.

APPENDIX A

TABLE A.1 DESIGN LOADS FOR ROADS IN EUROPE

Country	Freeways and expressways		Main highways		Secondary highways	
	Design traffic (commercial vehicles/day)	Design axle or wheel load, kip (ton)	Design traffic (commercial vehicles/day)	Design axle or wheel load, kip (ton)	Design traffic (commercial vehicles/day)	Design axle or wheel load, kip (ton)
Austria	N[a]	Single axle 22·5 (10) Tandem axle 36 (16)	N	Single axle 22·5 (10) Tandem axle 36 (16)		
Belgium	N	Axle load 29·25 (13)	No design	Axle load 29·25 (13)	No design	Axle load 29·25 (13)
Czechoslovakia	1 000	Single axle 22·5 (10)	500–1 000	Single axle 22·5 (10)	100–500	Single axle 22·5 (10)
Denmark	>4 000 (sum in both directions)	Single axle 18 (8)				
France		Single axle 29·25 (13) Tandem axle 45 (20)		Single axle 29·25 (13) Tandem axle 45 (20)		
Great Britain	Design based on total commercial vehicles carried during total life	Single axle 22·5 (10)	Design based on total commercial vehicles carried during total life	Single axle 22·5 (10)	Design based on total commercial vehicles carried during total life	Single axle 22·5 (10)

Netherlands	Approx. 3 000–5 000 converted to 22·5 kip (10 ton) axles	Maximum wheel load 11·25 (5)	Approx. 2 000–3 000	Maximum wheel load 11·25 (5)	Up to 2 000 vehicles, also bicycles	Maximum wheel load 11·25 (5)
Italy	N	Single axle 22·5 (10) Tandem axle 32·6 (14·5)	N	Single axle 22·5 (10) Tandem axle 32·6 (14·5)	N	Single axle 22·5 (10) Tandem axle 32·6 (14·5)
Spain	N	Single axle 29·25 (13) Tandem axle 47·25 (21)	N	Single axle 29·25 (13) Tandem axle 47·25 (21)	Not used	Not used
Sweden	>900	Single axle 22·5 (10) Tandem axle 36 (16)	300–900	Single axle 22·5 (10) Tandem axle 36 (16)	Not used as a rule	
Switzerland	2 000–4 000 (sum in both directions)	Single axle 22·5 (10) Tandem axle 36 (16)	1 000–2 000 (sum in both directions)	Single axle 22·5 (10) Tandem axle 36 (16)	100–1 000 (sum in both directions)	Single axle 22·5 (10) Tandem axle 36 (16)
West Germany	>1 000	Single axle 22·5 (10)	500–1 000	Single axle 22·5 (10)	500–1 000	Single axle 22·5 (10)

[a] N = not included in specifications.

APPENDIX B

AIRCRAFT CHARACTERISTICS

This Appendix summarises aircraft characteristics for 72 current and planned aircraft of 60 000 lb (27 200 kg) or greater gross weight. Table B.1 lists the major characteristics of these 72 aircraft models. Each line entry of the table is divided into five major items of information, described below.

Aircraft identification. Column (a) of the table identifies, by aircraft company, the aircraft model. For each company, the models are arranged by decreasing gross weight.

Aircraft landing gear assembly type. Column (b) is a coded description identifying the configuration made between the nose gear and the truck assemblies of the main gear. Figure B.1 depicts the three aircraft assembly configurations used by the aircraft listed in Table B.1.

Aircraft gross weight. Column (c) indicates the maximum gross weight of each designated aircraft.

Nose gear characteristics. Columns (d)–(h) give the load and geometric features of the nose gear. Column (d) identifies the nose gear assembly type by reference to Fig. B.2. This figure depicts the various configurations for nose and main truck assemblies. Column (e) gives the spacing between tyres of the nose gear. Columns (f), (g) and (h) need no explanation.

Main truck characteristics. Columns (i)–(q) present data on the trucks comprising the main gear assembly of the aircraft. Columns (i)–(m) contain the same type of information as the columns describing nose gear characteristics. Columns (n)–(q) give the distances used to locate the main truck assembly relative to the nose gear.

Figure B.3 defines the *x*, *y* distances for the main truck assembly. The 0,0 ordinate is the centreline of the nose gear. All *x* distances are measured from the geometric centre of the main truck(s) to the aircraft centreline. Similarly, the *y* distances are measured from the geometric centre of the main truck(s) to the centreline of the nose gear.

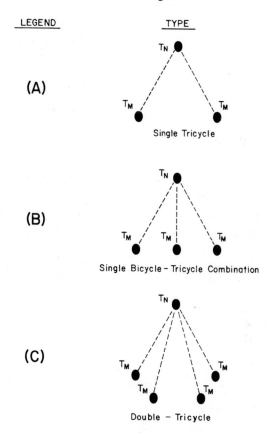

FIG. B.1. Various types of aircraft gear assemblies.

Several aircraft, such as the B-747, have more than one main truck assembly on each side of the aircraft. These aircraft are identified in the table by entries under 'x_2' and 'y_2', columns (o) and (q), or by the C legend in column (b).

Figure B.4 illustrates how to extract information from Table B.1 for L-500 aircraft (aircraft number 20 in the table).

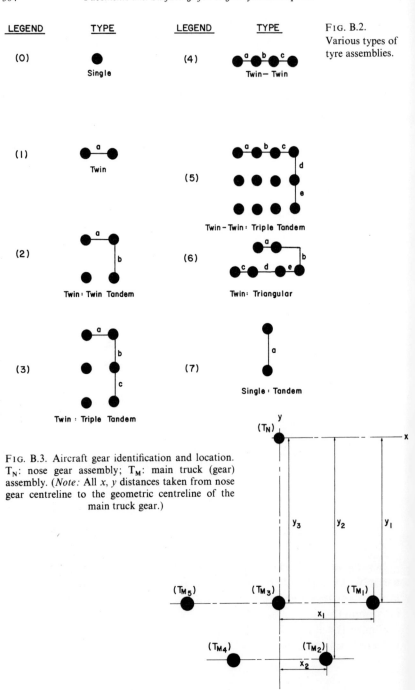

LEGEND   TYPE   LEGEND   TYPE

(0) Single

(4) Twin–Twin

(1) Twin

(5) Twin–Twin: Triple Tandem

(2) Twin: Twin Tandem

(6) Twin: Triangular

(3) Twin: Triple Tandem

(7) Single: Tandem

FIG. B.2. Various types of tyre assemblies.

FIG. B.3. Aircraft gear identification and location. T_N: nose gear assembly; T_M: main truck (gear) assembly. (*Note:* All x, y distances taken from nose gear centreline to the geometric centreline of the main truck gear.)

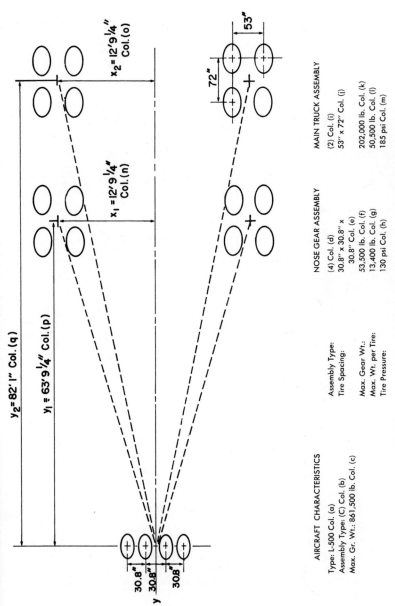

AIRCRAFT CHARACTERISTICS

Type: L-500 Col. (a)
Assembly Type: (C) Col. (b)
Max. Gr. Wt.: 861,500 lb. Col. (c)

NOSE GEAR ASSEMBLY

Assembly Type: (4) Col. (d)
Tire Spacing: 30.8″ x 30.8″ x
30.8″ Col. (e)
Max. Gear Wt.: 53,500 lb. Col. (f)
Max. Wt. per Tire: 13,400 lb. Col. (g)
Tire Pressure: 130 psi Col. (h)

MAIN TRUCK ASSEMBLY

(2) Col. (i)
53″ x 72″ Col. (j)

202,000 lb. Col. (k)
50,500 lb. Col. (l)
185 psi Col. (m)

FIG. B.4. L-500 aircraft characteristics summary. (*Note*: See Table B.1 for column references. See 'Notes', Table B.1 for metric unit conversion functions.)

	Col. (a)	Col. (b)	Col. (c)	Col. (d)	Col. (e)	Col. (f)	Col. (g)	Col. (h)
	Aircraft identification				*Nose gear assembly*			
	Aircraft company/model	*Landing gear assembly type[a]*	*Max. gross weight (lb × 10³)*	*Assembly type[b]*	*Tyre spacing (in)[d]*	*Max. gear weight[e] (lb × 10³)*	*Max. weight per tyre (lb × 10³)*	*Tyre pressure (psi)*
I. Boeing								
1.	B-747-F	(C)	778·0	(1)	36	50·8	25·4	180
2.	B-747-C	(C)	778·0	(1)	36	50·8	25·4	180
3.	B-747-B	(C)	778·0	(1)	36	50·8	25·4	180
4.	B-747	(C)	713·0	(1)	36	47·0	23·5	165
5.	B-707-320C	(A)	336·0	(1)	22	22·0	11·0	115
6.	B-707-320B	(A)	328·0	(1)	22	26·0	13·0	115
7.	B-707-320/420	(A)	316·0	(1)	22	25·0	12·5	115
8.	B-707-120B	(A)	258·0	(1)	22	16·8	8·4	90
9.	B-720-B	(A)	230·0	(1)	22	12·0	6·0	100
10.	B-720	(A)	230·0	(1)	22	12·0	6·0	100
11.	B-727-200	(A)	173·0	(1)	24	13·2	6·6	100
12.	B-727-100C	(A)	170·0	(1)	24	17·2	8·6	100
13.	B-727-100	(A)	170·0	(1)	24	16·2	8·1	100
14.	B-737-200C	(A)	111·0	(1)	15	8·0	4·0	135
15.	B-737-200	(A)	111·0	(1)	15	8·0	4·0	135
16.	B-737-100	(A)	111·0	(1)	15	8·0	4·0	135
II. Convair								
17.	Cv-990	(A)	255·0	(1)	17	13·6	6·8	200
18.	Cv-880M	(A)	193·5	(1)	17	8·7	4·4	140
19.	Cv-880	(A)	185·0	(1)	17	11·8	5·9	136
III. Lockheed								
20.	L-500	(C)	861·5	(4)	30·8 × 30·8 × 30·8	53·5	13·4	130
21.	L-1011-3	(A)	598·0	(1)	25	30·0	15·0	182
22.	L-1011-1	(A)	411·0	(1)	24	21·0	10·5	179
23.	L-1649A	(A)	160·0	(1)	15	8·0	4·0	75

[a] See Fig. B.1 for aircraft assembly type legend.

[b] See Fig. B.2 for gear assembly type legend.

[c] See Fig. B.3 for explanation of x, y values of aircraft gear layout.

[d] Dimensions shown for tyre spacings in Table B.1 are centre-to-centre of the tyres or axles, depending on direction. Dimensions are listed in the alphabetical order noted in Fig. B.2 for the respective assembly type.

[e] Nose and main gear weights are based on *static* condition. Maximum nose gear weights during braking conditions generally are in excess of those noted in Table B.1.

Col. (i)	Col. (j)	Col. (k)	Col. (l)	Col. (m)	Col. (n)		Col. (o)		Col. (p)		Col. (q)	
				Main truck assembly								
Assembly type[b]	Tyre spacing (in)[d]	Max. truck weight[e] (lb×10³)	Max. weight per tyre (lb×10³)	Tyre pressure (psi)	x_1[c] ft	in	x_2 ft	in	y_1 ft	in	y_2 ft	in
(2)	44×58	181·8	45·5	185	18	½	6	3	78	11½	89	½
(2)	44×58	181·8	45·5	185	18	½	6	3	78	11½	89	½
(2)	44×58	181·8	45·5	185	18	½	6	3	78	11½	89	½
(2)	44×58	166·5	41·6	204	18	½	6	3	78	11½	89	½
(2)	34·5×56	157·0	39·3	180	11	½	NA		59	0	NA	
(2)	34·5×56	151·0	37·8	180	11	½	NA		59	0	NA	
(2)	34·5×56	145·5	36·4	180	11	½	NA		59	0	NA	
(2)	34×56	120·6	30·2	170	11	½	NA		52	4	NA	
(2)	32×49	109·0	27·3	145	10	11½	NA		50	8	NA	
(2)	32×49	109·0	27·3	145	10	11½	NA		50	8	NA	
(1)	34	79·9	39·9	168	9	4½	NA		63	3	NA	
(1)	34	76·4	38·2	166	9	4½	NA		53	3	NA	
(1)	34	76·9	38·5	166	9	4½	NA		53	3	NA	
(1)	30·5	51·5	25·8	148	8	7	NA		37	4	NA	
(1)	30·5	51·5	25·8	148	8	7	NA		37	4	NA	
(1)	30·5	51·5	25·8	148	8	7	NA		37	4	NA	
(2)	24×46·5	120·7	30·2	186	9	11½	NA		57	3	NA	
(2)	21·5×45	92·4	23·1	155	9	5½	NA		53	1	NA	
(2)	21·5×45	87·1	21·8	150	9	5½	NA		53	1	NA	
(2)	53×72	202·0	50·5	185	12	9¼	12	9¼	63	9¼	82	1
(3)	42×56×56	284·0	47·3	196	16	9	NA		71	½	NA	
(2)	52×70	195·0	48·8	175	18	0	NA		70	0	NA	
(1)	30	76·0	38·0	140	19	2½	NA		54	4	NA	

Metric unit conversion factors applicable to this table are:

pound × 0·453 59 = kilogram (kg)

inch × 0·025 4 = metre (m)

foot × 0·304 8 = metre (m)

psi × 6·894 757 = kilonewton per square metre (kN/m²)

	Col. (a)	Col. (b)	Col. (c)	Col. (d)	Col. (e)	Col. (f)	Col. (g)	Col. (h)
	Aircraft identification				*Nose gear assembly*			
	Aircraft company/model	*Landing gear assembly type[a]*	*Max. gross weight (lb × 10³)*	*Assembly type[b]*	*Tyre spacing (in)[d]*	*Max. gear weight[e] (lb × 10³)*	*Max. weight per tyre (lb × 10³)*	*Tyre pressure (psi)*
24.	L-100-30	(A)	155·0	(1)	22	9·8	4·9	60
25.	L-100-20	(A)	155·0	(1)	22	9·5	4·8	60
26.	L-100-10	(A)	155·0	(1)	22	11·0	5·5	60
27.	L-1049H	(A)	142·0	(1)	15	7·1	3·6	70
28.	L-1049G	(A)	137·5	(1)	15	6·9	3·4	70
29.	L-1049C/D	(A)	133·0	(1)	15	6·7	3·3	70
30.	L-1049	(A)	120·0	(1)	15	6·0	3·0	70
31.	L-188C	(A)	116·0	(1)	16	5·8	2·9	90
32.	L-188A	(A)	113·0	(1)	16	5·7	2·8	90
33.	L-749A	(A)	107·0	(1)	15	5·4	2·7	80
34.	L-749	(A)	102·0	(1)	15	5·1	2·6	80
35.	L-149	(A)	100·0	(1)	15	5·0	2·5	80
36.	L-649A	(A)	98·0	(1)	15	4·9	2·5	80
37.	L-649	(A)	94·0	(1)	15	4·7	2·4	75
38.	L-49E	(A)	98·0	(1)	15	4·9	2·5	80
39.	L-49D	(A)	96·0	(1)	15	4·8	2·4	80
40.	L-49B/C	(A)	93·0	(1)	15	4·7	2·3	80
41.	L-49A	(A)	90·0	(1)	15	4·5	2·3	80
42.	L-49	(A)	86·3	(1)	15	4·3	2·2	80
	IV. McDonnell-Douglas							
43.	DC-10-30CF	(B)	533·0	(1)	25	42·8	21·4	180
					(This line entry refers to TM₃)[c]			
44.	DC-10-30	(B)	533·0	(1)	25	42·8	21·4	175
					(This line entry refers to TM₃)[c]			
45.	DC-10-20CF	(B)	533·0	(1)	25	42·8	21·4	180
					(This line entry refers to TM₃)[c]			
46.	DC-10-20	(B)	533·0	(1)	25	42·8	21·4	175
					(This line entry refers to TM₃)[c]			
47.	DC-10-10	(A)	413·0	(1)	24	24·8	12·4	155
48.	DC-8-63F	(A)	358·0	(1)	18·5	13·8	6·9	143

[a] See Fig. B.1 for aircraft assembly type legend.

[b] See Fig. B.2 for gear assembly type legend.

[c] See Fig. B.3 for explanation of x, y values of aircraft gear layout.

[d] Dimensions shown for tyre spacings in Table B.1 are centre-to-centre of the tyres or axles, depending on direction. Dimensions are listed in the alphabetical order noted in Fig. B.2 for the respective assembly type.

[e] Nose and main gear weights are based on *static* condition. Maximum nose gear weights during braking conditions generally are in excess of those noted in Table B.1.

Col. (i)	Col. (j)	Col. (k)	Col. (l)	Col. (m)	Col. (n)		Col. (o)		Col. (p)		Col. (q)	
				Main truck assembly								
Assembly type[b]	Tyre spacing $(in)^d$	Max. truck weight^e $(lb \times 10^3)$	Max. weight per tyre $(lb \times 10^3)$	Tyre pressure *(psi)*	$x_1{}^c$ ft	in	x_2 ft	in	y_1 ft	in	y_2 ft	in
(7)	60·5	75·3	37·7	110	7	$1\frac{1}{2}$	NA		40	$4\frac{1}{2}$	NA	
(7)	60·5	75·1	37·5	110	7	$1\frac{1}{2}$	NA		37	$\frac{3}{4}$	NA	
(7)	60·5	74·3	37·1	110	7	$1\frac{1}{2}$	NA		32	$\frac{3}{4}$	NA	
(1)	28	67·5	33·7	132	14	0	NA		49	11	NA	
(1)	28	65·3	32·7	128	14	0	NA		49	11	NA	
(1)	28	63·2	31·6	123	14	0	NA		49	11	NA	
(1)	28	57·0	28·5	110	14	0	NA		49	11	NA	
(1)	26	55·1	27·6	145	15	7	NA		48	3	NA	
(1)	26	53·7	26·8	140	15	7	NA		48	3	NA	
(1)	28	50·8	25·4	85	14	0	NA		39	3	NA	
(1)	28	48·5	24·2	85	14	0	NA		39	3	NA	
(1)	28	47·5	23·8	85	14	0	NA		39	5	NA	
(1)	28	46·6	23·3	85	14	0	NA		39	3	NA	
(1)	28	44·7	22·4	80	14	0	NA		39	3	NA	
(1)	28	46·6	23·3	85	14	0	NA		39	5	NA	
(1)	28	45·6	22·8	85	14	0	NA		39	5	NA	
(1)	28	44·2	22·1	85	14	0	NA		39	5	NA	
(1)	28	42·6	21·3	85	14	0	NA		39	5	NA	
(1)	28	41·0	20·5	85	14	0	NA		39	5	NA	
(2)	54 × 64	201·2	50·3	185	17	6	NA		72	5	NA	
(1)	38	87·8	43·9	170	0	0	NA		74	11	NA	
(2)	54 × 64	201·2	50·3	185	17	6	NA		72	5	NA	
(1)	38	87·8	43·9	170	0	0	NA		74	11	NA	
(2)	54 × 64	201·2	50·3	185	17	6	NA		72	5	NA	
(1)	38	87·8	43·9	170	0	0	NA		74	11	NA	
(2)	54 × 64	201·2	50·3	185	17	6	NA		72	5	NA	
(1)	38	87·8	43·9	170	0	0	NA		74	11	NA	
(2)	54 × 64	194·1	48·5	175	17	6	NA		72	5	NA	
(2)	32 × 55	172·1	43·0	196	10	5	NA		77	6	NA	

Metric unit conversion factors applicable to this table are:

pound × 0·453 59 = kilogram (kg)

inch × 0·025 4 = metre (m)

foot × 0·304 8 = metre (m)

psi × 6·894 757 = kilonewton per square metre (kN/m²)

	Col. (a)	Col. (b)	Col. (c)	Col. (d)	Col. (e)	Col. (f)	Col. (g)	Col. (h)
	Aircraft identification			*Nose gear assembly*				
	Aircraft company/model	*Landing gear assembly type[a]*	*Max. gross weight (lb × 10³)*	*Assembly type[b]*	*Tyre spacing (in)[d]*	*Max. gear weight[e] (lb × 10³)*	*Max. weight per tyre (lb × 10³)*	*Tyre pressure (psi)*
49.	DC-8-63	(A)	358·0	(1)	18·5	13·8	6·9	143
50.	DC-8-62F	(A)	353·0	(1)	18·5	17·4	8·7	168
51.	DC-8-62	(A)	353·0	(1)	18·5	23·4	11·7	168
52.	DC-8-61F	(A)	331·0	(1)	18·5	13·2	6·6	118
53.	DC-8-61	(A)	328·0	(1)	18·5	12·8	6·4	117
54.	DC-8-55F	(A)	328·0	(1)	18·5	17·4	8·7	171
55.	DC-8-55	(A)	328·0	(1)	18·5	17·8	8·9	171
56.	DC-8-43	(A)	318·0	(1)	18·5	22·0	11·0	162
57.	DC-7C	(A)	140·0	(0)	NA	14·2	14·2	70
58.	DC-7B	(A)	126·0	(0)	NA	14·3	14·3	85
59.	DC-7	(A)	122·2	(0)	NA	14·3	14·3	85
60.	DC-9-41	(A)	115·0	(1)	14	7·4	3·7	130
61.	DC-9-32	(A)	109·0	(1)	14	8·2	4·1	130
62.	DC-9-21	(A)	101·0	(1)	14	5·6	2·8	122
63.	DC-9-15	(A)	91·5	(1)	14	6·7	3·4	105
64.	DC-6A/B	(A)	107·0	(0)	NA	12·8	12·8	70
65.	DC-6	(A)	97·2	(0)	NA	12·1	12·1	70
66.	DC-4	(A)	73·0	(0)	NA	10·4	10·4	60
	V. SAF–BAC							
67.	Concorde	(A)	388·0	(1)	21·1	19·4	9·7	174
	VI. BAC							
68.	BAC 1-11-500	(A)	100·0	(1)	13·7	5·0	2·5	115
69.	Viscount 810	(A)	72·5	(1)	13·7	3·6	1·8	95
70.	Viscount 745D	(A)	64·5	(1)	13·7	3·2	1·6	92
	VII. SUD							
71.	Caravelle SE-210-6R[f]	(A)	110·2	(1)	16	7·4	3·7	90
72.	Caravelle SE-210-1	(A)	95·9	(1)	16	6·4	3·2	90

[a] See Fig. B.1 for aircraft assembly type legend.

[b] See Fig. B.2 for gear assembly type legend.

[c] See Fig. B.3 for explanation of x, y values of aircraft gear layout.

[d] Dimensions shown for tyre spacings in Table B.1 are centre-to-centre of the tyres or axles, depending on direction. Dimensions are listed in the alphabetical order noted in Fig. B.2 for the respective assembly type.

[e] Nose and main gear weights are based on *static* condition. Maximum nose gear weights during braking conditions generally are in excess of those noted in Table B.1.

Col. (i)	Col. (j)	Col. (k)	Col. (l)	Col. (m)	Col. (n)		Col. (o)	Col. (p)		Col. (q)
				Main truck assembly						
Assembly type[b]	Tyre spacing $(in)^d$	Max. truck weight[e] $(lb \times 10^3)$	Max. weight per tyre $(lb \times 10^3)$	Tyre pressure (psi)	$x_1{}^c$ ft	in	x_2 ft in	y_1 ft	in	y_2 ft in
(2)	32×55	172·1	43·0	196	10	5	NA	77	6	NA
(2)	32×55	167·8	42·0	191	10	5	NA	60	10	NA
(2)	32×55	164·8	41·2	187	10	5	NA	60	10	NA
(2)	30×55	158·9	39·7	190	10	5	NA	77	6	NA
(2)	30×55	157·6	39·4	188	10	5	NA	77	6	NA
(2)	30×55	155·3	38·8	186	10	5	NA	57	6	NA
(2)	30×55	155·1	38·8	186	10	5	NA	57	6	NA
(2)	30×55	148·0	37·0	177	10	5	NA	57	6	NA
(1)	29·8	67·0	33·5	126	17	4	NA	48	1	NA
(1)	29·8	60·0	30·0	127	12	4	NA	44	9	NA
(1)	29·8	58·4	29·2	123	12	4	NA	44	9	NA
(1)	26	53·8	26·9	163	8	$2\frac{1}{2}$	NA	56	2	NA
(1)	25	50·4	25·2	152	8	$2\frac{1}{2}$	NA	53	2	NA
(1)	25	47·7	23·8	144	8	$2\frac{1}{2}$	NA	43	8	NA
(1)	24	42·4	21·2	127	8	$2\frac{1}{2}$	NA	43	8	NA
(1)	30·8	51·4	25·7	106	12	4	NA	36	2	NA
(1)	30·8	45·8	22·9	95	12	4	NA	36	2	NA
(1)	30	35·4	17·7	75	12	4	NA	36	0	NA
(2)	$26·4 \times 66$	184·3	46·1	184	12	8	NA	59	8	NA
(1)	21	47·5	23·8	174	7	$1\frac{1}{2}$	NA	48	5	NA
(1)	19	34·4	17·2	138	11	11	NA	39	6	NA
(1)	19	30·6	15·3	122	11	11	NA	35	8	NA
(2)	$16·8 \times 48·7$	51·4	12·9	155	8	$6\frac{1}{2}$	NA	55	6	NA
(2)	$16·8 \times 48·7$	44·7	11·2	155	8	$6\frac{1}{2}$	NA	55	6	NA

[f] Nominal tyre spacing values used for Caravelle are due to different fore and aft lateral dimensions in dual tandem main gear. Maximum tyre pressure is used for all tyres of main gear. Metric unit conversion factors applicable to this table are:

    pound $\times 0·453\,59$ = kilogram (kg)

    inch $\times 0·025\,4$ = metre (m)

    foot $\times 0·304\,8$ = metre (m)

    psi $\times 6·894\,757$ = kilonewton per square metre (kN/m^2)

TESTS USED IN DETERMINING SUBGRADE STRENGTH

1. CALIFORNIA BEARING RATIO (CBR)

The CBR test consists of measuring the load required to cause a plunger of standard size to penetrate a soil specimen at a specified rate. The CBR is the ratio between the load, in lb/in^2, required to force a piston of a certain area into the soil a certain depth, and that required to force the piston the same depth into a standard sample of crushed stone. Usually, penetration depths of either 0·1 or 0·2 in are used. The penetration loads for the standard sample of crushed stone are 1000 and 1500 psi for 0·1 and 0·2 in penetration, respectively. The CBR test is probably the most widely used test to provide the relative bearing value of subgrade, sub-base and base materials. Tests on laboratory-compacted specimens are usually performed to give the information needed for the design. The in-place field test can be used to check that the compacted soil has reached the load-carrying capacity considered in the design. When the in-place field test is performed on materials which may later, during the life of the pavement, undergo changes of moisture content, undisturbed samples of the field-compacted materials are tested in the laboratory for conditions of moisture content representing those expected in the field.

Details regarding the equipment required for remoulded, undisturbed or field in-place tests, procedure of compaction of cohesive, cohesionless and swelling soils, and preparation of test specimens can be found in soil mechanics textbooks (such as the *Soil Manual No. 10* of the Asphalt Institute). The general procedure of the test is summarised below, and a sketch of the apparatus that is generally used is shown in Fig. C.1.

The test is done in a 6 in diameter cylindrical mould which is 7 in high, provided with a base plate and a collar, 2 in high, which will fit on either end.

During the compaction process a spacer disc, $2\frac{1}{2}$ in high, is placed inside the cylinder so that the compacted height of the sample, after removal of the disc and collar and levelling off with the edge of the cylinder, is about $4\frac{1}{2}$ in. The specimen is compacted in five 1 in layers with a compaction hammer consisting of a 2 in diameter steel tamping foot, a $\frac{5}{8}$ in steel rod, a weight and a handle. The maximum weight of the assembled compaction hammer is $17\frac{1}{2}$ lb. The number of blows of the hammer per layer is either 55 (modified AASHO T180 Method D), 26 or 12. A sufficient number of test specimens over a range of water contents that will definitely establish the optimum water content and the maximum density should be compacted.

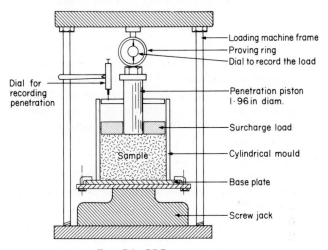

FIG. C.1. CBR test apparatus.

In the case of swelling soils, the proper moisture content and unit weight are not necessarily the optimum values. Generally, the minimum swell and highest soaked CBR will occur at a moulding moisture content slightly wetter than optimum. For this reason, it may be necessary when testing highly swelling soils, to prepare samples for a wide range of moisture contents to establish the relationship between moisture content, density, swell and CBR for a given soil. Where it is desirable to limit swell by the addition of overburden load, additional specimens should be prepared for soaked CBR tests, using various added amounts of surcharge during soaking.

To determine the percentage of swell of swelling soils, the mould, with the sample, is immersed in water to within $\frac{1}{2}$ in of the top, while the top surface of the soil is exposed to an intensity of surcharge loading equal to the weight of the base material and pavement, within ± 5 lb, but not less than 10 lb. Water is also placed inside the mould to the same level as water on the outside of the mould. The level of the soil surface, when determined, will be the initial measurement for swell. The specimen is then allowed to soak for 4 days, maintaining constant water level outside and inside the mould, and the final swell measurement is determined:

$$\text{swell } (\%) = \frac{\text{swell}}{\text{height of specimen}} \times 100 =$$

$$100 \times \frac{\text{final swell measurement} - \text{initial swell measurement}}{5} \quad \text{(C.1)}$$

If the mould is then removed from the water, inside free water poured off, surcharge weights removed and the specimen allowed to drain for 15 min, the specimen will be ready for the penetration test.

The procedure for penetration testing is the same for all types of re-moulded specimens and is also applicable for undisturbed and field in-place tests after the testing surface has been prepared. The steps are as follows:

1. Place a 5 lb annular disc surcharge weight on the soil surface.
2. Place the mould in the loading frame, and adjust its position until the piston is centred on the specimen.
3. Set both the load dial and strain dials to zero when the load on the penetration piston is 10 lb. This initial load is required to ensure satisfactory seating of the piston and is the zero load when determining stress–penetration relations.
4. Add penetration surcharge weights that produce an intensity of loading equal to the weight of the base material and pavement (within ± 5 lb) but not less than 10 lb. If the sample has been previously soaked, the surcharge weight should be equal to the weight of the soaking surcharge.
5. Apply the load to the piston at a uniform rate of 0·05 in of penetration per minute, using an electric motor, if possible, to achieve this uniform rate.
6. Record the total load readings corresponding to 0·025, 0·050, 0·075, 0·100, 0·125, 0·150, 0·175, 0·200, 0·250 and 0·300 in penetration. Calculate the stress corresponding to each of the above-mentioned

penetration values by dividing the load by 3, which is the area of the penetration piston in in².

7. Plot the stress–penetration curve on cross-section paper, as shown in Fig. C.2, after completion of the test. If the curve has a concave upward shape, as shown in curve 2 of Fig. C.2, draw a line tangent to the steepest point of the curve to meet the base at a point which

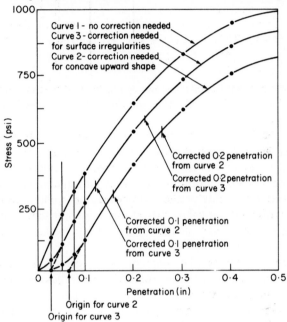

FIG. C.2. Correction of stress–penetration curves.

represents the corrected origin or zero point. For surface irregularities, as shown in curve 3 of Fig. C.2, extend the straight line portion of the curve to the base to obtain the corrected origin.

8. The corrected load values are determined at 0·1 in and 0·2 in penetration and the CBR values are determined using the following formula:

$$CBR (\%) = \frac{P_{soil}}{P_{crushed\ stone}} \times 100 \qquad (C.2)$$

where P_{soil} = soil resistance or the unit load on the piston (psi), for the specified penetration.

$P_{crushed\ stone}$ = standard unit load (psi), for the standard well-graded crushed stone.

The CBR is determined by dividing the corrected values of the unit loads at 0·1 and 0·2 in penetration by the standard unit loads of 1000 and 1500 psi, respectively, and multiplying by 100. The CBR is usually obtained at 0·1 in penetration. However, if the CBR at 0·2 in penetration is greater, the test should be rerun. If the results remain unchanged, the CBR at 0·2 in penetration should be used.

2. DETERMINATION OF BEARING VALUE (PLATE-BEARING TEST)

This test can be used to measure the strength of any layer in an asphalt pavement structure: surface of the subgrade, top of the sub-base or base course, or surface of the finished pavement.

In order that this test can provide a representative measurement of subgrade strength, the tested soil must be in the same condition as that expected after equilibrium with the environmental influences of moisture, density, frost, drainage and traffic. This may be achieved by:

1. performing the plate-bearing test beneath an existing asphalt pavement having similar subgrade soil, and where the pavement has been in place long enough for the subgrade to have reached equilibrium with its environment; or
2. performing the plate-bearing test on a specially constructed test section of the soil with an adequate depth processed to duplicate the conditions expected after it reaches equilibrium with its environment following the paving operation.

The criterion for plate-bearing test for highways is a 12 in diameter bearing plate, 0·2 in deflection and 10 repetitions of load; for airfields it is a 30 in diameter bearing plate, 0·5 in deflection and 10 repetitions of load.

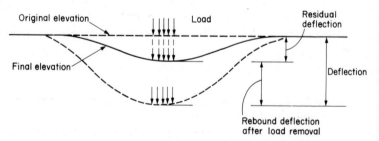

FIG. C.3. Identification of deflection, residual deflection and rebound deflection.

The difference between deflection, residual deflection and rebound deflection is illustrated in Fig. C.3.

The equipment needed for the plate-bearing test includes a loading device (a loaded truck or similar), an hydraulic jack assembly, bearing plates of different diameters, and two or more dial gauges mounted on a beam at least 18 ft long.

The test procedure is as follows:

1. Set the proper bearing plate, level in a thin bed of a mixture of fine sand and plaster of Paris, and centred under the jack assembly. Place the other plates with smaller diameter on top of the bearing plate and concentric with it.

2. Place the two dials 1 in from the ends of a diameter of the top plate and with their stems resting on the upper surface of the next plate. The dials extend from and are supported on the beam, which should rest on supports located at least 8 ft from the circumference of the bearing plate or the nearest wheel.

3. After properly arranging the equipment, seat the bearing plate assembly by a quick application and release of a load that produces a deflection between 0·01 and 0·02 in. When the needles of the dial gauges come to rest after releasing the load, reseat the plate by applying one-half of the recorded load producing the 0·01–0·02 in deflection. When the needles have come to rest, set each dial accurately at its zero mark.

4. Apply a load giving a 0·04 in deflection, start a stop watch, and maintain the same load constantly until the rate of deflection is less than or equal to 0·001 in/min for 3 successive minutes. Then release the load, and observe the rebound until the rate of recovery is less than or equal to 0·001 in for 3 successive minutes. Apply and release the same load in this manner 10 times, and record the readings of the two dial gauges at the end of each minute. The readings of other dial gauges, which may be set beyond the perimeter of the bearing plate (Fig. C.4), are recorded just before the application as well as the release of the load, for each repetition.

5. Increase the load to give a deflection of about 0·2 in, and proceed as explained in step 4. Also, increase the load to give a deflection of about 0·4 in, and proceed as before. The deflection for a given load at any time is determined by averaging the readings of the dial assembly on the bearing block.

6. Record the air temperature in the shade near the bearing plate at half-hour intervals.

7. For each repetition of each load, determine the deflection at which the rate of increase of deflection is 0·001 in/min. This deflection, which is

important for the calculations, is termed 'end point deflection' and can be determined from visual inspection of the deflection data for each repetition of load recorded.

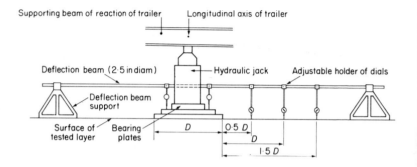

F IG. C.4. Arrangement of equipment for plate-bearing test.

8. Correct the applied loads, as read from the pressure gauge of the jack, if the jack calibration indicates a difference between the values read from the pressure gauge and the true loads. Also, adjust the applied load for the load of bearing plates, jack, etc.

9. Tabulate corrected total load versus measured deflection at which the rate of deflection is exactly 0·001 in/min, for each repetition of each load.

10. Correct the measured deflections by plotting the corrected total loads versus measured deflections for the fifth repetition of load. This curve meets the abscissa (the deflection scale) at a point which generally represents a negative value for the deflection. This value should be added to each of the measured deflections to obtain the corrected values. (If a positive value is obtained, then it should be subtracted from the measured deflections.)

11. Plot the corrected deflection data versus number of repetitions of each corrected total load on a semi-logarithmic graph and extend straight lines through the plotted points as shown in Fig. C.5.

Similar graphs may be plotted in which either corrected residual deflection or corrected rebound deflection are plotted versus the number of repetitions of each corrected total load.

12. Plot corrected total load versus corrected deflection for 10 repetitions of load, as illustrated by Fig. C.6.

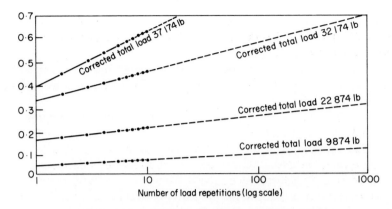

FIG. C.5. Influence of load repetitions on the deflection.

The value of the load at 0·5 in deflection for 10 load repetitions (using 30 in diameter bearing plate) is the bearing value, in the case of airfields. Similarly, if a 12 in diameter bearing plate is used, then the value of the load at 0·2 in deflection for 10 load repetitions will represent the bearing value for highways.

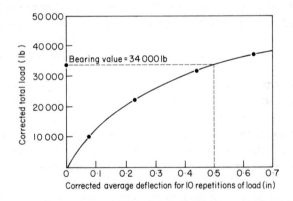

FIG. C.6. Determination of bearing value for airfields.

13. If the bearing value corresponding to a greater number of load repetitions is required, then either more repetitions of each load may be applied to the plate or the curves of Fig. C.5 may be extrapolated, as shown, to obtain the values of the deflection for the desired number of repetitions.

3. HVEEM'S RESISTANCE VALUE (*R*) METHOD

This method of evaluating treated and untreated materials for bases, sub-bases and subgrades for pavement thickness design, is based on two separate measurements:

1. The *R* value, which determines the thickness of cover required to prevent plastic deformation of the soil under imposed wheel loads.

2. The expansion pressure test, which determines the thickness or weight of cover necessary to maintain the compaction of the soil.

The design *R* value is determined from the moisture content and density at which the above-mentioned two thicknesses are equal. The testing method for determining the *R* value consists of the following parts, and should be done in the same sequence:

(a) sample preparation
(b) test specimen compaction
(c) determination of exudation pressure of test specimens
(d) determination of the expansion pressures of the test specimens
(e) determination of the *R* value.

The *R* value test requires the preparation of four test briquettes at different moisture contents, with the first one serving as a pilot specimen. The size of the compacted specimens is of 4 in diameter and $2\frac{1}{2}$ in high after compaction. The exudation pressure for one of the other three samples should be above 300 psi, and for the other two below 300 psi; or two above and one below 300 psi. All samples should exude moisture between 100 and 800 psi. When very high expansive pressures are expected, wetter specimens are sometimes necessary to get expansion pressures low enough to provide an intersection with the *R* value curve.

After compacting the specimens in 4 in inside diameter × 5 in high steel moulds, using the mechanical kneading compactor, as developed by the California Division of Highways, the compressive stress necessary to exude water from the specimen should be determined.

The procedure for determining this stress is as follows.

1. Place a bronze disc on top of the tamped specimen in the mould, and one piece of filter paper on the disc.

2. Place the mould with sample in an inverted position on the contact plate of a moisture exudation indicator. Then put the assembly in a centred position on the plate of the compression-testing machine and turn on the moisture exudation indicator switch.

3. Place a metal follower ramp, of 4 in outside diameter and 6 in high,

on top of the sample, and force the sample down in the mould to the contact plate.

4. Apply an increasing load at the rate of 2000 lb/min, using the testing machine, until five of the six outer lights of the indicating device are on. The total load at that instant, converted to psi, is recorded as the exudation pressure. However, if free moisture becomes visible around the bottom of the mould and at least three outer lights are on, the load in psi at that moment is recorded as the exudation pressure. The specimens should then be left in a covered mould for at least one-half hour before starting the next part of this test method.

5. Specimens with exudation pressure lower than 100 or higher than 800 psi should be discarded except in the case of very expansive materials.

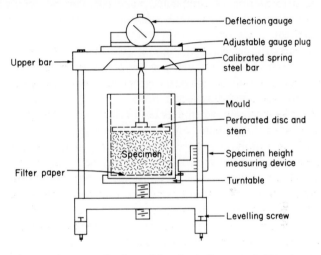

FIG. C.7. Apparatus for determining the specimen expansion pressure.

The next step is to determine the expansion pressure that a soil will develop in the presence of free water. This can be done using the calibrated expansion pressure device, shown in Fig. C.7, as follows.

1. Determine the gross weight of the specimen (mould weight should be deducted) for density calculations.

2. Raise or lower the adjustment plug until the deflection gauge is on −0·0010 in, i.e. the deflection gauge will read 0·0090.

3. Put the mould containing the specimen, with the perforated disc and stem placed firmly into position on the face of the compacted specimen, on the turntable in the expansion pressure device.

4. Turn up the turntable until a surcharge deflection of 0·0010 in is applied to the specimen. The dial gauge should record a zero reading.

5. Add about 200 ml of water to the specimen in the mould, allow expansion pressure to develop for 16–20 h, and read the deflection of the calibrated spring steel bar to 0·0001 in.

6. Determine the expansion pressure from the following equation:

$$p = Kd \qquad (C.3)$$

where p = expansion pressure of the soil (psi);
K = spring steel bar constant (psi per 0·0001 in deflection); and
d = deflection indicated by the dial gauge.

The resistance (R) value, which is a measure of the stability or resistance to plastic deformation of compacted materials, can be determined by using the Hveem stabilometer. This stabilometer is a triaxial-type testing device consisting of a rubber sleeve in a metal cylinder with a liquid between the sleeve and cylinder wall. Any lateral deformation of a specimen subjected to a vertical load will cause a horizontal pressure which will be transferred to the liquid. This pressure is registered on a horizontal pressure gauge. The testing procedure is as follows.

1. Force the specimen, which has been previously tested for expansion pressure, into the stabilometer, and place the follower on top of it with the stabilometer assembly centred under the testing machine head. Lower the testing machine head until it just engages the follower without producing any load in the specimen. Adjust the stabilometer pump to give a 5 psi horizontal pressure and apply a vertical load to the test specimen at a speed of 0·05 in/min.

2. Record the stabilometer gauge readings when the vertical pressures are 80 and 160 psi. These correspond to applied vertical loads of 1000 and 2000 lb, respectively.

3. Stop the vertical loading at 2000 lb and reduce it immediately to 1000 lb. Then turn the stabilometer pump to reduce the horizontal pressure to 5 psi. Set the turns displacement dial indicator to zero and turn the handle of the pump, at approximately two turns per second, until the stabilometer gauge reads 100 psi.

4. Record the number of turns. Each 0·10 in reading on the turns indicator dial is equal to one turn.

5. Use the following formula to calculate the stabilometer R value:

$$R = 100 - \frac{100}{\dfrac{2 \cdot 5}{D} \left(\dfrac{P_v}{P_H} - 1 \right) + 1} \qquad (C.4)$$

where P_v = vertical pressure = 160 psi;

P_H = stabilometer gauge reading representing the horizontal pressure for 160 psi vertical pressure; and

D = number of turns.

To determine the *design R* value additional computations must be made using the results of the exudation pressure test, the expansion pressure test and the stabilometer test. The procedure is as follows.

1. Use the chart for the design of flexible pavements given in Fig. 6.4 of Chapter 6, on Design of Flexible Pavements, to determine the thickness of the flexible pavement corresponding to the *R* value for each specimen as obtained by using the stabilometer. Write the values of the exudation pressure and expansion pressure beside the thickness for each specimen.

2. Plot the thickness in inches for each specimen against the corresponding exudation pressure. Determine the thickness at the intersection of the curve with the 300 psi line and convert to *R* value using the design chart of Fig. 6.4. This value is the *R* value by exudation pressure.

3. Calculate the thickness of cover required by expansion pressure for each specimen by dividing the expansion pressure by the unit weight of cover in lb/in³. The chart of Fig. C.8 can be used for this purpose.

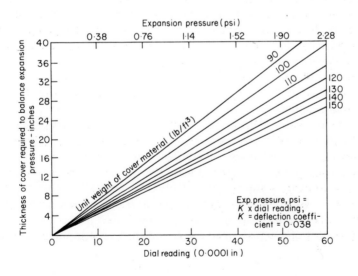

FIG. C.8. Determination of thickness of cover corresponding to expansion pressure.

4. Plot the thickness indicated by the stabilometer against the thickness indicated by expansion pressure, for each specimen, using the same scale for the abscissa and ordinate. Draw a 45° line through the origin and determine the thickness value at which the curve connecting the points crosses this 45° balance line.

5. Use again the design chart of Fig. 6.4 to determine the *R* value by expansion pressure.

6. The design *R* value (*R* value of equilibrium) is the lower value of the *R* value by exudation and the *R* value by expansion.

To use Fig. 6.4 in determining the thickness of the flexible pavement which corresponds to a certain *R* value, the design traffic number (DTN) must be known. The value of DTN should be determined from the traffic data provided for the highway for which these tests are to be carried out, following the procedure given in Chapter 6.

4. DETERMINATION OF MODULUS OF SUBGRADE REACTION

The modulus of subgrade reaction, *K*, is a measure of the strength of the supporting soil, which may be the sub-base or the subgrade. Its value is given in pounds per square inch per inch deflection.

In general, soil strength depends upon density, moisture content and texture of the soil. Increase in density is usually accompanied by increase in strength, whereas increase in moisture content above a certain limit is usually accompanied by decrease in soil strength. However, moisture and density are interrelated and in some cases it is possible to cause a decrease in soil strength by increasing the density. For this reason, the value of the modulus of subgrade reaction should be evaluated for the density and moisture content conditions expected to be approximated beneath the slab under service conditions.

The modulus of subgrade reaction *K* is generally determined by a plate loading test carried out in the field on the compacted soil at its natural moisture content. A plate of 30 in diameter is generally employed in the test and the subgrade is subjected to pressures at a predetermined rate of speed. The value of the pressure generally used in the test is 10 psi. The modulus of subgrade reaction *K* can then be determined from the relationship:

$$K = \frac{p}{\Delta} \tag{C.5}$$

where p = pressure on the plate (psi), usually 10 psi; and
  Δ = deflection of plate (in), corresponding to the assigned pressure.
  Sometimes it is not practical to conduct tests in the field on representative subgrade soils at the various densities and water contents which will

TABLE C.1 VALUE OF K FOR DIFFERENT SOIL GROUPS

Soil groups	Soil description	Range of K value (psi/in)
Gravel and gravelly soils	Well-graded gravel and gravel sand mixtures, little or no fines	500–700
	Well-graded gravel–sand–clay mixtures, excellent binder	400–700
	Poorly graded gravel and gravel–sand mixtures, little or no fines	300–500
	Gravel with fines, very silty gravel, clayey gravel, poorly graded gravel–sand–clay mixtures	250–500
Sand and sandy soils	Well-graded sands and gravelly sands, little or no fines	250–580
	Well-graded sand–clay mixtures, excellent binder	250–550
	Poorly graded sands, little or no fines	200–300
	Sands with fines, very silty sands, clayey sands, poorly graded sand–clay mixtures	175–300
Low to medium compressible fine-grained soils	Inorganic silts and very fine sands, silty or clayey fine sands with slight plasticity	150–300
	Inorganic clays of low to medium plasticity, sandy clays, silt clays, lean clays	125–200
	Organic silts and organic silt clays of low plasticity	100–150
Highly compressible fine-grained soils	Micaceous or diatomaceous fine sandy and silty soils, elastic silts	50–150
	Inorganic clays of high plasticity, fat clays	50–125
	Organic clays of medium to high plasticity	50–100

approximate expected service conditions. For this reason, when the field K has been determined, for existing conditions, it is necessary to adjust the value to the most unfavourable subgrade condition that can be expected. This is accomplished by first running in the laboratory a consolidation or simple compression test on a subgrade soil sample at the moisture content and density prevailing during the field test. The deformation produced in the laboratory sample by a 10 psi load is determined. A second sample, which is compacted in a similar way, is then saturated and loaded by a 10 psi load. It is assumed that the ratio of the deflections for the unsaturated and saturated tests in the laboratory will be approximately the same as that in the field. A modified K for pavement design can then be obtained by using the formula:

$$K_{\text{modified}} = \frac{d}{d_s} K \qquad (C.6)$$

where d and d_s are the deflections obtained in the laboratory for the field sample and saturated sample, respectively.

Table C.1 gives ranges of K values for different soil groups.

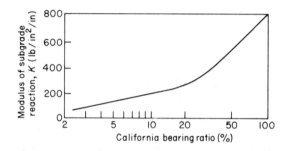

FIG. C.9. Relation between California bearing ratio and modulus of subgrade reaction.

The relationship between the California Bearing Ratio (CBR) and the modulus of subgrade reaction (K) is shown in Fig. C.9.

APPENDIX D

AIRCRAFT EQUIVALENCY DATA

Figures D.2–D.65 summarise the aircraft equivalency values for 22 of the major present and future jet aircraft listed in Table B.1 of Appendix B. These diagrams are used to find the number of equivalent DC-8-63F strain repetitions at the critical design location for both the subgrade vertical compressive strain criterion and the asphalt concrete horizontal tensile strain criterion.

The diagrams are separated into two major groups by type of strain criterion. Figures D.2–D.30 are used for analysis of subgrade vertical compressive strain, while Figs. D.31–D.65 are used for analysis of asphalt concrete horizontal tensile strain. For each criterion, the equivalency diagrams are developed for a particular set of assumed thicknesses, T_A— 10 in (25 cm), 20 in (51 cm), 30 in (76 cm) and 40 in (102 cm) for subgrade strain; and 10 in (25 cm), 30 in (76 cm) and 50 in (127 cm) for asphalt concrete strain. Inherent in the equivalency results is the consideration of aircraft wander from centreline along a straight taxiway, assumed to be a normally distributed variable with a standard deviation of 3·5 ft (1·1 m). This value can be considered a close estimate of aircraft wander while taxiing. Each diagram has several curves for specific distances from the centreline, and yields estimates of the equivalent DC-8-63F strain repetitions at a specific critical design location for a specific aircraft type, assumed thickness of asphalt concrete (T_A), and the total number of movements of the aircraft along the taxiway. Equivalency values for aircraft that are not included in this appendix may be approximated as follows.

In Table B.1, Appendix B, locate an aircraft (listed also in Table D.1) with characteristics that most nearly match those of the aircraft under consideration particularly, gross weight, main gear configuration, dimension and tyre pressure. For example, a DC-9-32 aircraft, which is not listed in this Appendix, has very nearly the same characteristics as the DC-9-41,

527

which is listed in the Appendix. Accordingly, the diagrams in this chapter for the similar aircraft can be used to find the equivalency values.

Figure D.1 shows how to use the equivalency diagrams, while Table D.1 indexes the figure number for each diagram by aircraft type, T_A value and type of strain criterion.

TABLE D.1 INDEX OF AIRCRAFT EQUIVALENCY DIAGRAMS

Thickness, T_A	Subgrade vertical compressive strain, ε_c				AC horizontal tensile strain, ε_t		
	10 in (25 cm)	20 in (51 cm)	30 in (76 cm)	40 in (102 cm)	10 in (25 cm)	30 in (76 cm)	50 in (127 cm)
Aircraft	Figure				Figure		
1. B-747F	D-2	D-2	D-2	D-2	D-31	D-31	D-31
2. B-747	D-3	D-3	D-3	D-3	D-32	D-32	D-32
3. B-707-320C	D-4	D-4	D-4	D-4	D-33	D-33	D-33
4. B-707-120B	D-5	D-5	D-5	D-5	D-34	D-34	D-34
5. B-720B	D-6	D-6	D-6	D-6	D-35	D-36	D-36
6. B-727-200	D-7	D-7	D-7	D-7	D-37	D-37	D-37
7. B-737-200C	D-8	D-8	D-8	D-8	D-38	D-38	D-39
8. CV-990	D-9	D-9	D-9	D-9	D-40	D-40	D-41
9. CV-880M	D-10	D-10	D-10	D-10	D-42	D-42	D-43
10. L-500	D-11	D-11	D-12	D-12	D-44	D-44	D-45
11. L-1011-8	D-13	D-13	D-14	D-14	D-46	D-46	D-47
12. L-1011-1	D-15	D-15	D-16	D-16	D-48	D-48	D-49
13. DC-10-30CF	D-17	D-17	D-18	D-18	D-50	D-50	D-51
14. DC-10-10	D-19	D-19	D-20	D-20	D-52	D-52	D-53
15. DC-8-63F	D-21	D-21	D-21	D-21	D-54	D-54	D-54
16. DC-8-61	D-22	D-22	D-23	D-23	D-55	D-55	D-56
17. DC-9-41	D-24	D-24	D-24	D-24	D-57	D-57	D-57
18. DC-9-15	D-25	D-25	D-25	D-25	D-58	D-58	D-59
19. Concorde	D-26	D-26	D-27	D-27	D-60	D-61	D-61
20. BAC-1-11-500	D-28	D-28	D-28	D-28	D-62	D-62	D-62
21. VIS 810	D-29	D-29	D-29	D-29	D-63	D-63	D-63
22. SE-210-6R	D-30	D-30	D-30	D-30	D-64	D-64	D-65

Figures D.2–D.30 are the aircraft equivalency diagrams relevant to subgrade compressive strain ε_c.

Figures D.31–D.65 are the aircraft equivalency diagrams relevant to a.c. horizontal tensile strain, ε_t.

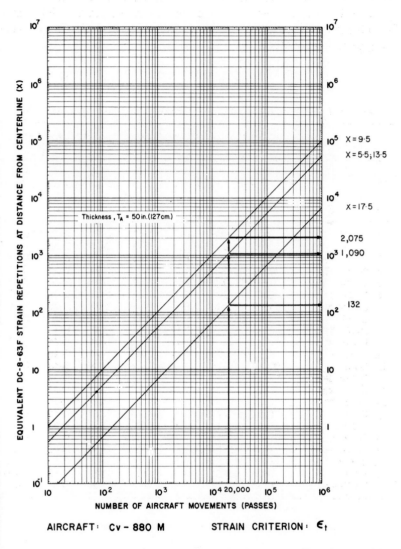

FIG. D.1. How to use equivalency diagrams (see 64 diagrams in succeeding pages).

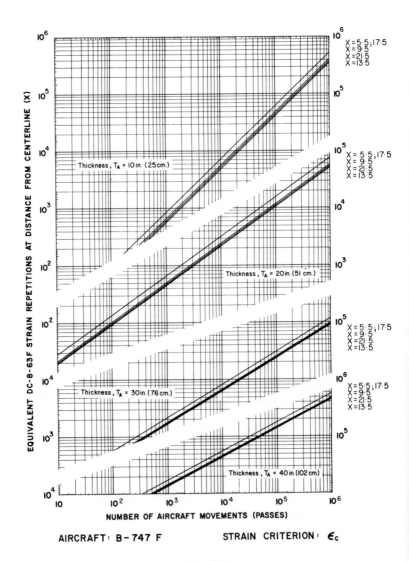

AIRCRAFT: B-747 F   STRAIN CRITERION: ϵ_c

FIG. D.2

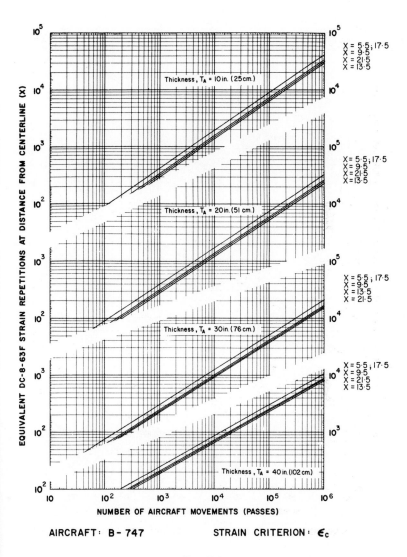

AIRCRAFT: **B- 747** STRAIN CRITERION: $\boldsymbol{\epsilon}_c$

FIG. D.3

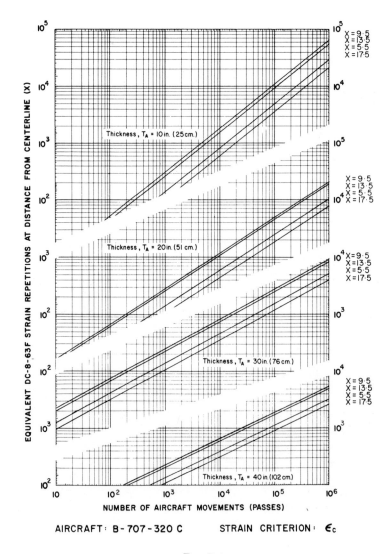

FIG. D.4

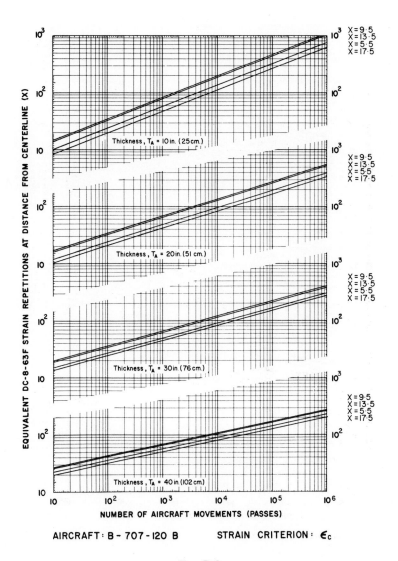

FIG. D.5

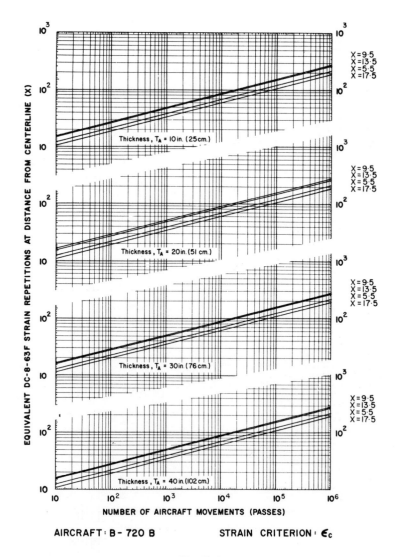

AIRCRAFT: B-720 B   STRAIN CRITERION: ϵ_c

FIG. D.6

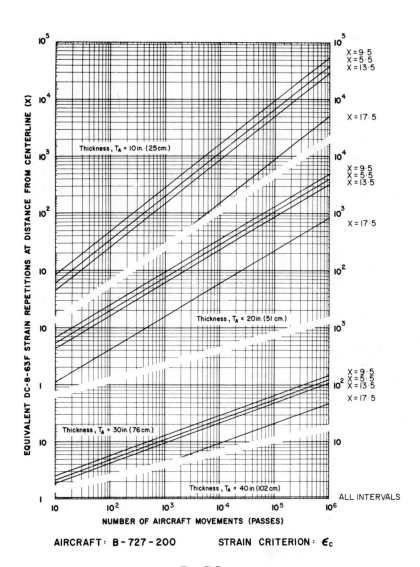

FIG. D.7

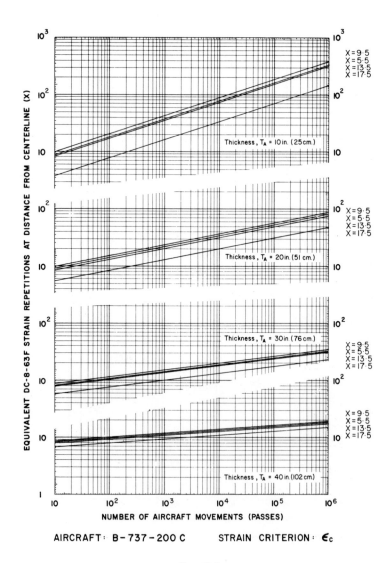

AIRCRAFT: B-737-200 C   STRAIN CRITERION: ϵ_c

FIG. D.8

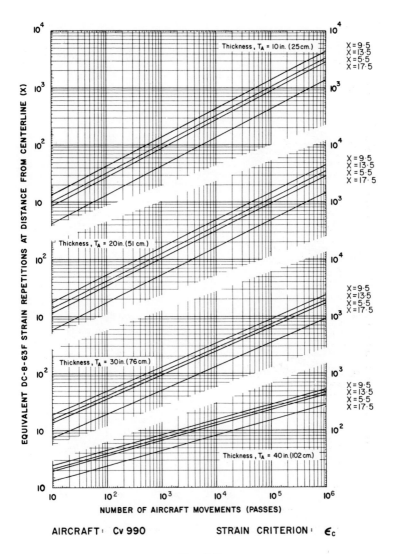

FIG. D.9

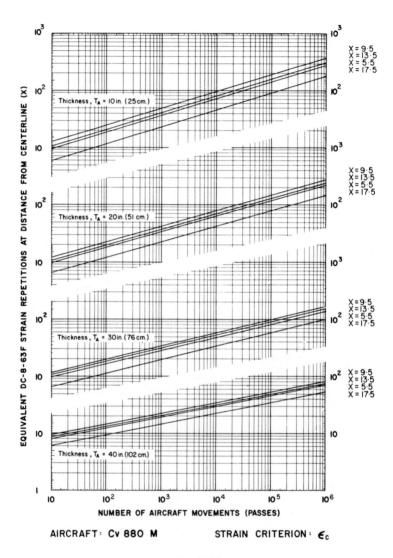

F ɪ ɢ. D.10

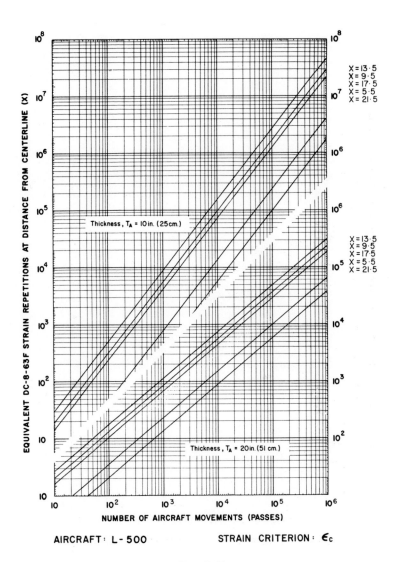

AIRCRAFT: L-500   STRAIN CRITERION: ϵ_c

F<small>IG</small>. D.11

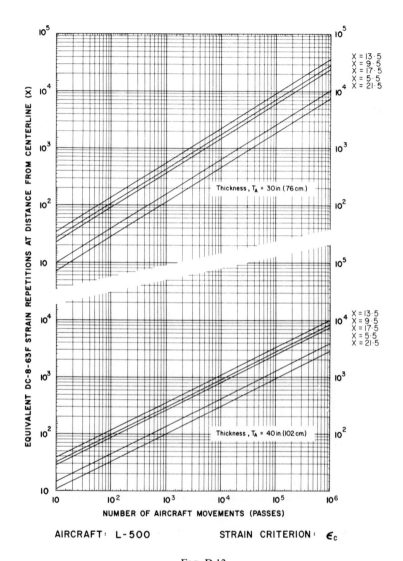

FIG. D.12

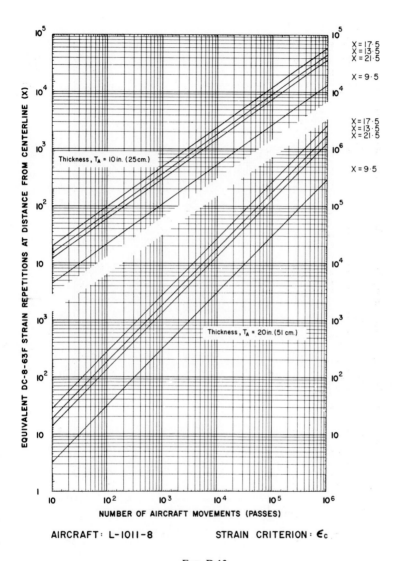

F<small>IG</small>. D.13

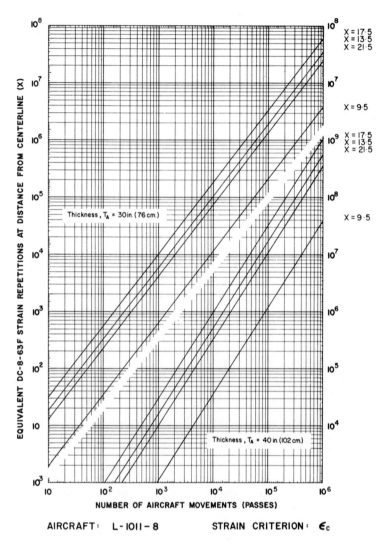

FIG. D.14

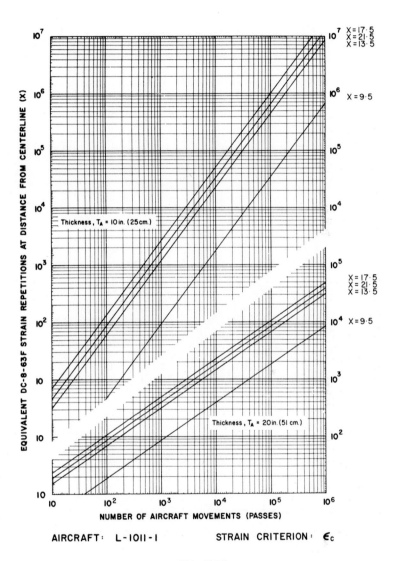

FIG. D.15

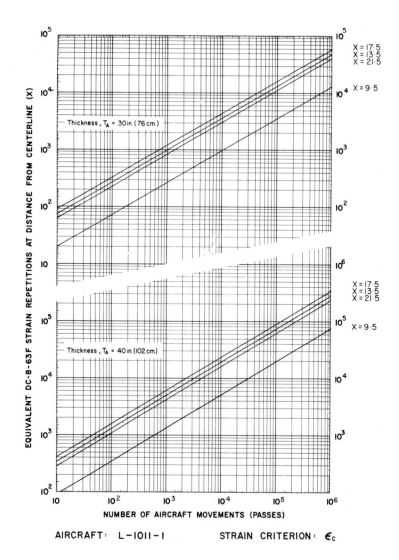

FIG. D.16

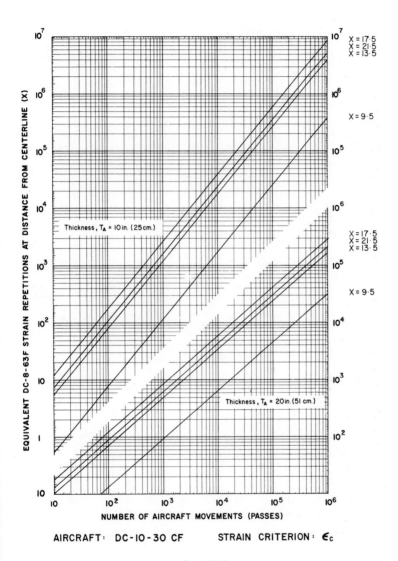

FIG. D.17

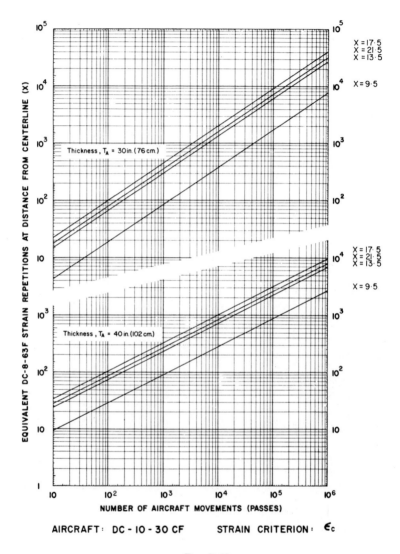

AIRCRAFT : DC - 10 - 30 CF   STRAIN CRITERION : ϵ_c

FIG. D.18

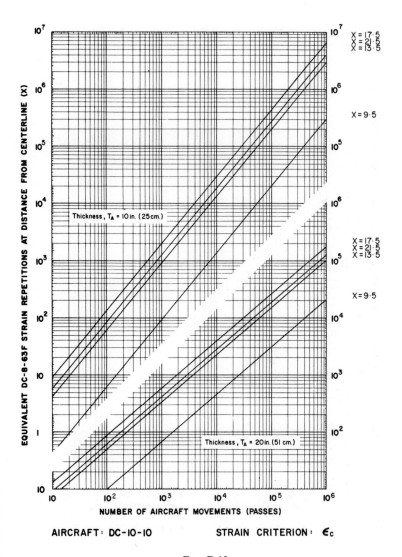

FIG. D.19

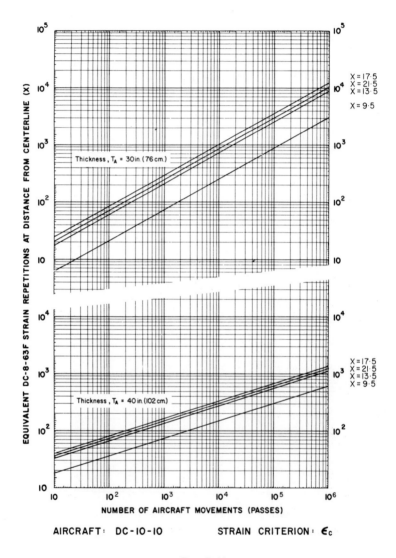

AIRCRAFT: DC-10-10   STRAIN CRITERION: ϵ_c

FIG. D.20

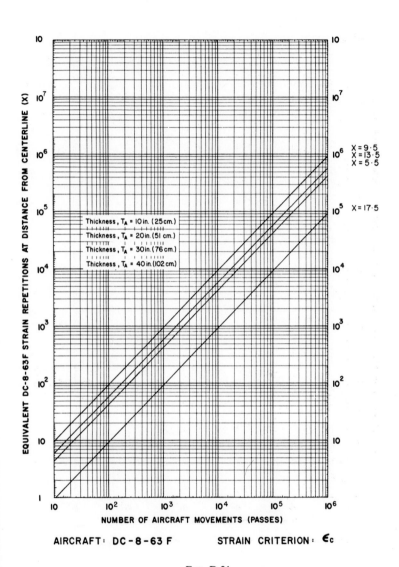

AIRCRAFT: DC-8-63 F   STRAIN CRITERION: ϵ_c

FIG. D.21

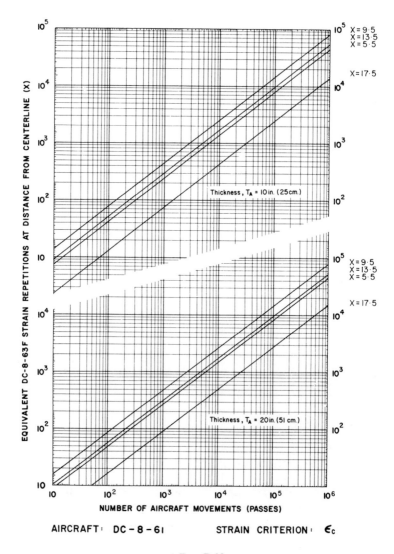

AIRCRAFT: DC - 8 - 61   STRAIN CRITERION: ϵ_c

FIG. D.22

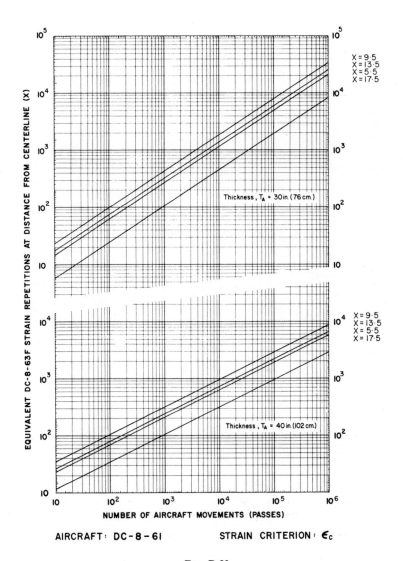

FIG. D.23

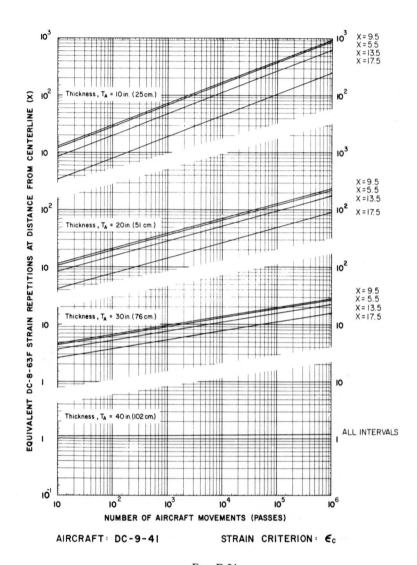

FIG. D.24

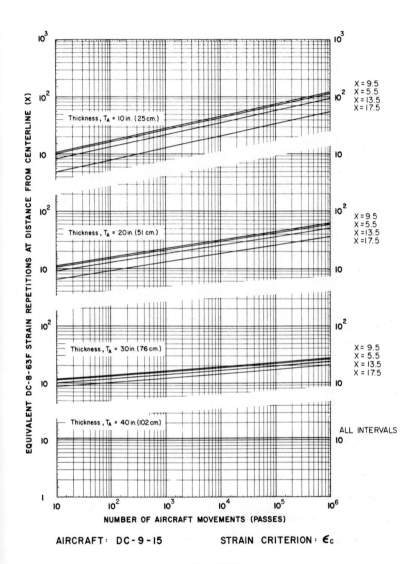

AIRCRAFT: DC-9-15   STRAIN CRITERION: ϵ_c

FIG. D.25

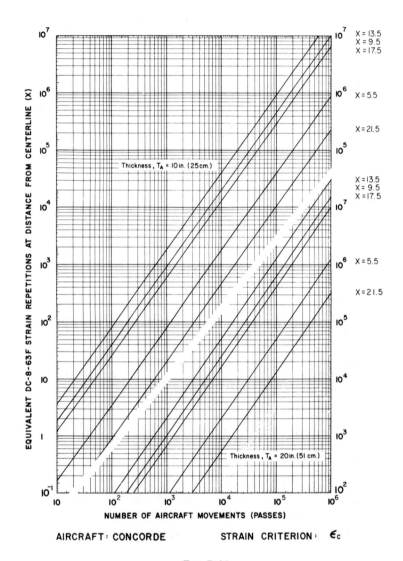

AIRCRAFT: CONCORDE   STRAIN CRITERION: ϵ_c

FIG. D.26

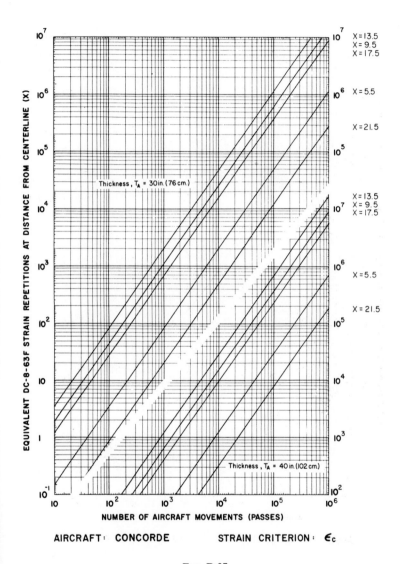

AIRCRAFT: CONCORDE   STRAIN CRITERION: ϵ_c

FIG. D.27

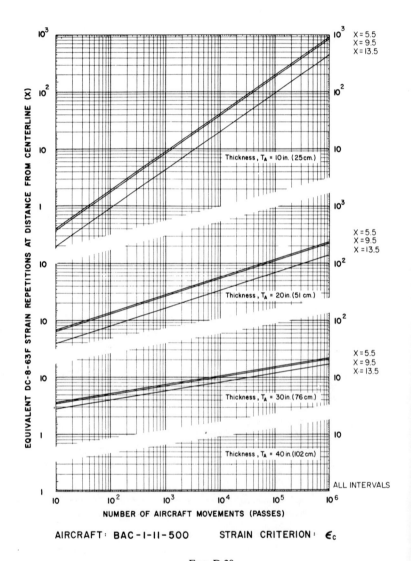

AIRCRAFT: BAC-1-11-500   STRAIN CRITERION: ϵ_c

FIG. D.28

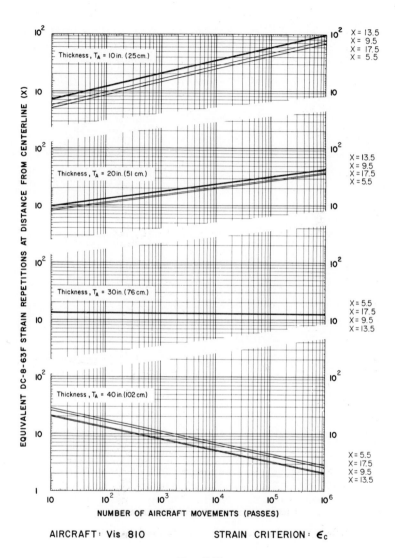

FIG. D.29

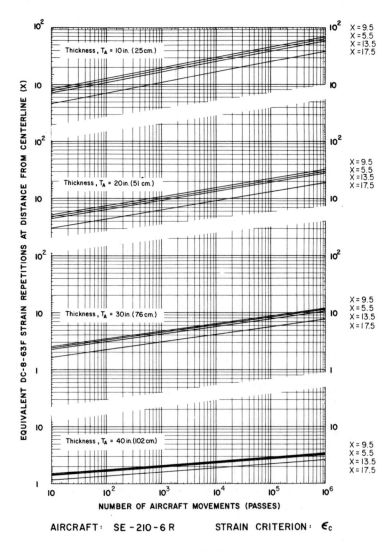

F IG. D.30

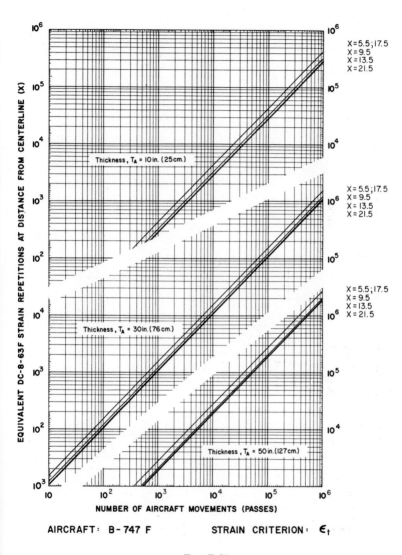

AIRCRAFT: B-747 F   STRAIN CRITERION: ϵ_t

F\ıG. D.31

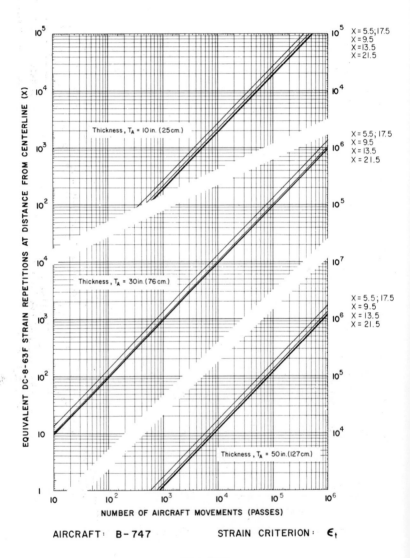

FIG. D.32

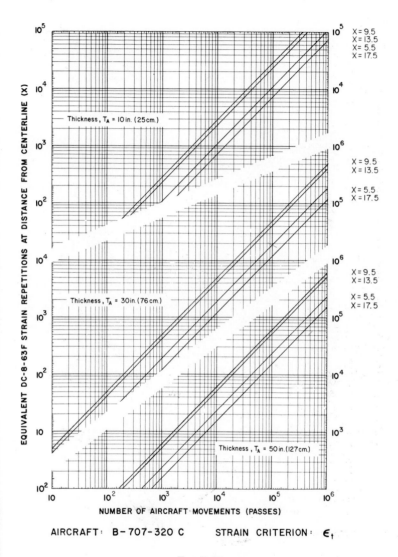

FIG. D.33

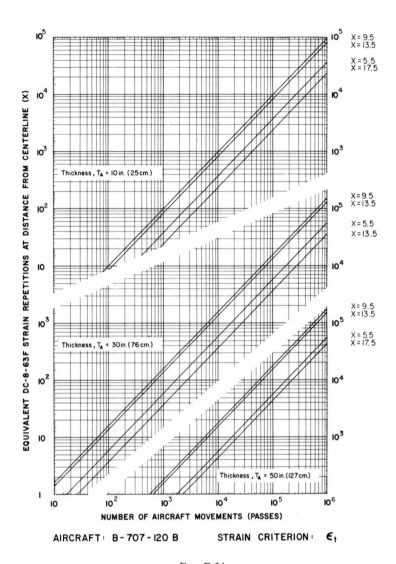

AIRCRAFT: B-707-120 B   STRAIN CRITERION: ϵ_t

FIG. D.34

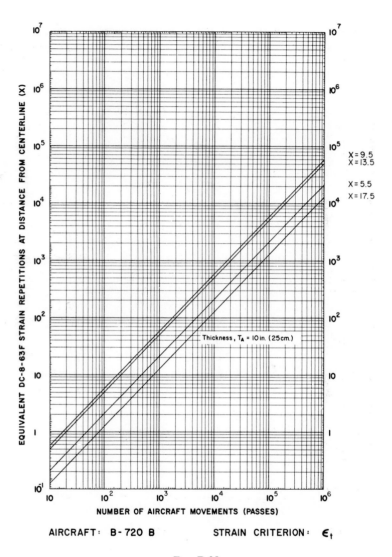

FIG. D.35

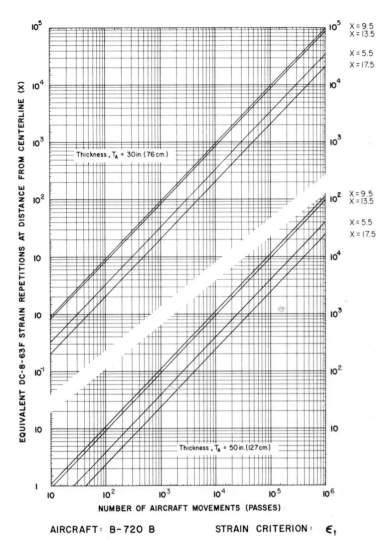

AIRCRAFT: B-720 B   STRAIN CRITERION: ϵ_t

F<small>IG</small>. D.36

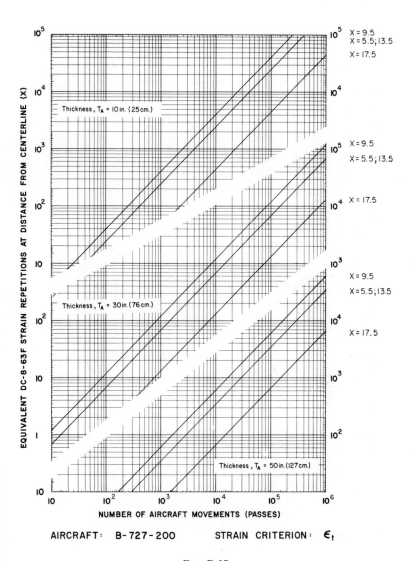

FIG. D.37

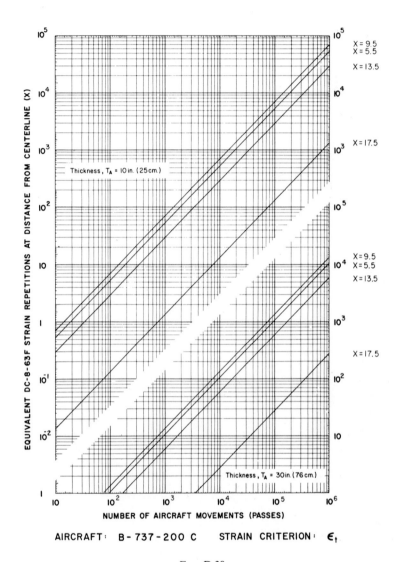

AIRCRAFT: B-737-200 C   STRAIN CRITERION: ϵ_t

FIG. D.38

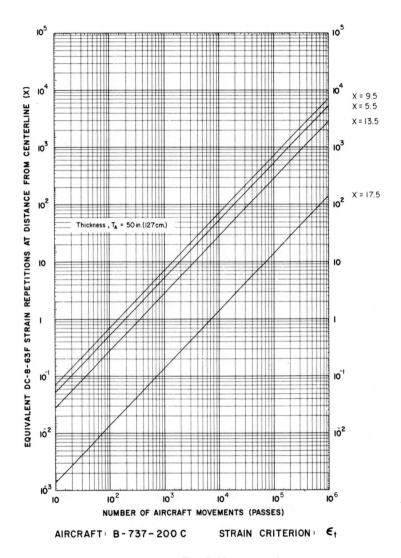

AIRCRAFT: B-737-200C   STRAIN CRITERION: ϵ_t

FIG. D.39

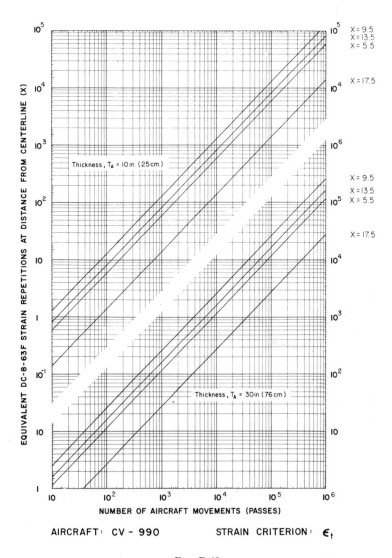

AIRCRAFT: CV - 990   STRAIN CRITERION: ϵ_t

FIG. D.40

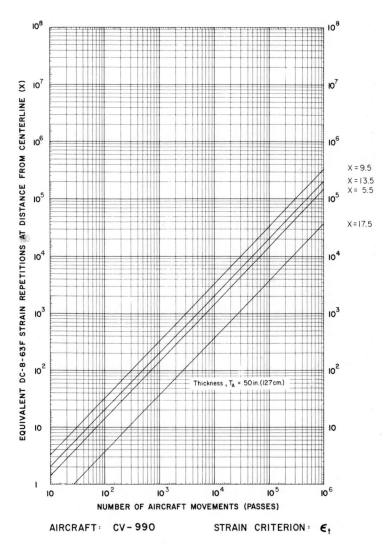

FIG. D.41

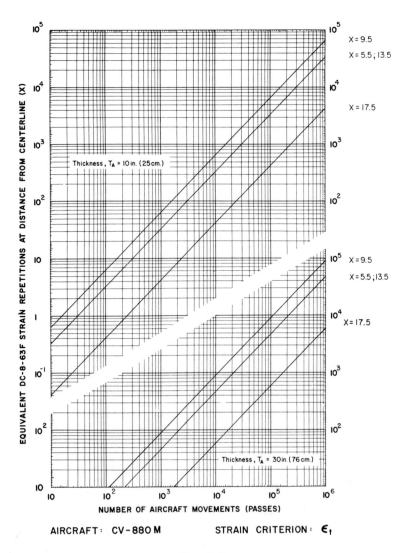

AIRCRAFT: CV-880M   STRAIN CRITERION: ϵ_t

FIG. D.42

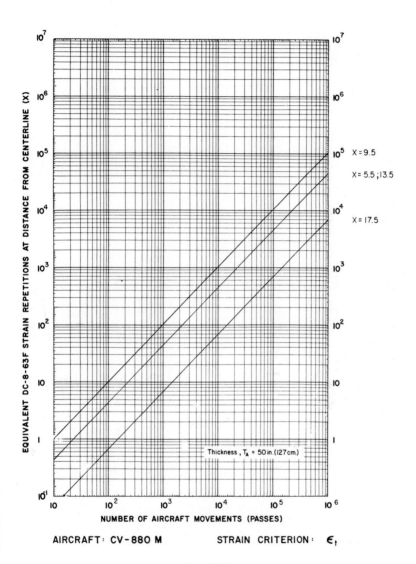

AIRCRAFT: CV-880 M   STRAIN CRITERION: ϵ_t

FIG. D.43

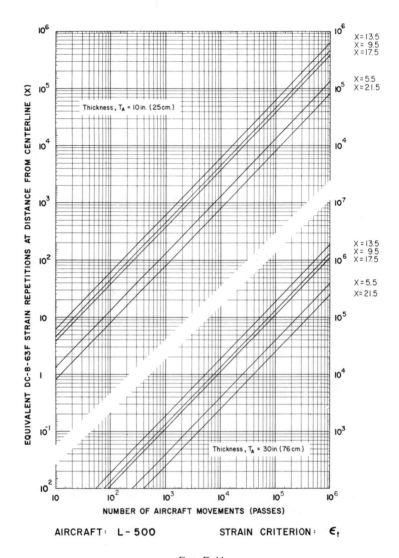

AIRCRAFT: **L - 500**   STRAIN CRITERION: ϵ_t

FIG. D.44

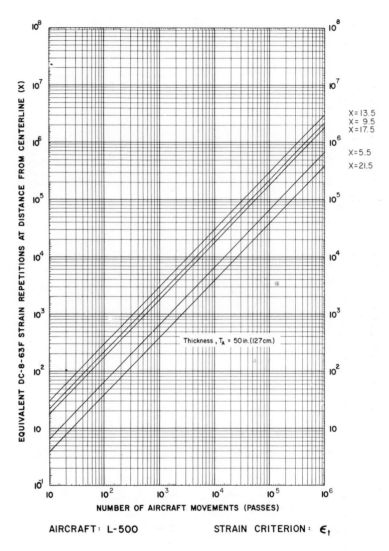

AIRCRAFT: **L-500**   STRAIN CRITERION: ϵ_t

FIG. D.45

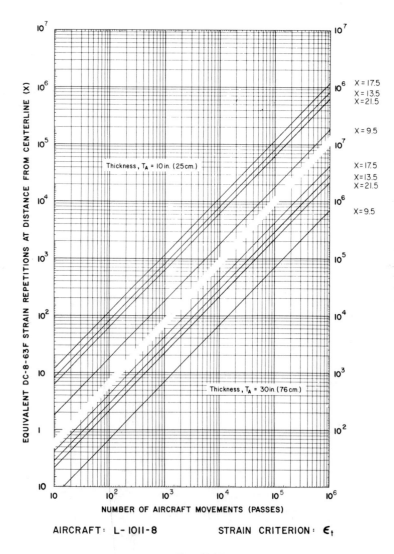

AIRCRAFT : L- 1011-8   STRAIN CRITERION : ϵ_t

FIG. D.46

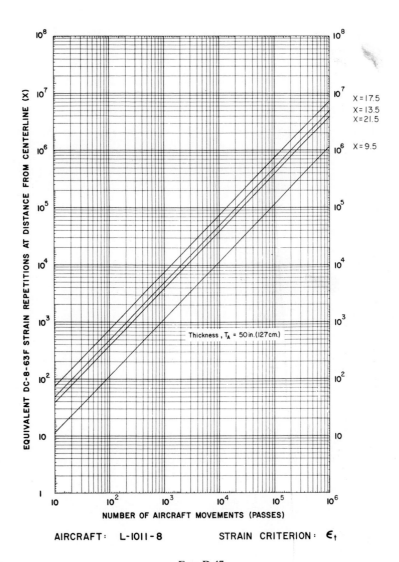

AIRCRAFT: **L-1011-8**   STRAIN CRITERION: ϵ_t

FIG. D.47

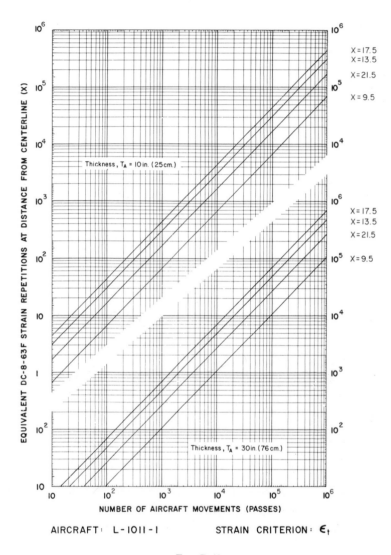

AIRCRAFT: L-1011-1   STRAIN CRITERION: ϵ_t

Fig. D.48

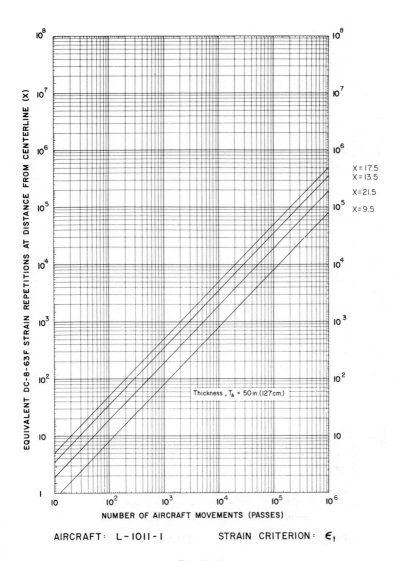

AIRCRAFT: L-1011-1   STRAIN CRITERION: ϵ_t

FIG. D.49

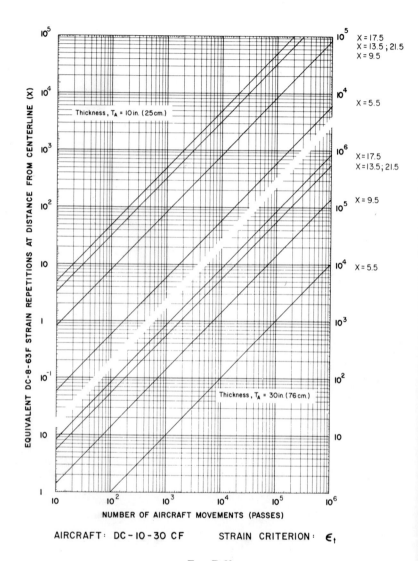

AIRCRAFT: DC-10-30 CF   STRAIN CRITERION: ϵ_t

FIG. D.50

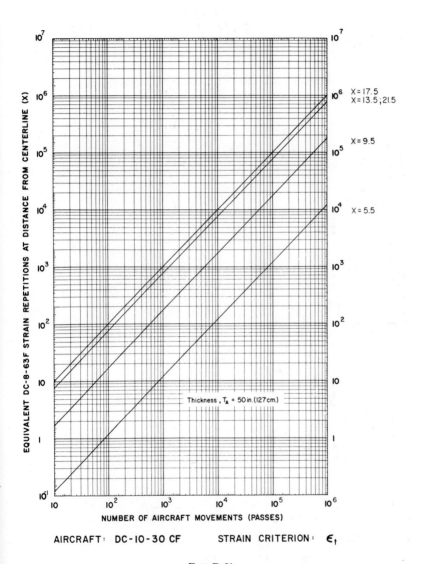

AIRCRAFT: DC-10-30 CF   STRAIN CRITERION: ϵ_t

FIG. D.51

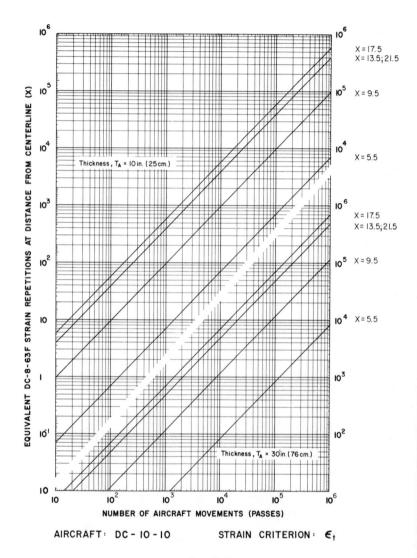

AIRCRAFT: DC - 10 - 10 STRAIN CRITERION: ϵ_t

F<small>IG</small>. D.52

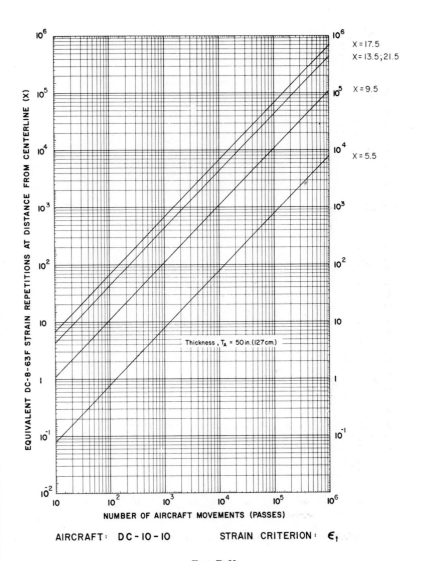

AIRCRAFT: DC-IO-IO   STRAIN CRITERION: ϵ_t

FIG. D.53

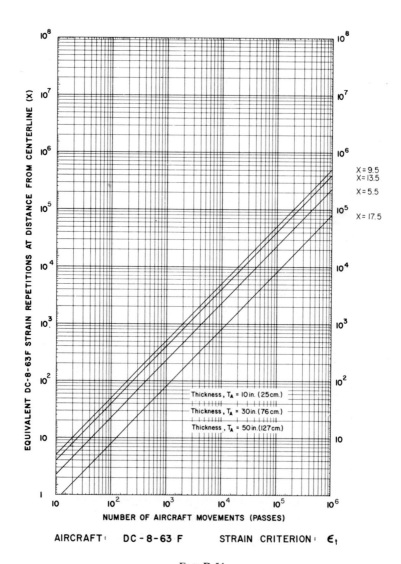

AIRCRAFT: DC - 8-63 F   STRAIN CRITERION: ϵ_t

F<small>IG</small>. D.54

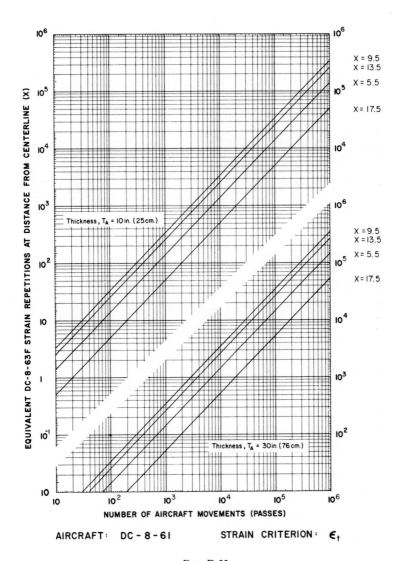

AIRCRAFT: DC - 8 - 61   STRAIN CRITERION: ϵ_t

F IG. D.55

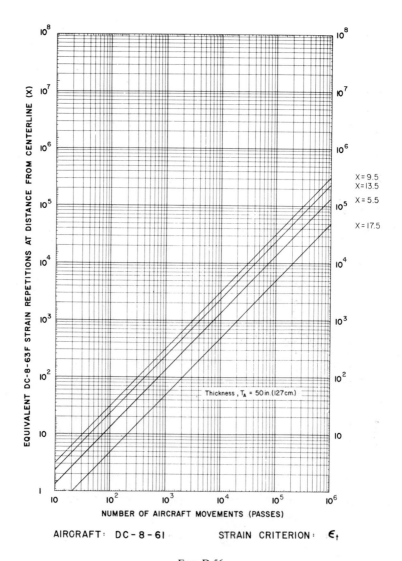

AIRCRAFT: DC‑8‑61   STRAIN CRITERION: ϵ_t

Fig. D.56

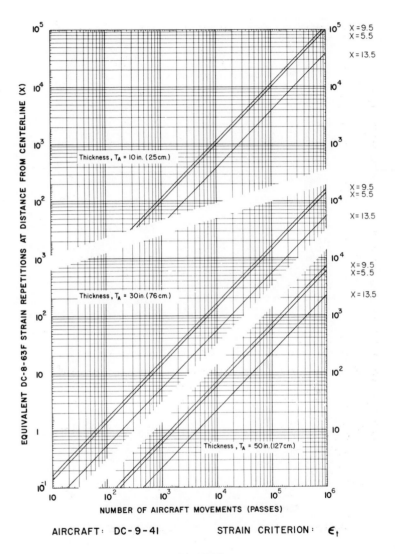

AIRCRAFT: DC-9-41   STRAIN CRITERION: ϵ_t

FIG. D.57

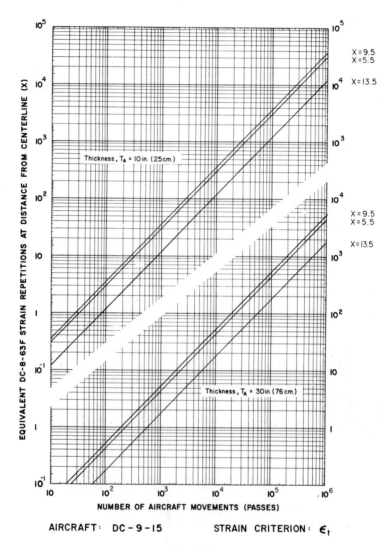

AIRCRAFT: DC-9-15   STRAIN CRITERION: ϵ_t

FIG. D.58

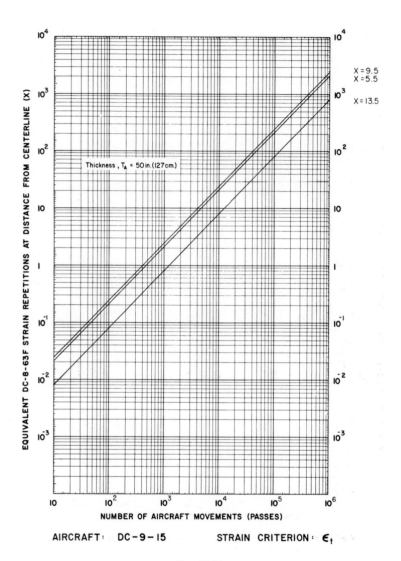

AIRCRAFT: DC-9-15   STRAIN CRITERION: ϵ_t

FIG. D.59

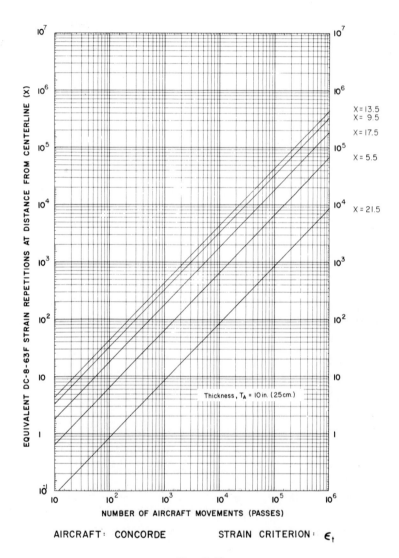

AIRCRAFT: CONCORDE   STRAIN CRITERION: ϵ_t

FIG. D.60

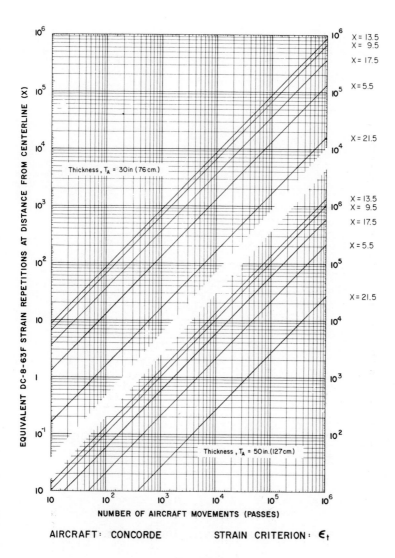

EQUIVALENT DC-8-63F STRAIN REPETITIONS AT DISTANCE FROM CENTERLINE (X)

NUMBER OF AIRCRAFT MOVEMENTS (PASSES)

AIRCRAFT : CONCORDE   STRAIN CRITERION : ϵ_t

FIG. D.61

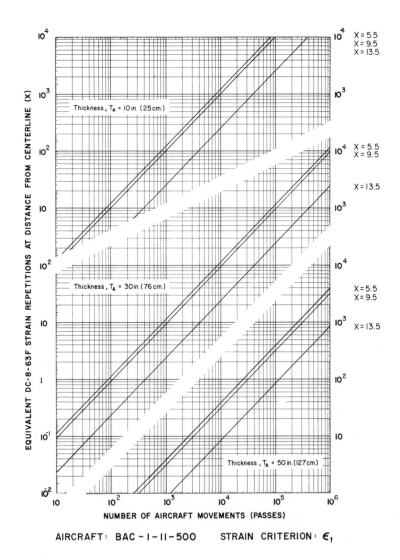

AIRCRAFT: BAC-1-11-500   STRAIN CRITERION: ϵ_t

FIG. D.62

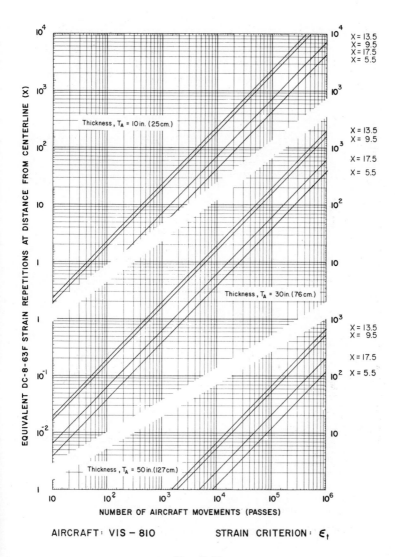

AIRCRAFT: VIS − 810   STRAIN CRITERION: ϵ_t

FIG. D.63

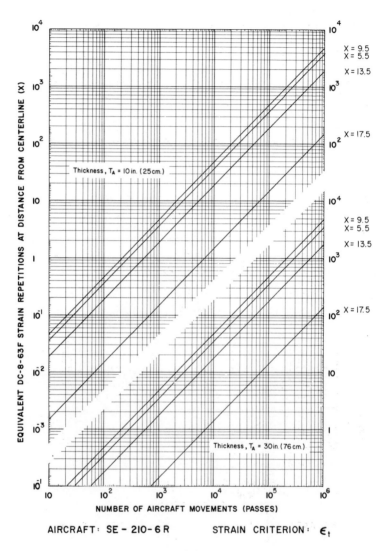

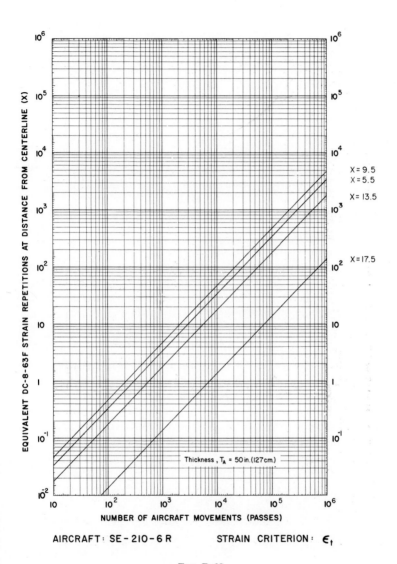

AIRCRAFT : SE - 210 - 6 R   STRAIN CRITERION : ϵ_t

FIG. D.65

RANDOM SAMPLING

Table E.1 contains random numbers for the general procedure of selecting sampling locations at pavement site. To use this table, the following steps are necessary.

1. For subgrade sampling locations, break down a highway or each runway, taxiway and apron of an airport into sections whose boundaries are defined by changes in major soil types.

2. Determine the number of sampling locations within a section by selecting the maximum average longitudinal distance desired between samples and dividing the length of the section by the maximum average longitudinal distance.

3. Select a column of random numbers in Table E.1 by placing 28 1 in (2·5 cm) square pieces of cardboard, numbered 1 to 28, into a container, shaking them to get them thoroughly mixed and drawing out one.

4. Go to the column of random numbers identified with the number drawn from the container. Locate all numbers, in sub-column A, which are equal to and less than the number of sampling locations per section desired.

5. Multiply the total length of the section by the decimal values in sub-column B, found opposite the numbers located in sub-column A. Add the result to the station number at the beginning of the section to get the station of the sampling location.

6. Multiply the total width of the proposed pavement in the section by the decimal values in sub-column C, found opposite the numbers located in sub-column A. Subtract one-half of the total width of the proposed pavement from the result to obtain the offset distance from the centreline to the sampling location. A positive number indicates the distance to the right of centreline and a negative number will be the distance to the left of the centreline.

TABLE E.1 RANDOM NUMBERS FOR GENERAL SAMPLING PROCEDURE

Column No. 1			Column No. 2			Column No. 3			Column No. 4			Column No. 5			Column No. 6			Column No. 7		
A	B	C	A	B	C	A	B	C	A	B	C	A	B	C	A	B	C	A	B	C
15	0·033	0·576	05	0·048	0·879	21	0·013	0·220	18	0·089	0·716	17	0·024	0·863	30	0·030	0·901	12	0·029	0·386
21	0·101	0·300	17	0·074	0·156	30	0·036	0·853	10	0·102	0·330	24	0·060	0·032	21	0·096	0·198	18	0·112	0·284
23	0·129	0·916	18	0·102	0·191	10	0·052	0·746	14	0·111	0·925	26	0·074	0·639	10	0·100	0·161	20	0·114	0·848
30	0·158	0·434	06	0·105	0·257	25	0·061	0·954	28	0·127	0·840	07	0·167	0·512	29	0·133	0·388	03	0·121	0·656
24	0·177	0·397	28	0·179	0·447	29	0·062	0·507	24	0·132	0·271	28	0·194	0·776	24	0·138	0·062	13	0·178	0·640
11	0·202	0·271	26	0·187	0·844	18	0·087	0·887	19	0·285	0·899	03	0·219	0·166	20	0·168	0·564	22	0·209	0·421
16	0·204	0·012	04	0·188	0·482	24	0·105	0·849	01	0·326	0·037	29	0·264	0·284	22	0·232	0·953	16	0·221	0·311
08	0·208	0·418	02	0·208	0·577	07	0·139	0·159	30	0·334	0·938	11	0·282	0·262	14	0·259	0·217	29	0·235	0·356
19	0·211	0·798	03	0·214	0·402	01	0·175	0·641	22	0·405	0·295	14	0·379	0·994	01	0·275	0·195	28	0·264	0·941
29	0·233	0·070	07	0·245	0·080	23	0·196	0·873	05	0·421	0·282	13	0·394	0·405	06	0·277	0·475	11	0·287	0·199
07	0·260	0·073	15	0·248	0·831	26	0·240	0·981	13	0·451	0·212	06	0·410	0·157	02	0·296	0·497	02	0·336	0·992
17	0·262	0·308	29	0·261	0·087	14	0·255	0·374	02	0·461	0·023	15	0·438	0·700	26	0·311	0·144	15	0·393	0·488
25	0·271	0·180	30	0·302	0·883	06	0·310	0·043	06	0·487	0·539	22	0·453	0·635	05	0·351	0·141	19	0·437	0·655
06	0·302	0·672	21	0·318	0·088	11	0·316	0·653	08	0·497	0·396	21	0·472	0·824	17	0·370	0·811	24	0·466	0·773
01	0·409	0·406	11	0·376	0·936	13	0·324	0·585	25	0·503	0·893	05	0·488	0·118	09	0·388	0·484	14	0·531	0·014
13	0·507	0·693	14	0·430	0·814	12	0·351	0·275	15	0·594	0·603	01	0·525	0·222	04	0·410	0·073	09	0·562	0·678
02	0·575	0·654	27	0·438	0·676	20	0·371	0·535	27	0·620	0·894	12	0·561	0·980	25	0·471	0·530	06	0·601	0·675
18	0·591	0·318	08	0·467	0·205	08	0·409	0·495	21	0·629	0·841	08	0·652	0·508	13	0·486	0·779	10	0·612	0·859
20	0·610	0·821	09	0·474	0·138	16	0·445	0·740	17	0·691	0·583	18	0·668	0·271	15	0·515	0·867	26	0·673	0·112
12	0·631	0·597	10	0·492	0·474	03	0·494	0·929	09	0·708	0·689	30	0·736	0·634	23	0·567	0·798	23	0·738	0·770
27	0·651	0·281	13	0·499	0·892	27	0·543	0·387	07	0·709	0·012	02	0·763	0·253	11	0·618	0·502	21	0·753	0·614
04	0·661	0·953	19	0·511	0·520	17	0·625	0·171	11	0·714	0·049	23	0·804	0·140	28	0·636	0·148	30	0·758	0·851
22	0·692	0·089	23	0·591	0·770	02	0·699	0·073	23	0·720	0·695	25	0·828	0·425	27	0·650	0·741	27	0·765	0·563
05	0·779	0·346	20	0·604	0·730	19	0·702	0·934	03	0·748	0·413	10	0·843	0·627	16	0·711	0·508	07	0·780	0·534
09	0·787	0·173	24	0·654	0·330	22	0·816	0·802	20	0·781	0·603	16	0·858	0·849	19	0·778	0·812	04	0·818	0·187
10	0·818	0·837	12	0·728	0·523	04	0·838	0·166	26	0·830	0·384	04	0·903	0·327	07	0·804	0·675	17	0·837	0·353
14	0·895	0·631	16	0·753	0·344	15	0·904	0·116	04	0·843	0·022	09	0·912	0·382	08	0·806	0·952	05	0·854	0·818
26	0·912	0·376	01	0·806	0·134	28	0·969	0·742	12	0·884	0·582	27	0·935	0·162	18	0·841	0·414	01	0·867	0·133
28	0·920	0·163	22	0·878	0·884	09	0·974	0·046	29	0·926	0·700	20	0·970	0·582	12	0·918	0·114	08	0·915	0·538
03	0·945	0·140	25	0·939	0·162	05	0·977	0·494	16	0·951	0·601	19	0·975	0·327	03	0·992	0·399	25	0·975	0·584

Column No. 8			Column No. 9			Column No. 10			Column No. 11			Column No. 12			Column No. 13			Column No. 14		
A	B	C	A	B	C	A	B	C	A	B	C	A	B	C	A	B	C	A	B	C
09	0·042	0·071	14	0·061	0·935	26	0·038	0·023	27	0·074	0·779	16	0·073	0·987	03	0·033	0·091	26	0·035	0·175
17	0·141	0·411	02	0·065	0·097	30	0·066	0·371	06	0·084	0·396	23	0·078	0·056	07	0·047	0·391	17	0·089	0·363
02	0·143	0·221	03	0·094	0·228	27	0·073	0·876	24	0·098	0·524	17	0·096	0·076	28	0·064	0·113	10	0·149	0·681
05	0·162	0·899	16	0·122	0·945	09	0·095	0·568	10	0·133	0·919	04	0·153	0·163	12	0·066	0·360	28	0·238	0·075
03	0·285	0·016	18	0·158	0·430	05	0·180	0·741	15	0·187	0·079	10	0·254	0·834	26	0·076	0·552	13	0·244	0·767
28	0·291	0·034	25	0·193	0·469	12	0·200	0·851	17	0·227	0·767	06	0·284	0·628	30	0·087	0·101	24	0·262	0·366
08	0·369	0·557	24	0·224	0·572	13	0·259	0·327	20	0·236	0·571	12	0·305	0·616	02	0·127	0·187	08	0·264	0·651
01	0·436	0·386	10	0·225	0·223	21	0·264	0·681	01	0·245	0·988	25	0·319	0·901	06	0·144	0·068	18	0·285	0·311
20	0·450	0·289	09	0·233	0·838	17	0·283	0·645	04	0·317	0·291	01	0·320	0·212	25	0·202	0·674	02	0·340	0·131
18	0·455	0·789	20	0·290	0·120	23	0·363	0·063	29	0·350	0·911	08	0·416	0·372	01	0·247	0·025	29	0·353	0·478
23	0·488	0·715	01	0·297	0·242	20	0·364	0·366	26	0·380	0·104	13	0·432	0·556	23	0·253	0·323	06	0·359	0·270
14	0·496	0·276	11	0·337	0·760	16	0·395	0·363	28	0·425	0·864	02	0·489	0·827	24	0·320	0·651	20	0·387	0·248
15	0·503	0·342	19	0·389	0·064	02	0·423	0·540	22	0·487	0·526	29	0·503	0·787	10	0·328	0·365	14	0·392	0·694
04	0·515	0·693	13	0·411	0·474	08	0·432	0·736	05	0·552	0·511	15	0·518	0·717	27	0·338	0·412	03	0·408	0·077
16	0·532	0·112	20	0·447	0·893	10	0·476	0·468	14	0·564	0·357	28	0·524	0·998	13	0·356	0·991	27	0·440	0·280
22	0·557	0·357	22	0·478	0·321	03	0·508	0·774	11	0·572	0·306	16	0·542	0·352	16	0·401	0·792	22	0·461	0·830
11	0·559	0·620	29	0·481	0·993	01	0·601	0·417	21	0·594	0·197	19	0·585	0·462	17	0·423	0·117	16	0·527	0·003
12	0·650	0·216	27	0·562	0·403	22	0·687	0·917	09	0·607	0·524	05	0·695	0·111	21	0·481	0·838	30	0·531	0·486
21	0·672	0·320	04	0·566	0·179	29	0·697	0·862	19	0·650	0·572	07	0·733	0·838	08	0·560	0·401	25	0·678	0·360
13	0·709	0·273	08	0·603	0·758	11	0·701	0·605	18	0·664	0·101	11	0·744	0·948	19	0·564	0·190	21	0·725	0·014
07	0·745	0·687	15	0·632	0·927	07	0·728	0·498	25	0·674	0·428	18	0·793	0·748	05	0·571	0·054	05	0·797	0·595
30	0·780	0·285	06	0·707	0·107	14	0·745	0·679	02	0·697	0·674	27	0·802	0·967	18	0·587	0·584	15	0·801	0·927
19	0·845	0·097	28	0·737	0·161	24	0·819	0·444	03	0·767	0·928	21	0·826	0·487	15	0·604	0·145	12	0·836	0·294
26	0·846	0·366	17	0·846	0·130	15	0·840	0·823	16	0·809	0·529	24	0·835	0·832	11	0·641	0·298	04	0·854	0·982
29	0·861	0·307	07	0·874	0·491	25	0·863	0·568	30	0·838	0·294	26	0·855	0·142	22	0·672	0·156	11	0·884	0·928
25	0·906	0·874	05	0·880	0·828	06	0·878	0·215	13	0·845	0·470	14	0·861	0·462	20	0·674	0·887	19	0·886	0·832
24	0·919	0·809	23	0·931	0·659	18	0·930	0·601	08	0·855	0·524	20	0·874	0·625	14	0·752	0·881	07	0·929	0·932
10	0·952	0·555	26	0·960	0·365	04	0·954	0·827	07	0·867	0·718	30	0·929	0·056	09	0·774	0·560	09	0·932	0·206
06	0·961	0·504	21	0·978	0·194	28	0·963	0·004	12	0·881	0·722	09	0·935	0·582	29	0·921	0·752	01	0·970	0·692
27	0·969	0·811	12	0·982	0·183	19	0·988	0·020	23	0·937	0·872	22	0·947	0·797	04	0·959	0·099	23	0·973	0·082

Column No. 15			Column No. 16			Column No. 17			Column No. 18			Column No. 19			Column No. 20			Column No. 21		
A	B	C	A	B	C	A	B	C	A	B	C	A	B	C	A	B	C	A	B	C
15	0·023	0·979	19	0·062	0·588	13	0·045	0·004	25	0·027	0·290	12	0·052	0·075	20	0·030	0·881	01	0·010	0·946
11	0·118	0·465	25	0·080	0·218	18	0·086	0·878	06	0·057	0·571	30	0·075	0·493	12	0·034	0·291	10	0·014	0·939
07	0·134	0·172	09	0·131	0·295	26	0·126	0·990	26	0·059	0·026	28	0·120	0·341	22	0·043	0·893	09	0·032	0·346
01	0·139	0·230	18	0·136	0·381	12	0·128	0·661	07	0·105	0·176	27	0·145	0·689	28	0·143	0·073	06	0·093	0·180
16	0·145	0·122	05	0·147	0·864	30	0·146	0·337	18	0·107	0·358	02	0·209	0·957	03	0·150	0·937	15	0·151	0·012
20	0·165	0·520	12	0·158	0·365	05	0·169	0·470	22	0·128	0·827	26	0·272	0·818	04	0·154	0·867	16	0·185	0·455
06	0·185	0·481	28	0·214	0·184	21	0·244	0·433	23	0·156	0·440	22	0·299	0·317	19	0·158	0·359	07	0·227	0·277
09	0·211	0·316	14	0·215	0·757	23	0·270	0·849	15	0·171	0·157	18	0·306	0·475	29	0·304	0·615	02	0·304	0·400
14	0·248	0·348	13	0·224	0·846	25	0·274	0·407	08	0·220	0·097	20	0·311	0·653	06	0·369	0·633	30	0·316	0·074
25	0·249	0·890	15	0·227	0·809	10	0·290	0·925	20	0·252	0·066	04	0·348	0·156	18	0·390	0·536	18	0·328	0·799
13	0·252	0·577	11	0·280	0·898	01	0·323	0·490	04	0·268	0·576	16	0·381	0·710	17	0·403	0·392	20	0·352	0·288
30	0·273	0·088	01	0·331	0·925	24	0·352	0·291	14	0·275	0·302	01	0·411	0·607	23	0·404	0·182	26	0·371	0·216
18	0·277	0·689	10	0·399	0·992	15	0·361	0·155	11	0·297	0·589	13	0·417	0·715	01	0·415	0·457	19	0·448	0·754
22	0·372	0·958	30	0·417	0·787	29	0·374	0·882	01	0·358	0·305	21	0·472	0·484	07	0·437	0·696	13	0·487	0·598
10	0·461	0·075	08	0·439	0·921	08	0·432	0·139	09	0·412	0·089	04	0·478	0·885	24	0·446	0·546	12	0·546	0·640
28	0·519	0·536	20	0·472	0·484	04	0·467	0·266	16	0·429	0·834	25	0·479	0·080	26	0·485	0·768	24	0·550	0·038
17	0·520	0·090	24	0·498	0·712	22	0·508	0·880	10	0·491	0·203	11	0·566	0·104	15	0·511	0·313	03	0·604	0·780
03	0·523	0·519	04	0·516	0·396	27	0·632	0·191	28	0·542	0·306	10	0·576	0·659	10	0·517	0·290	22	0·621	0·930
26	0·573	0·502	03	0·548	0·688	16	0·661	0·836	12	0·563	0·091	29	0·665	0·397	30	0·556	0·853	21	0·629	0·154
19	0·634	0·206	23	0·597	0·508	19	0·675	0·629	02	0·593	0·321	19	0·739	0·298	25	0·561	0·837	11	0·634	0·908
24	0·635	0·810	21	0·681	0·114	14	0·680	0·890	30	0·692	0·198	14	0·749	0·759	09	0·574	0·599	05	0·696	0·459
21	0·679	0·841	02	0·739	0·298	28	0·714	0·508	19	0·705	0·445	08	0·756	0·919	13	0·613	0·762	23	0·710	0·078
27	0·712	0·366	29	0·792	0·038	06	0·719	0·441	24	0·709	0·717	07	0·798	0·183	11	0·698	0·783	29	0·726	0·585
05	0·780	0·497	22	0·829	0·324	09	0·735	0·040	13	0·820	0·739	23	0·834	0·647	14	0·715	0·179	17	0·749	0·916
23	0·861	0·106	17	0·834	0·647	17	0·741	0·906	05	0·848	0·866	06	0·837	0·978	16	0·770	0·128	04	0·802	0·186
12	0·865	0·377	16	0·909	0·608	11	0·747	0·205	27	0·867	0·633	03	0·849	0·964	08	0·815	0·385	14	0·835	0·319
29	0·882	0·635	06	0·914	0·420	20	0·850	0·047	03	0·883	0·333	24	0·851	0·109	05	0·872	0·490	08	0·870	0·546
08	0·902	0·020	27	0·958	0·856	02	0·859	0·356	17	0·900	0·443	05	0·859	0·935	21	0·885	0·999	28	0·871	0·539
04	0·951	0·482	26	0·981	0·976	07	0·870	0·612	21	0·914	0·483	17	0·863	0·220	02	0·958	0·177	25	0·971	0·369
02	0·977	0·172	07	0·983	0·624	03	0·916	0·463	29	0·950	0·753	09	0·863	0·147	27	0·961	0·980	27	0·984	0·252

TABLE E.1 (continued)

Column No. 22			Column No. 23			Column No. 24			Column No. 25			Column No. 26			Column No. 27			Column No. 28		
A	B	C	A	B	C	A	B	C	A	B	C	A	B	C	A	B	C	A	B	C
12	0·051	0·032	26	0·051	0·187	08	0·015	0·521	02	0·039	0·005	16	0·026	0·102	21	0·050	0·952	29	0·042	0·039
11	0·068	0·980	03	0·053	0·256	16	0·068	0·994	16	0·061	0·599	01	0·033	0·886	17	0·085	0·403	07	0·105	0·293
17	0·089	0·309	29	0·100	0·159	11	0·118	0·400	26	0·068	0·054	04	0·088	0·686	10	0·141	0·624	25	0·115	0·420
01	0·091	0·371	13	0·102	0·465	21	0·124	0·565	11	0·073	0·812	22	0·090	0·602	05	0·154	0·157	09	0·126	0·612
10	0·100	0·709	24	0·110	0·316	18	0·153	0·158	07	0·123	0·649	13	0·114	0·614	06	0·164	0·841	10	0·205	0·144
30	0·121	0·744	18	0·114	0·300	17	0·190	0·159	05	0·126	0·658	20	0·136	0·576	07	0·197	0·013	03	0·210	0·054
02	0·166	0·056	11	0·123	0·208	26	0·192	0·676	14	0·161	0·189	05	0·138	0·228	16	0·215	0·363	23	0·234	0·533
23	0·179	0·529	09	0·138	0·182	01	0·237	0·030	18	0·166	0·040	10	0·216	0·565	08	0·222	0·520	13	0·266	0·799
21	0·187	0·051	06	0·194	0·115	12	0·283	0·077	28	0·248	0·171	02	0·233	0·610	13	0·269	0·477	20	0·305	0·603
22	0·205	0·543	22	0·234	0·480	03	0·286	0·318	06	0·255	0·117	07	0·278	0·357	02	0·288	0·012	05	0·372	0·223
28	0·230	0·688	20	0·274	0·107	10	0·317	0·734	15	0·261	0·928	30	0·405	0·273	25	0·333	0·633	26	0·385	0·111
19	0·243	0·001	21	0·331	0·292	05	0·337	0·844	10	0·301	0·811	06	0·421	0·807	28	0·348	0·710	30	0·422	0·315
27	0·267	0·990	08	0·346	0·085	25	0·441	0·336	24	0·363	0·025	12	0·426	0·583	20	0·362	0·961	17	0·453	0·783
15	0·283	0·440	27	0·382	0·979	27	0·469	0·786	22	0·378	0·792	08	0·471	0·708	14	0·511	0·989	02	0·460	0·916
16	0·352	0·089	07	0·387	0·865	24	0·473	0·237	27	0·379	0·959	18	0·473	0·738	26	0·540	0·903	27	0·461	0·841
03	0·377	0·648	28	0·411	0·776	20	0·475	0·761	19	0·420	0·557	19	0·510	0·207	27	0·587	0·643	14	0·483	0·095
06	0·397	0·769	16	0·444	0·999	06	0·557	0·001	21	0·467	0·943	03	0·512	0·329	12	0·603	0·745	12	0·507	0·375
09	0·409	0·428	04	0·515	0·993	07	0·610	0·238	17	0·494	0·225	15	0·640	0·329	29	0·619	0·895	28	0·509	0·748
14	0·465	0·406	17	0·518	0·827	09	0·617	0·041	09	0·620	0·081	09	0·665	0·354	23	0·623	0·333	21	0·583	0·804
13	0·499	0·651	05	0·539	0·620	13	0·641	0·648	30	0·623	0·106	14	0·680	0·884	22	0·624	0·076	22	0·587	0·993
04	0·539	0·972	02	0·623	0·271	22	0·664	0·291	03	0·625	0·777	26	0·703	0·622	18	0·670	0·904	16	0·689	0·339
18	0·560	0·747	30	0·637	0·374	04	0·668	0·856	08	0·651	0·790	29	0·739	0·394	11	0·711	0·253	06	0·727	0·298
26	0·575	0·892	14	0·714	0·364	19	0·717	0·232	12	0·715	0·599	25	0·759	0·386	01	0·790	0·392	04	0·731	0·814
29	0·756	0·712	15	0·730	0·107	02	0·776	0·504	23	0·782	0·093	24	0·803	0·602	04	0·813	0·611	08	0·807	0·983
20	0·760	0·920	19	0·771	0·552	29	0·777	0·548	20	0·810	0·371	27	0·842	0·491	19	0·843	0·732	15	0·833	0·757
05	0·847	0·925	23	0·780	0·662	14	0·823	0·223	01	0·841	0·726	21	0·870	0·435	03	0·844	0·511	19	0·896	0·464
25	0·872	0·891	10	0·924	0·888	23	0·848	0·264	29	0·862	0·009	28	0·906	0·367	30	0·858	0·299	18	0·916	0·384
24	0·874	0·135	12	0·929	0·204	30	0·892	0·817	25	0·891	0·873	23	0·948	0·367	09	0·929	0·199	01	0·948	0·610
08	0·911	0·215	01	0·937	0·714	28	0·943	0·190	04	0·917	0·264	11	0·956	0·142	24	0·931	0·263	11	0·976	0·799
07	0·946	0·065	25	0·974	0·398	15	0·975	0·962	13	0·958	0·990	17	0·993	0·989	15	0·939	0·947	24	0·978	0·633

Example

Select subgrade sampling locations for a new runway pavement 6000 ft (1829 m) long, beginning at station 20+00, and 150 ft (46 m) wide. The runway passes through two major soil areas, the boundary between them being at station 39+50 on the centreline, station 36+00 at the right edge of the proposed pavement and station 43+00 at its left edge. An average centreline interval of subgrade soil samples of 250 ft (76 m) is desired.

Solution

1. For sampling purposes, divide the proposed runway into two sections whose boundaries on the centreline are:
   Section 1: Station 20+00 to station 39+50 (1950 ft (594 m) from the beginning of the runway).
   Section 2: Station 39+50 to station 80+00 (4050 ft (1235 m) from the beginning of the runway).

2. Calculate the number of sampling locations in each section.

   $$\text{Section 1}: \frac{1950 \text{ ft}}{250 \text{ ft}} = 7 \cdot 8, \text{ i.e. 8 locations}$$

   $$\text{Section 2}: \frac{4050 \text{ ft}}{250 \text{ ft}} = 16 \cdot 2, \text{ i.e. 16 locations}$$

3. Assume, in the example, that the numbers 6 and 20, drawn from a container, identify the columns of random numbers in the table to use for the two sections.

4. For Section 1, the numbers selected from column 6 are shown in the table below.

Column A	Column B	Column C
01	0·275	0·195
06	0·277	0·475
02	0·296	0·497
05	0·351	0·141
04	0·410	0·073
07	0·804	0·675
08	0·806	0·952
03	0·992	0·399

Section 2: The numbers selected for column 20 are given in the table below.

Column A	Column B	Column C
12	0·034	0·291
03	0·150	0·937
04	0·154	0·867
06	0·369	0·633
01	0·415	0·457
07	0·437	0·696
15	0·511	0·313
10	0·517	0·290
09	0·574	0·599
13	0·613	0·762
11	0·698	0·783
14	0·715	0·179
16	0·770	0·128
08	0·815	0·385
05	0·872	0·490
02	0·958	0·177

5. Station numbers of sampling locations are as given below.
Section 1 : Length of section = 1950 ft (594 m).

Length of section[a]	×	Column B	=	Distance from beginning of section (ft)[a]	+	Station at beginning of section[a]	=	Station number of sampling location[a]
1 950		0·275		536		20 + 00		25 + 36
1 950		0·277		540		20 + 00		25 + 40
1 950		0·296		577		20 + 00		25 + 77
1 950		0·351		685		20 + 00		26 + 85
1 950		0·410		800		20 + 00		28 + 00
1 950		0·804		1 568		20 + 00		35 + 68
1 950		0·806		1 572		20 + 00		35 + 72
1 950		0·992		1 934		20 + 00		39 + 34

[a] Metres = feet × 0·304 8.

Section 2: Length of section = 4050 ft (1235 m).

Length of section[a]	×	Column B	=	Distance from beginning of section (ft)[a]	+	Station at beginning of section[a]	=	Station number of sampling location[a]
4 050		0·034		138		39 + 50		40 + 88
4 050		0·150		607		39 + 50		45 + 57
4 050		0·154		624		39 + 50		45 + 74
4 050		0·369		1 494		39 + 50		54 + 44
4 050		0·415		1 680		39 + 50		56 + 30
4 050		0·437		1 770		39 + 50		57 + 20
4 050		0·511		2 070		39 + 50		60 + 20
4 050		0·517		2 095		39 + 50		60 + 45
4 050		0·574		2 321		39 + 50		62 + 71
4 050		0·613		2 480		39 + 50		64 + 30
4 050		0·698		2 825		39 + 50		67 + 75
4 050		0·715		2 896		39 + 50		68 + 46
4 050		0·770		3 119		39 + 50		70 + 69
4 050		0·815		3 301		39 + 50		72 + 51
4 050		0·872		3 532		39 + 50		74 + 82
4 050		0·958		3 880		39 + 50		78 + 30

[a] Metres = feet × 0·304 8.

6. Offset distance from runway centreline to sample locations are shown in the two tables below.

Width of pavement[a]	×	Column C	=	Distance from left edge of pavement (ft)[a]	=	$\frac{1}{2}$ width of pavement[a]	=	Offset distance from centreline to sampling location[a]
150		0·195		29		75		−46
150		0·475		71		75		−4
150		0·497		75		75		0
150		0·141		21		75		−54
150		0·073		11		75		−64
150		0·675		101		75		26
150		0·952		143		75		68
150		0·399		60		75		−15

[a] Metres = feet × 0·304 8.

Section 1: Total width of proposed pavement = 150 ft (46 m).

Width of pavement[a]	×	Column C	=	Distance from left edge of pavement (ft)[a]	=	$\frac{1}{2}$ width of pavement[a]	=	Offset distance from centreline to sampling location[a]
150		0·291		45		75		−30
150		0·937		141		75		66
150		0·867		130		75		55
150		0·633		95		75		20
150		0·457		69		75		−6
150		0·696		105		75		30
150		0·313		47		75		−28
150		0·290		44		75		−31
150		0·599		90		75		15
150		0·762		114		75		39
150		0·783		117		75		42
150		0·179		27		75		−48
150		0·128		19		75		−56
150		0·385		58		75		−17
150		0·490		74		75		−1
150		0·177		27		75		−48

[a] Metres = feet × 0·304 8.

Section 2: Total width of proposed pavement = 150 ft (46 m).

7. Sampling locations are given in the table below as well as in Fig. E.1.

	Station number[a]	Distance (ft) from centreline to sampling location[a]	
		Left	Right
Section 1	25 + 36	46	
	25 + 40	4	
	25 + 77	0	
	26 + 85	54	
	28 + 00	64	
	35 + 68		26
	35 + 72		68
	39 + 34	15	
Section 2	40 + 88	30	
	45 + 57		66
	45 + 74		55
	54 + 44		20
	56 + 30	6	
	57 + 20		30
	60 + 20	28	
	60 + 45	31	
	62 + 71		15
	64 + 30		39
	67 + 75		42
	68 + 46	48	
	70 + 69	56	
	72 + 51	17	
	74 + 82	1	
	78 + 30	48	

[a] Metres = feet × 0·304 8.

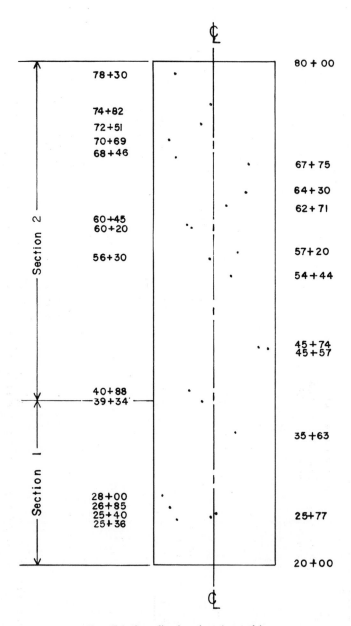

FIG. E.1. Sampling locations (example).

APPENDIX F

TABLE F.1 THICKNESS OF PLAIN CONCRETE PAVEMENTS AND SUB-BASES FOR ROADS IN EUROPE

| | Pavement thickness, in (cm) | | | Sub-base | | | |
| | | | | Cement treated | | Other materials | |
	Freeways and expressways	Main highways	Secondary highways	Material used	Thickness, in (cm)	Material used	Thickness, in (cm)
Austria	R 8·70 (22)	R 8·70 (22)				R Hot bitumen gravel	
Belgium	Unreinforced 9 (23) Continuously reinforced 8 (20)	Unreinforced 9 (23) Continuously reinforced 8 (20)	8–9 (20–30)	A Crushed stone, gravel, lean concrete	A 8–14 (20–35) 7·2–8 (18–20)	Dense bitumen binder	2·4 (6) Isolation layer
Czechoslovakia	8·7–9·5 (22–24)	8·0–8·7 (20–22)	7·2 (18)	Gravel and sand natural granular 0–45 mm		(a) Hot bitumen gravel (b) Penetrated macadam	
Denmark	8·0 (20)			Cement bound granular	6 (15)		
France	A 10–11 (25–28)	N 9–10 (23–25)		R Gravel and sand 0–20 mm Cement content 3·5–4·0%	A 6 (15)	R 15–20% granulated slag 1% line	R 6 (15)
Great Britain[c]	A 9·5–12 (24–30·5)	A 8·7–10·7 (22–27)	A 8–10 (20–25)	Soil-cement, cement bound granular lean concrete	0–6 (0–15) Normal 6 (15)	Crushed rock Slag or concrete	0–6 (0–15)

TABLE F.1—contd.

| | Pavement thickness, in (cm) | | | Sub-base | | | |
| | Freeways and expressways | Main highways | Secondary highways | Cement treated | | Other materials | |
				Material used	Thickness, in (cm)	Material used	Thickness, in (cm)
Netherlands[a]	10 (25) concrete + 2 (5) asphalt + 16 (40) sand/cement	10 (25) concrete + 6 (15) tar–lime stabilisation	5–7·2 (12–18) No stabilisation	Sand–cement	6–16 (15–40)	Tar–lime stabilised sand	6 (15)
Italy	N 9·5 (24)	N 8·7 (22)	N 8 (20)	R Crushed stone, sand and gravel	R 8 (20)	E Granular pozzolano	E 10 (25)
Spain[c]	10 (25)	Normal practice 10 (25)	Not used	R Crushed stone 3·5–4·0 Cement content	6–8 (15–20)	Crushed stone	6–8 (15–20)
Sweden[a]	8 (20)	8 (20)		Poor-quality gravel	6 (15)	Well-graded good-quality gravel	
Switzerland	A 8–8·7 (20–22)	7·2–8 (18–20)	A 6·4–7·2 (16–18)	Natural granular	A 6–12 (15–30)	Natural granular	8–24 (20–60)
West Germany[b]	R 8·7 (22)	R 8 (20)	R 8·7 (22)	Gravelly sand	6 (15)	R Lime, tar or bitumen	6 (15) for lime and tar 3·2–4 (8–10) for bitumen

A: As shown on plans. R: Definitely required by specifications. E: Where authorised by engineer.

[a] Light reinforcement is definitely required by specifications for the concrete pavement slabs.
[b] Light reinforcement is definitely required for slabs over 17 ft (5 m) and for differential settlement areas.
[c] Light reinforcement can be permitted under certain conditions.

APPENDIX G

The US Customary units have been used in the text, followed, wherever possible, by the SI (International System) units in parentheses. However, in some cases, the reader may need to convert values given in US units to corresponding values in SI units. Table G.1 can be used for this purpose.

TABLE G.1 LIST OF US UNITS, SI EQUIVALENTS AND CONVERSION FACTORS

US units to convert from	SI units to	Multiply by
foot	metre (m)	0·304 8
inch	centimetre (cm)	2·54
yard	metre (m)	0·914 4
mile	metre (m)	1 609·344
pound-mass (lbm)	kilogram (kg)	0·453 6
ton (short, 2 000 lbm)	kilogram (kg)	907·185
pound-force (lbf)	newton (N)	4·448
kip (1 000 lbf)	newton (N)	4 448·222
pound-force/inch	newton/metre (N/m)	175·127
pound-force/foot	newton/metre (N/m)	14·594
pound-force/inch^2 (psi)	newton/metre^2 (N/m^2)	6 894·760
pound-force/foot^2 (psf)	newton/metre^2 (N/m^2)	47·88

AUTHOR INDEX

609

SUBJECT INDEX

Bond breaker, 335, 396, 478
Bottom tension crack, 362
Boundary conditions, 247
Buckling failure, 379
Bulk specific gravity, 210
Butt joint, 81

California Bearing Ratio, 75, 130, 180, 516
    method, 144, 168
Capacity operations, 268
Capital recovery factor, 228, 422
Car road meter, 224
Cathode ray, 103
Cement
    bound insulation layers, 95
    bound materials, 115
    grout, 383
    stabilised soil, 80
    types, 311, 399
Central load, 241
Channelised traffic, 18, 281
Chevron computer program, 153
Chloe profilometer, 226, 466
Civil aircraft, 157
Clay, 60, 68
Climate, 44
Coefficient of thermal expansion, 255
Cohesiometer, 146
Cohesion, 343
Commerical vehicles, 8
Compaction, 69, 421
    factor, 400
Complex modulus, 102
Component analysis method, 469
Concrete
    compacting, 407
    mixing, 405
    spreading, 406
Concrete airport pavement, 277
Constraint conditions, 233
Construction
    cost, 5
    methods, 56, 404
Contact
    area, 165
    pressure, 40

Continuous reinforcement, 329
Continuous slabs, 378, 379
Continuously reinforced concrete, 7, 301, 325, 335, 420
Cores, 308
Cork strips, 289
Corner loading, 239, 264, 303
Corps of Engineers, 42, 49, 171
Corrosion-resistant, 291
Corrugations, 440
Cost
    analysis, 226
    coefficients, 423
    variables, 5
Coverage, 171, 392
Crack starter, 331
Cracking, 44, 111, 305, 309, 331, 355, 384, 396, 412, 417, 432, 446
    frequency, 113
    index, 113
    spacing, 309, 310
    width, 309
Critical areas, 271
Cumulative damage theory, 172
Curing, 81, 310, 311
Curling stresses, 254, 256, 356
Cycle ratio, 153

Defects, 431
Deflection, 27, 248, 316, 417, 467
Deformation, 44
Deformed bars, 306, 307
Degree day, 47
De-icing agents, 416
Dense-graded asphalt, 181
Density, 71, 210, 524
Density-voids analysis, 210
Design
    aircraft, 21, 170
    lane, 123, 126
    load, 500
    location, 178, 192
    period, 123
    subgrade value, 180
    thickness, 130, 137, 203, 258, 261, 268, 282, 317, 321, 325, 369, 388